Cloud Scale Analytics with Azure Data Services

Build modern data warehouses on Microsoft Azure

Patrik Borosch

Packt>

BIRMINGHAM—MUMBAI

Cloud Scale Analytics with Azure Data Services

Group Product Manager: Kunal Parikh

Publishing Product Manager: Reshma Raman

Senior Editor: David Sugarman

Content Development Editor: Joseph Sunil

Technical Editor: Manikandan Kurup

Copy Editor: Safis Editing

Project Coordinator: Aparna Nair

Proofreader: Safis Editing

Indexer: Rekha Nair

Production Designer: Vijay Kamble

First published: July 2021

Production reference: 1250621

Published by Packt Publishing Ltd.
Livery Place
35 Livery Street
Birmingham
B3 2PB, UK.

ISBN 978-1-80056-293-6

www.packt.com

A big thank you to Packt Publishing: Reshma Raman, Gebin George, David Sugarman, Aishwarya Mohan and, of course, Joseph Sunil. It was a ride, and your support was vital to finish this book. I am very proud, humbled, and thankful that you approached me and gave me the chance to write this book.

Special thanks go to my manager Zoran Draganic, who encouraged me to take on this challenge and supported me throughout the writing process.

I want to say another huge thank you to my CSA colleague Meinrad Weiss. Your technical expertise, your honest (and always to the point) feedback, and our reviews helped me to learn even more and to improve the quality of the book.

A fourth big thank you to Liviana Zürcher, another CSA alongside Meinrad and myself. Your reviews and your support, especially when I started the writing, were more than important, as you gave me faith and kept me going during the critical starting phase.

Finally, I need to thank my wife, Simone, from the deepest bottom of my heart. She was with me in this challenge and stayed patient during all those night sessions, and all the times I told her, "I need to finish the next chapter!" Without your support I wouldn't have been able to research, experiment, and finish this book.

Contributors

About the author

Patrik Borosch is a Cloud Solution Architect for Data and AI at Microsoft Switzerland GmbH. He has more than 25 years of BI and analytics development, engineering, and architecture experience and is a Microsoft Certified Data Engineer and a Microsoft Certified AI Engineer. Patrik has worked on numerous significant international Data Warehouse, Data Integration and Big Data projects. There, he has built and extended his experience in all facets from requirement engineering over data modelling and ETL all the way to reporting and dashboarding. At Microsoft Switzerland, he supports customers in their journey into the analytical world of Azure Cloud.

About the reviewers

Pradeep Menon is a seasoned data analytics professional with more than 18 years of experience in data and AI. Currently, Pradeep works as a data and AI strategist with Microsoft. In this role, he is responsible for helping Microsoft's strategic customers across Asia to be more data-driven by using cloud, big data, and AI technologies. He is also a distinguished speaker and blogger and has given numerous keynotes on cloud technologies, data, and AI. He has previously worked at Microsoft, Alibaba Group, and IBM.

Liviana Zürcher started her career in Romania as Technical Sales Consultant for Business Intelligence and Data Warehouse at Oracle. She has been working a few years as a Big Data Warehouse and Business Intelligence consultant and trainer, having projects all over the world. Finally, she has joined Microsoft Switzerland team as Cloud Solution Architect for Data and Artificial Intelligence at Microsoft, where she is currently working since 2018.

First of all, I would like to thank my parents that have always helped me and encouraged me to go out from my comfort zone. Special thank you to my wonderful husband and to our two amazing children for their support. I thank Patrik for this opportunity, I was flattered by your request.

Meinrad Weiss works as a Senior Cloud Solution Architect in the Data and AI Team at Microsoft Switzerland. He is a very experienced Database expert and comes with a long and successful track record in Data, BI and analytical projects. Meinrad's expertise spans from RDMS on-prem and in the cloud up to the most complex analytical architectures with the Azure Data Services and the field of IoT at Microsoft. He joined Microsoft in 2017 and became a vital and reliable pillar of the Data and AI architects team.

Table of Contents

2
Connecting Requirements and Technology

Section 2: The Storage Layer

3
Understanding the Data Lake Storage Layer

4
Understanding Synapse SQL Pools and SQL Options

Section 3: Cloud-Scale Data Integration and Data Transformation

5
Integrating Data into Your Modern Data Warehouse

6
Using Synapse Spark Pools

4
Understanding Synapse SQL Pools and SQL Options

Section 3: Cloud-Scale Data Integration and Data Transformation

5
Integrating Data into Your Modern Data Warehouse

6
Using Synapse Spark Pools

7
Using Databricks Spark Clusters

8
Streaming Data into Your MDWH

9

Integrating Azure Cognitive Services and Machine Learning

10

Loading the Presentation Layer

Section 4:
Data Presentation, Dashboarding,
and Distribution

11
Developing and Maintaining the Presentation Layer

12

Distributing Data

13

Introducing Industry Data Models

14

Establishing Data Governance

Other Books You May Enjoy

Index

Preface

Azure Data Lake, the modern data warehouse architecture, and related data services on Azure enable organizations to build their own customized analytical platform to fit any analytical requirements in terms of volume, speed, and quality.

This book is your guide to learning all the features and capabilities of Azure data services for storing, processing, and analyzing data (structured, unstructured, and semi-structured) of any size. You will explore key techniques for ingesting and storing data and perform batch, streaming, and interactive analytics. The book also shows you how to overcome various challenges and complexities relating to productivity and scaling. You will be able to develop and run massive data workloads to perform different actions. Using a cloud-based big data modern data warehouse analytics setup, you will also be able to build secure, scalable data estates for enterprises. Finally, you will not only learn how to develop a data warehouse but also how to create enterprise-grade security and auditing big data programs.

By the end of this Azure book, you will have learned how to develop a powerful and efficient analytical platform to meet enterprise needs.

Who this book is for

This book is for data architects, ETL developers, or anyone who wants to get well-versed in Azure data services to implement an analytical data estate for their enterprise. The book will also appeal to data scientists and data analysts who want to explore all the capabilities of Azure data services, which can be used to store, process, and analyze any kind of data. A beginner-level understanding of data analysis and streaming will be required.

What this book covers

Chapter 1, Balancing the Benefits of Data Lakes over Data Warehouses, explores the evolution of data lakes in the analytical world, and also helps us understand the value of data warehouses.

Chapter 2, Connecting Requirements and Technology, focuses on the architecture of the modern data warehouse and introduces various Azure services, and guides you in choosing the right ones for your needs.

Chapter 3, Understanding the Data Lake Storage Layer, examines the setup and organization of the Data Lake Gen2 storage. You'll learn how to access data and monitor your storage account. You will also learn about backups and disaster recovery, and examine various security and networking options for your storage.

Chapter 4, Understanding Synapse SQL Pools and SQL Options, explores MPP in a cloud PaaS service. You'll also explore the replication and distribution of data in a database. You'll learn about various evolutionary steps of SQL pools in Synapse and other components. You'll also check out various alternative SQL database services in Azure and how you can use them.

Chapter 5, Integrating Data into Your Modern Data Warehouse, shows how to implement ETL/ELT pipelines with Synapse pipelines, or Azure Factory. You'll examine various source connectors and work on integration jobs. You'll also learn how to monitor your integration environment.

Chapter 6, Using Synapse Spark Pools, discusses Synapse Spark pools and how to implement them on Azure. You will examine how to implement notebooks and Spark jobs and integrate additional libraries with your clusters. Finally, we will examine security features and see how to monitor our environment.

Chapter 7, Using Databricks Spark Clusters, examines Azure Databricks. We will learn how to work with it and perform various operations. We'll also learn how to create and use dashboards and run ETL jobs. Finally, you'll learn how to set up Databricks with VNets and implement access control within the workspace.

Chapter 8, Streaming Data into Your MDWH, explores Azure Stream Analytics and how it can be used for analysis. You'll learn how to set up and use the service, and you'll learn about various SQL queries with windowing functions and pattern recognition to detect and highlight various events. You'll also build an online dashboard with Power BI that monitors data streaming in real time.

Chapter 9, Integrating Azure Cognitive Services and Machine Learning, examines various machine learning models that you can use as services in Azure. You'll then explore the Azure Machine Learning service and learn how to implement your own model using the graphical user interface there.

Chapter 10, Loading the Presentation Layer, shows you how to load data into your presentation layer using various tools, such as PolyBase, the COPY command, and Synapse pipelines. We'll also check out how to implement SQL in your data lake. Lastly, we'll explore some options for exchanging metadata between various compute components to improve efficiency.

Chapter 11, Developing and Maintaining the Presentation Layer, examines how to use Azure Synapse, and particularly Synapse Studio, when you implement your presentation layer. You will see how to integrate Azure Synapse with Azure DevOps and how you can automate your deployments. In your role as an modern data warehouse developer, you will also enjoy the developer productivity features that Synapse Studio offers. You'll also dive into disaster recovery and some security aspects of your environment.

Chapter 12, Distributing Data, shows you ways to create data marts to distribute insights in your modern data warehouse with Power BI. You will see how to use Power BI data models and the options to visualize and publish their content and even use the data with other tools. We will also examine Azure Data Share as another option to provide datasets to others.

Chapter 13, Introducing Industry Data Models, showcases various industry data models that you can utilize in your projects using Microsoft's CDM tool. We'll also explore a service in Azure called Industry Data Workbench.

Chapter 14, Establishing Data Governance, takes you through the options that the Azure Purview preview offers for scanning your data and qualifying it. You will see how you can benefit from predefined and custom search patterns and how Purview helps you to find information in your data estate. You will also see how to integrate with other Azure services such as Azure Synapse Analytics or Data Factory.

To get the most out of this book

You will need a system with a good internet connection and an Azure account.

Please note: all the services that you might use during the exercises of this book will cause cost within your Azure subscription.

Try to always scale down the services where possible or even delete them after you have finished going through the exercises.

Download the example code files

You can download the example code files for this book from GitHub at `https://github.com/PacktPublishing/Cloud-Scale-Analytics-with-Azure-Data-Services`. In case there's an update to the code, it will be updated on the existing GitHub repository.

We also have other code bundles from our rich catalog of books and videos available at `https://github.com/PacktPublishing/`. Check them out!

Download the color images

We also provide a PDF file that has color images of the screenshots/diagrams used in this book. You can download it here: `https://static.packt-cdn.com/downloads/9781800562936_ColorImages.pdf`.

Conventions used

There are a number of text conventions used throughout this book.

`Code in text`: Indicates code words in text, database table names, folder names, filenames, file extensions, pathnames, dummy URLs, user input, and Twitter handles. Here is an example: "You can now start entering an alias for your input connection. Name it something such as `airdelaystreaminginput`."

A block of code is set as follows:

```
SELECT
    t1.Cartype,
    SUM(t2.mgNOx/60) as SumNOx
FROM
    Cartraffic as t1 TIMESTAMPED BY ObservedT
JOIN
    CarStats as t2
ON
    t1.Cartype = t2.Cartype
GROUP BY
    t1.Cartype,
    TUMBLINGWINDOW(minute, 10)
```

When we wish to draw your attention to a particular part of a code block, the relevant lines or items are set in bold:

```
SELECT
    CensusStation,
    COUNT(*) as Amount
FROM
    Cartraffic
TIMESTAMP BY
    ObservedT
GROUP BY
    CensusStation,
    System.Timestamp()
```

Bold: Indicates a new term, an important word, or words that you see onscreen. For example, words in menus or dialog boxes appear in the text like this. Here is an example: "Please click **Create** to start the provisioning of your configuration."

> **Tips or important notes**
> Appear like this.

Get in touch

Feedback from our readers is always welcome.

General feedback: If you have questions about any aspect of this book, mention the book title in the subject of your message and email us at customercare@packtpub.com.

Errata: Although we have taken every care to ensure the accuracy of our content, mistakes do happen. If you have found a mistake in this book, we would be grateful if you would report this to us. Please visit www.packtpub.com/support/errata, selecting your book, clicking on the Errata Submission Form link, and entering the details.

Piracy: If you come across any illegal copies of our works in any form on the Internet, we would be grateful if you would provide us with the location address or website name. Please contact us at copyright@packt.com with a link to the material.

If you are interested in becoming an author: If there is a topic that you have expertise in and you are interested in either writing or contributing to a book, please visit authors.packtpub.com.

Reviews

Please leave a review. Once you have read and used this book, why not leave a review on the site that you purchased it from? Potential readers can then see and use your unbiased opinion to make purchase decisions, we at Packt can understand what you think about our products, and our authors can see your feedback on their book. Thank you!

For more information about Packt, please visit packt.com.

Share Your Thoughts

Once you've read *Cloud Scale Analytics with Azure Data Services*, we'd love to hear your thoughts! Scan the QR code below to go straight to the Amazon review page for this book and share your feedback.

https://packt.link/r/1-800-56293-4

Your review is important to us and the tech community and will help us make sure we're delivering excellent quality content.

Section 1: Data Warehousing and Considerations Regarding Cloud Computing

This section will examine the question of whether data warehouses are still required given the rise of the enterprise data lake and provides a brief overview of the trends and development on the market of data and AI. As cloud computing adds flexible and scalable services to AI, there are no more limits in terms of source formats and volumes that can be processed for AI requirements, and given that AI and machine learning are on everybody's mind at the moment, the book attempts to ask what all this entails and where we are heading. In addition, we'll take a technology-agnostic look at the components that make up a successful analytical system. From an agnostic viewpoint, we will try to find the right Azure services to build a modern data warehouse.

This section comprises the following chapters:

- *Chapter 1, Balancing the Benefits of Data Lakes over Data Warehouses*
- *Chapter 2, Connecting Requirements and Technology*

1

Balancing the Benefits of Data Lakes Over Data Warehouses

Is the Data Warehouse dead with the advent of Data Lakes? There is disagreement everywhere about the need for Data Warehousing in a modern data estate. With the rise of Data Lakes and Big Data technology, many people use other, newer technologies compared to databases for their analytical efforts. Establishing a data-driven company seems to be possible without all those narrow definitions and planned structures, the ETL/ELT, and all the indexing for performance. But when we examine the technology carefully, when we compare the requirements that are formulated in analytical projects, free of prejudice to the functionality that the chosen services or software packages can deliver, we often find gaps on both ends. This chapter discusses the capabilities of Data Warehousing and Data Lakes and introduces the concept of the Modern Data Warehouse.

With all the innovations that have been brought to us in the last few years, such as faster hardware, new technologies, and new dogmas such as the Data Lake, older concepts and methods are being questioned and challenged. In this chapter, I would like to explore the evolution of the analytical world and try to answer the question, *is the Data Warehouse really obsolete?*

We'll find out by covering the following topics:

- Distinguishing between Data Warehouses and Data Lakes
- Understanding the opportunities of modern cloud computing
- Exploring the benefits of AI and ML
- Answering the question

Distinguishing between Data Warehouses and Data Lakes

There are several definitions of **Data Warehousing** on the internet. The narrower ones characterize a warehouse as the database and the model used in the database; the wider descriptions look at the term as a method, and a suitable collection of all organizational and technological components that make up a BI solution. They talk about everything from the **Extract Transform Load tool** (**ETL tool**) to the database, the model, and, of course, the reporting and dashboarding solution.

Understanding Data Warehouse patterns

When we look at the Data Warehousing method in general, at its heart, we find a database that offers a certain table structure. We almost always find two main types of artifacts in the database: **Facts** and **Dimensions**.

Facts provide all the measurable information that we want to analyze; for example, the quantities of products sold per customer, per region, per sales representative, and per time. Facts are normally quite narrow objects, but with a lot of rows stored.

In the Dimensions, we will find all the descriptive information that can be linked to the Facts for analysis. Every piece of information that a user puts on their report or dashboard to aggregate and group the fact data, filter it, and view it is collected in the Dimensions. All the data related to customer information, such as Product, Contract, Address, and so on, that might need to be analyzed and correlated is stored here. Typically, these objects are stored as tables in the database and are joined using their key columns. Dimensions are normally wide objects, sometimes with controlled redundancy, that look at the given modeling method.

Three main methods for modeling the Facts and Dimensions within a Data Warehouse database have crystalized over the years of its evolution:

- **Star-Join/Snowflake**: This is probably the most famous method for Data Warehouse modeling. Fact tables are put in the center of the model, while Dimension tables are arranged around them, inheriting their Primary Key into the Fact table. In the Star-Join method, we find a lot of redundancy in the tables since as all the Dimension data, including all hierarchical information (such as Product Group -> Product SubCategory -> Product Category) regarding a certain artifact (Product, Customer, and so on), is stored in one table. In a Snowflake schema, hierarchies are spread in additional tables per hierarchy level and are linked over relationships with each other. This, when expressed in a graph, turns out to show a kind of snowflake pattern.

- **Data Vault**: This is a newer method that reflects the rising structural volatility of data sources, which offers higher flexibility and speed for developing. Entities that need to be analyzed are stored over **Hubs**, **Satellites**, and **Links**. Hubs simply reflect the presence of an entity by storing its ID and some audit information such as its data source, create times, and so on. Each hub can have one or more Satellite(s). These Satellites store all the descriptive information about the entity. If we need to change the system and add new information about an entity, we can add another Satellite to the model, reflecting just the new data. This will bring the benefit of non-destructive deployments to the productive system in the rollout. In the Data Vault, the Customer data will be stored in one Hub (the CustomerID and audit columns) and one or more Satellites (the rest of the customer information). The structure in the model is finally brought by Links. They are provided with all the Primary Keys of the Hubs and, again, some metadata. Additionally, Links can have Satellites of their own that describe the relationships of the Link content. Therefore, the connection between a Customer and the Products they bought will be reflected in a Link, where the Customer-Hub-Key and the Product-Hub-Key are stored together with audit columns. A Link Satellite can, for example, reflect some characteristics of the relationship, such as the amount bought, the date, or a discount. Finally, we can even add a Star-Join-View schema to abstract all the tables of the Data Vault and make it easier to understand for the users.

- **3rd Normal form**: This is the typical database modeling technique that is also used in (and first created for) so-called **Online Transactional Processing (OLTP)** databases. Artifacts are broken up into their atomic information and are spread over several tables so that no redundancy is stored in either table. The Product information of a system might be split in separate tables for the product's name, color, size, price information, and many more. To derive all the information of a 3rd Normal Form model, a lot of joining is necessary.

Investigating ETL/ELT

But how does data finally land in the Data Warehouse database? The process and the related tools are named **Extract, Transform, Load** (**ETL**), but depending on the sequence we implement it in, it may be referred to as **ELT**. You'll find several possible ways to implement data loading into a Data Warehouse. This can be done by implementing specialized ETL tools such as **Azure Data Factory** in the cloud, **SQL Server Integration services (SSIS)**, **Informatica**, **Talend**, or **IBM Data Stage**, for example.

The biggest advantage of these tools is the availability of wide catalogues of ready-to-use source and target connectors. They can connect directly to a source, query the needed data, and even transform it while being transported to the target. In the end, data is loaded into the Data Warehouse database. Other advantages include its graphical interfaces, where complex logic can be implemented on a "point-and-click" basis, which is very easy to understand and maintain.

There are other options as well. Data is often pushed by source applications and a direct connection for the data extraction process is not wanted at all. Many times, files are provided that are stored somewhere near the Data Warehouse database and then need to be imported. Maybe there is no ETL tool available. Since nearly every database nowadays provides loader tools, the import can be accomplished using those tools in a scripted environment. Once the data has made its way to the database tables, the transformational steps are done using Stored Procedures that then move the data through different DWH stages or layers to the final **Core Data Warehouse**.

Understanding Data Warehouse layers

Talking about the Data Warehouse layers, we nearly always find several steps that are processed before the data is provided for reporting or dashboarding. Typically, there are at least the following stages or layers:

- **Landing or staging area**: Data is imported into this layer in its rawest format. Nothing is changed on its way to this area and only audit information is added to track down all the loads.

- **QS, transient, or cleansing area**: This is where the work is done. You will only find a few projects where data is consistent. Values, sometimes even mandatory ones from the source, may be missing, content might be formatted incorrectly or even corrupted, and so on. In this zone, all the issues with the source data are taken care of, and the data is curated and cleansed.

- **Core Data Warehouse**: In this part of the database, the clean data is brought into the data model of choice. The model takes care of data historization, if required. This is also one possible point of security – or, even better, the *central* point of security. Data is secured so that all participating users can only see what they can see. The Core Data Warehouse is also the place where performance is boosted. By using all the available database techniques, such as indexes, partitioning, compression, and so on, the Data Warehouse will be tuned to suit the performance needs of the reporting users. The Core Data Warehouse is often also seen as the **presentation layer** of the system, where all kinds of reporting and dashboarding and self-service BI tools can connect and create their analysis.

Additionally, Data Warehouses have also brought in other types of layers. There might be sublayers for the cleansing area, for example, or the **Data Marts**, which are used to slice data semantically for the needs of certain user groups, or the **Operational Data Store (ODS)**, which has different definitions, depending on who you ask. Sometimes, it is used as an intermediary data store that can be used for reporting. Other definitions speak of a transient zone that stores data for integration and further transport to the DWH. Sometimes, it is also used as a kind of archived data pool that the Data Warehouse can always be reproduced from. The definitions vary.

Implementing reporting and dashboarding

In terms of the reporting/dashboarding solution, we will also find several approaches to visualize the information that's provided by the Data Warehouse. There are the typical reporting tools, such as **SQL Server Reporting Service** or **Crystal Reports**, for example. These are page-oriented **report generators** that access the underlying database, get the data, and render it according to the template being used.

The more modern approach, however, has resulted in tools that can access a certain dataset, store it in an internal caching database, and allow interactive reporting and dashboarding based on that data. **Self-Service BI** tools such as **Power BI**, **QLIK**, and **Tableau** allow you to access the data, create your own visuals, and put them together in dashboards. Nowadays, these tools can even correlate data from different data sources and allow you to analyze data that might not have made it to the Data Warehouse database yet.

Loading bigger amounts of data

You can scale databases with newer, faster hardware, more memory, and faster disks. You can also go from **Symmetric Multi-Processing (SMP)** to **Massively Parallel Processing (MPP)** databases. However, the usual restrictions still apply: the more data we need to process, the longer it will take to do so. And there are also workloads that databases will not support, such as image processing.

Starting with Data Lakes

It's funny that we find a similar mix or maybe even call it confusion when we examine the term **Data Lake**. Many people refer to a Data Lake as a **Hadoop Big Data** implementation that delivers one or more clusters of computers, where a **distributed filesystem** and computational software is installed. It can deal with a distributed **Input and Output (I/O)** on the one hand but can also do distributed and parallel computation. Such a system adds specialized services for all kinds of workloads, be it just SQL queries against the stored data, in-memory computation, streaming analytics – you name it. Interestingly, as we mentioned while discussing Data Warehouses discussion, the narrower definition of the Data Lake only refers to the storage solution, and less to the method and the collection of services. To be honest, I like the wider definition far better, as it describes the method more holistically.

With the **Hadoop Distributed File System (HDFS)** from the **Apache Software Foundation**, a system is delivered that is even capable of storing data distributed over cheap legacy hardware clusters. HDFS splits the files into blocks and will also replicate those blocks within the system. This not only delivers a failsafe environment, but it also generates the biggest advantage of HDFS: parallel access to the data blocks for the consuming compute components. This means that a Spark cluster, for example, can read several data blocks in parallel, and with increased speed, because it can access several blocks of a file with several parallel threads.

With the possibility to hand files to a filesystem and start analyzing them just where they are, we enter another dogma.

MapReduce, the programming paradigm, supports parallel processing many files in this distributed environment. Every participating node runs calculations over a certain subset of all the files that must be analyzed. In the end, the results are aggregated by a driver node and returned to the querying instance. Different to the Data Warehouse, we can start analyzing data on a **Schema-On-Read** basis. This means we decide on the structure of the files that are processed, right when they are being processed. The Data Warehouse, in comparison, is based on a **Schema-On-Write** strategy, where the tables must be defined before the data is loaded into the database.

In the Data Lake world, you, as the developer, do not need to plan structures weeks ahead, as you would in the Data Warehouse. Imagine you have stored a vast quantity of files, where each row in a file will reflect an event, and each row also consists of, let's say, 100 columns. Using Schema-On-Read, you can just decide to only count the rows of all the files. You won't need to cut the rows into the columns for this process. For other purposes, you might need to split the rows during the reading process into their own columns to access the detailed information. You might need to predict a machine failure based on the content of the columns. Using a database, you can also mimic this behavior, but you would need to create different objects and selectively store data for each intention.

Understanding the Data Lake ecosystem

HDFS is the center of a Data Lake system. But we can't get further with just a filesystem, even one as sophisticated as this one. The Hadoop **open source** world around HDFS has therefore developed many services to interact and process the content that's kept in distributed storage. There are services such as **Hive**, a SQL-like Data Warehousing service; **Pig**, for highly parallel analysis jobs; **Spark** as a large-scale, in-memory analytical engine; **Storm** as a distributed real-time streaming analysis engine; **Mahout** for machine learning; **Hbase** as a database; **Oozie**, for workflows; and many more.

Spark clusters, with their ability to run distributed, in-memory analytical processes on distributed datasets, have influenced the market heavily in the recent years. **Scala**, one of the main programming languages used with Spark, is a Java-based programming language. It can be used to write high-performance routines for **ETL/ELT**, **data cleansing**, and **transformation**, but also for **machine learning** and **artificial intelligence** on massive datasets. **Python** also made it into this technology and is one of the go-to languages here. **R**, as the statistical programming language of choice for many data scientists, was also a must for the Spark offering. Dealing with data has been important for so many years and guess what – **SQL** is still a language that was not possible to skip in such an environment.

These services and languages are making the open source Hadoop ecosystem a rich, complex engine for processing and analyzing excessive amounts of data. They are available all over the clusters and can interact with the HDFS to make use of the distributed files and compute.

> Tip
>
> Jumping into a big data challenge should, just like a Data Warehouse project, always be a well-prepared and intensively examined project. Just starting it from an "I want that too!" position would be the worst driver you can have, and unfortunately happens far too often. When you start with purpose and find the right tool for that purpose, at least that selection might create the right environment for a successful project. Maybe we will find some alternatives for you throughout this book.

Comparing Data Lake zones

Very similar to the Data Warehouse layers, as we discussed previously, a Data Lake is also structured in different layers. These layers form the stages that the data must go through on its way to be formed into information. The three major zones are pretty much comparable. We can form a *landing zone* in the Data Lake, where data is written to "as-is" in the exact format as it comes from the source. The *transient zone* then compares to the cleansing area of the Data Warehouse. Data is transformed, wrangled, and massaged to suit the requirements of the analyzing users. As no one should access either the landing zone or the transient zone, the *curated zone* is where you're heading, and what you, as the developer, will allow your users to access. Other (similar) concepts talk about *Bronze*, *Silver*, and *Gold* areas, or use other terms. Just like in the Data Warehouse concept, the Data Lakes can include other additional zones. For example, zones for **master data**, user sandboxes, or logging areas for your routines. We can have several user- or group-related folder structures in the curated zone where data is aggregated, just like the Data Marts of the DWH.

Discovering caveats

So, yes, you can be flexible and agile. But in a Data Lake with Schema-On-Read, you'll need to decide which attributes you want to analyze or are needed by your machine learning models for training. This will, after all, force you to structure your sources and will therefore force you into a certain project life cycle. You will go through user stories, requirements analysis, structuring and development, versioning, and delivering artifacts.

If you're *only* analyzing tabular-oriented data, maybe it's worth checking if your hardware, your ETL-tool, and your company databases can scale to the needed volume.

This question, along with the nature of the source's data, complexity, and format, should be taken into account when you're deciding on the technology to use. If you need to process sound files, images, PDFs, or similar, then this is no job for a database. SQL does not offer language elements for this (you can add programming extensions to your database). But here, we have a definitive marker for the possible usage of technologies other than databases.

Once you have analyzed the so-called **unstructured data**, you will structure the results back into a tabular-oriented result set as an array (a tabular representation) of data. This needs to be delivered somehow to the recipients of your analysis. But how is that done in a Data Lake?

The typical reporting tools still require tables or table-like source data to work with. Often, they will import the source data into their internal database to be able to quickly answer reporting and dashboarding operations by their users. Experiences with these visualization tools have shown that it is not very performant to directly report from a vast number of files from a Data Lake. Data always needs to be condensed into digestible chunks for these purposes and should then also be stored in a suitable and accessible way.

Funnily enough, many of the services in the Hadoop ecosystem are equipped with similar functionality to the ones that databases have offered for ages now and are optimized for. Data Warehouse databases are more mature in many respects: when we look at Hive as the Data Warehouse service in Hadoop, for example, it still can't update the data and can't run queries with nested subqueries in Hive-QL. But over the years, it has been extended with all kinds of database-like functionality (views, indexes, and many more).

> **Important Note**
> The Data Lake approaches are still missing a fine-grained security model that can deliver centralized security mechanisms to control **Row-Level-Security** and **Column-Level-Security** over data while also allowing an easy-to-implement mechanism for **Data Masking**.

Understanding the opportunities of modern cloud computing

Hyperscale cloud vendors such as **Microsoft**, **AWS**, and **Google**, where you can just swipe your credit card and get storage, a web server, or provision a **VM** of your choice, are a real game changer in the IT world and started an unstoppable revolution. The speed at which new applications are developed and rolled out gets faster and faster, and these cloud technologies reduce time-to-market dramatically. And the cloud vendors don't stop there. Let's examine the different types of offerings they provide you as their customer.

Understanding Infrastructure-as-a-Service

Infrastructure-as-a-Service (IaaS) was only the start. Foremost, we are talking about delivering VMs You can set up a system of VMs just as you would within your own data center, with domain controllers, database and application servers, and so on. Another example for IaaS would be storage that can be integrated just like a virtual network drive into the customers' workspaces.

Sure, a VM with a pre-installed OS and maybe some other software components is a nice thing to have. And you even can scale it if you need more power. But still, you would need to take care of patching the operating system and all the software components that have been installed, the backup and disaster recovery measurements, and so on.

Nowadays, cloud vendors are making these activities easier and help automate this process. But still, this is not the end of the speed, automation, and convenience that a cloud offering could bring:

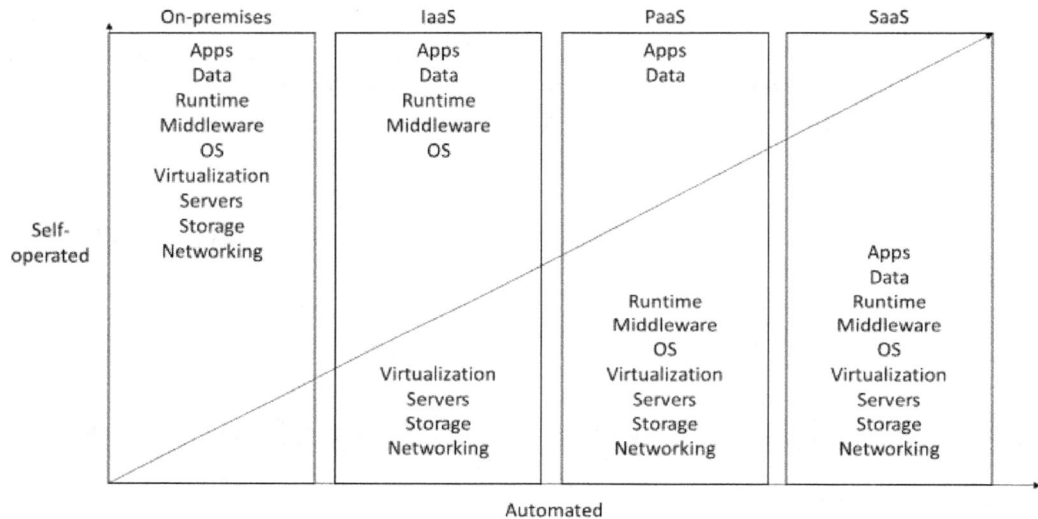

On-premises	IaaS	PaaS	SaaS
Apps	Apps	Apps	
Data	Data	Data	
Runtime	Runtime		
Middleware	Middleware		
OS	OS		
Virtualization			Apps
Servers			Data
Storage		Runtime	Runtime
Networking		Middleware	Middleware
		OS	OS
	Virtualization	Virtualization	Virtualization
	Servers	Servers	Servers
	Storage	Storage	Storage
	Networking	Networking	Networking

Self-operated ← → Automated

Figure 1.1 – On-premises, IaaS, PaaS, and SaaS compared

We'll now check out PaaS.

Understanding Platform-as-a-Service

Platform-as-a-Service (PaaS) soon gained the attention of cloud users. Need a database to store data from your web app? You go to your cloud portal, click the necessary link, fill in some configuration information and click *OK*, and there, you are done… your web app now runs on a service that can offer far better **Service-Level Agreements (SLAs)** than the tower PC under the desk at your home.

For example, databases, as one interesting representative of a PaaS, offer automated patching with new versions, backup, and disaster recovery deeply integrated into the service. This is also available for a queue service or the streaming component offered on the cloud.

For databases, there are even point-in-time restores from backups that go quite far back in time, and automated geo redundancy and replication are possible, among many more functions. Without PaaS support, an administrator would need to invest a lot of time into providing a similar level of maintenance and service.

No wonder that other use cases made it to the **Hyperscalers'** (the big cloud vendors, which can offer nearly unlimited resources) offerings. This is also why they all (with differing levels and completeness, of course) offer PaaS capabilities to implement a Modern Data Warehouse/Data Lakehouse in their data centers nowadays.

Some of these offerings are as follows:

- Storage components that can store nearly unlimited volumes of data.
- Data integration tools that can access on-premises, cloud, and third-party software that reaches compute components.
- The ability to scale and process data in a timely manner.
- Databases that can hold the amount of data the customer needs and answer the queries against that data with enough power to fulfill the time requirements set by users.
- Reporting/dashboarding tools, or the connectivity for tools of other vendors, to consume the data in an understandable way.
- **DevOps** life cycle support is available to help you develop, version, deploy, and run an application and gain even more speed with a higher quality.

Understanding Software-as-a-Service

The next evolutionary step was into **Software-as-a-Service (SaaS)**, where you can get an "all-you-can-eat" software service to fulfill your requirements, without implementing the software at all. Only the configuration, security, and connection need to be managed with a SaaS solution. Microsoft **Dynamics 365** is an example of such a system. You can use an **Enterprise Resource Planning** (**ERP**) suite from the cloud without the need to have it licensed and installed in your data center. Sure, the right configuration and the development of the ERP in the cloud still needs to be done, but the Hyperscaler cloud vendor and/or the SaaS provider relieves you of which server hardware, backup and restore, SLAs, and so on to use.

However, not all cloud users want to or can rely on predefined software to solve their problems. The need for individual software is still high and, as we discussed previously, the **virtualization** options that are available on the cloud platforms are very attractive in this case.

Examining the possibilities of virtual machines

Development can follow a few different options. **Microsoft's Azure**, for example, offers different capabilities to support application development on this platform. Starting with **VMs,** you can make use of any setup and configuration for this application that you can think of. Spin the VM, install and run the app, and off you go.

But don't forget – everything that you need to run the app needs to be installed and configured upfront. Maybe there are VM images available that come predefined in a certain configuration. Still, you as the developer are, after spinning it up, responsible for keeping the operating system and all the installed components up-to-date and secure.

The cloud vendor, however, supports you with **automation features** to help you keep up with changes – to back up the VM, for example – but like every individual server or PC, a VM is a complete computer and brings that level of complexity.

Let's assume you need to run a small piece of software to offer a function or one or more microservices; a cloud-based VM may be overkill. Even with all the cloud advantages of having a VM without the need of spinning up and maintaining your own hardware and data center, this can be quite costly and more complex than what is really needed.

Again, the Hyperscaler can help as it has different offerings to overcome this complexity. Looking at Microsoft's example, there are different approaches to solving this issue.

Understanding Serverless Functions

One of these approaches is **Azure Serverless Functions**, which you can use to implement functionality in different languages. You can choose between **C#**, **Java**, **JavaScript**, **PowerShell**, or **Python** to create the function. Then, you can deploy it to the environment and just rely on the runtime environment. Seamless integration with the rest of the services, especially the data services on the platform, is available to you.

Serverless, in this case, means that the underlying infrastructure will scale with increasing and decreasing requests against the function, without needing the developer or the administrator to manually scale the backing components. This can be configured using different consumption plans, with different limits to keep control of the cost.

Cost, besides implementation speed and flexibility, is one of the most important factors when moving to cloud platforms. With the example of Serverless Functions, we find a great template for the elasticity of cloud services and the possibilities of saving money in comparison to on-premises projects.

Using a seamless scaling service for your functionality gives you the possibility to start small and start experimenting, without spending too much money. Should you find out that your development trajectory points to a dead end, you can directly stop all efforts. You can then archive or even delete the developed artifacts and restart your development with a new approach, without spending a fortune purchasing new hardware and software licenses.

Time is the investment here. Maybe these possibilities will lead to more and better judgement about the status and viability of a project that is underway. And maybe, in the future, less time will be wasted because of investments that need to be justified and followed. However, this may lead to quick-started, half-baked artifacts that are rolled out too early just because they can be started easily and cheaply. This is a situation that might lead to expensive replacement efforts and a "running in circles" situation. Therefore, a cost-effective runtime environment and a quick and easy development and deployment situation should still be approached carefully, and designed and planned to a certain degree.

Looking at the importance of containers

Container technologies are another chance to benefit from the cloud's elastic capabilities. They come with some more detailed possibilities for development and configuration in comparison to Serverless Functions. Every Hyperscaler nowadays offers one or more container technologies.

Containers enable you to deploy **modularized** functionalities into a runtime environment. They abstract from the OS, support all the necessary libraries and dependencies, and can hold any routine or executable that you might need. In comparison to VMs, containers can easily be ported and deployed from one OS to another. The containers that are running in a certain environment use the kernel of the OS and share it with each other.

VMs run their own OSes, apart from each other's, so they will also need far more maintenance and care. Within a container, you can concentrate on the logic and the code you want to implement, without the need to organize the "computer" around that. This also leads to a far smaller footprint for a container compared to VMs. Containers also boot far quicker than VMs and can be available quickly in an on-demand fashion. This leads to systems that can react nearly instantly to all kinds of events.
A container failing, for example, can be mitigated by spinning up the image and redirecting the connection to it internally.

Therefore, in terms of their basic usage, containers are stateless to ensure quick. This adds up to the elasticity of cloud services. Developing and versioning of container modules eases their usage and increases the stability of applications based on this technology. You can roll back a buggy deployment easily, and deployments can be modularized down to small services (**microservices approach**).

Containers can be developed and implemented on your laptop and then deployed to a cloud repository, where they can be instantiated by the container runtime of choice.

In Microsoft Azure, the offering for containers comes from the **container registry**, where the typical container technologies can be placed and instantiated from. To run containers on Azure, you can use different offerings, such as **Azure Container Instances** or **Azure Kubernetes Service (AKS)**. **Azure Red Hat OpenShift** and **Azure Batch** can also be used to run containers in some situations. It is also worth mentioning **Azure Service Fabric** and **Azure App Service**, which can host, use, or orchestrate containers on Azure:

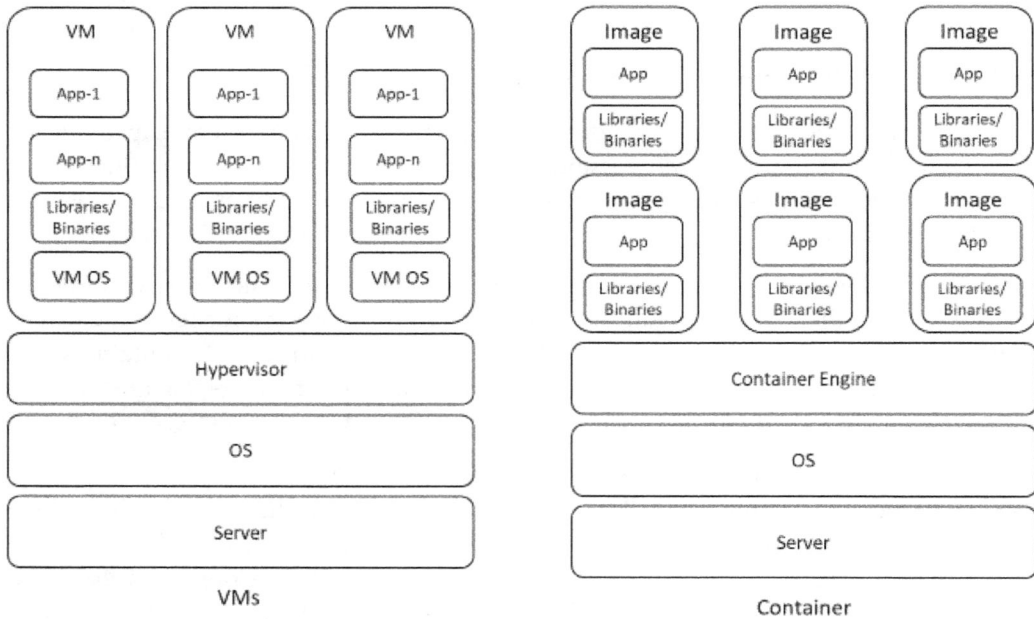

Figure 1.2 – Virtual machines versus containers

Exploring the advantages of scalable environments

Modern Data Warehouse requirements may reach beyond the functionalities of the typical out-of-the-box components of the ETL/ELT tools. As serverless functions and containers can run a wide variety of programming languages, they can be used to add nearly any function that is not available by default. A typical example is consuming streaming data as it pours into the platform to the queuing endpoints. As some of the streaming components might not offer extended programming functionality, serverless functions and/or containers can add the needed features, and can also add improved performance while keeping the overall solution simple and easy to implement.

Implementing elastic storage and compute

Talking about the possibilities created by cloud computing in the field of data, analytics, and AI, I need to mention **storage** and **compute** and the distinction between the two. To scale cloud components to fulfill volume or performance requirements, there are often services where the two are closely coupled. Looking at VMs, for example, higher performance measures such as **virtual CPUs** automatically require more local storage, which will always cause higher costs in both dimensions. This is also true for scaling many databases. Adding **vCores** to a database will also add memory and will scale diskspace ranges. Looking at the databases on Azure, for example, the diskspace can be influenced in certain ranges, but these ranges are still coupled to the amount of compute that the database consumes.

The development of serverless services in the data and AI sector is leading to new patterns in storage and compute. Many use cases benefit from the separate scaling possibilities that this distinction offers. There are also complex computations on smaller datasets whose internal complexity needs some significant computational power, be it for a high amount of iterations or wider join requirements. Elastically increasing compute for a particular calculation is an opportunity to save money. Why?

As you spin up or scale a computational cluster and take it back down once the computation is done, you only pay for what you consume. The rest of the time, you run the clusters in "keep-the-lights-on" mode at a lower level to fulfill routine requests. If you don't need a compute component, you can take it down and switch it off:

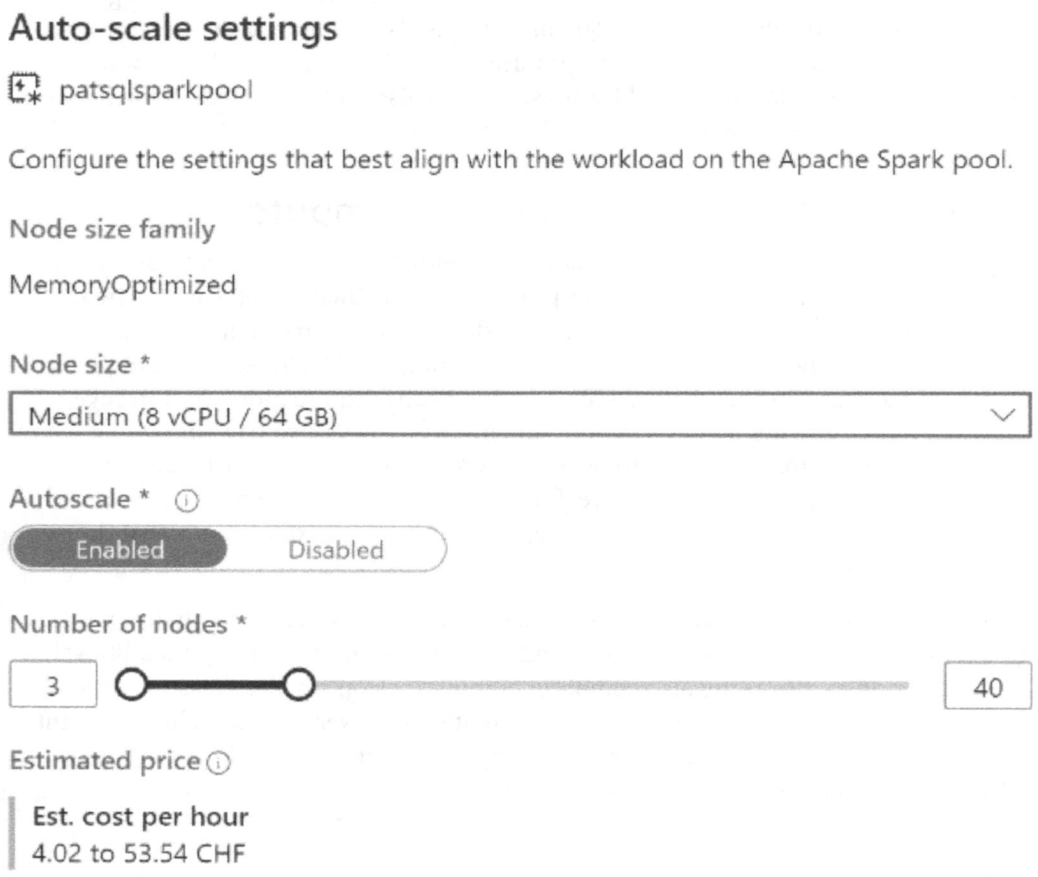

Auto-scale settings

patsqlsparkpool

Configure the settings that best align with the workload on the Apache Spark pool.

Node size family

MemoryOptimized

Node size *

Medium (8 vCPU / 64 GB)   ⌄

Autoscale *   ⓘ

(**Enabled**   Disabled)

Number of nodes *

[3] O━━━━━O━━━━━━━━━━━━━━━━━━━━━━━━━━━━━━━━━━━━ [40]

Estimated price ⓘ

Est. cost per hour
4.02 to 53.54 CHF

Figure 1.3 – Scaling options for Spark pools in Azure Synapse

More and more, we even see **automatic scaling** for compute components, such as in Spark clusters (as shown in the preceding screenshot). You can decide on lower and upper limits for the cluster to instantly react on computational peaks when they occur during a computation. The clusters will even go into **hibernation** mode if there is no computation going on. This can also be accomplished with the database technology that makes up the Modern Data Warehouse, for example, on Microsoft Azure.

As soon as a database is not experiencing high demand, you can scale it back down to a minimum to keep it working and answer random requests. Or, if it is not needed at all, you just switch it off and avoid any compute costs during that time. This comes in handy when you're working in development or test systems, or databases that are created for certain purposes and aren't needed 24/7. You can even do this with your production systems, which are only used during office times for an ELT/ETL load.

Hyperscaler cloud vendors don't limit you in terms of the amount and formats of data you can put into your storage service. They also often offer different **storage tiers**, where you can save money by deciding how much data needs to be in **hot** mode, a slightly more expensive mode for recurring read/write access, or "**cool**" mode, which is cheaper for longer periods without access. Even **archive** modes are available. They are the cheapest modes and intended for data that must be kept but will be accessed only in rare cases.

Cloud storage systems act just like a hard drive would in your laptop, but with far greater capacity. Looking at a Modern Data Warehouse that wants to access the storage, you would need sufficient performance when writing to or reading from such a storage. And performance is always worked on at the big cloud vendors.

Talking about Azure specifically, many measures are taken to improve performance, but also stability and reliance. One example is the fact that files are tripled and spread over different discs in the data center; they are also split into blocks, such as on HDFS, for parallel access and performance gains. This, with other, far more sophisticated techniques, increases the reading speed for analysis significantly and adds to the stability, availability, and reliability of the system.

The cloud vendors, with their storage, computational capabilities, and the elasticity of their service offerings, will lay the foundation for you to build successful and financial competitive systems. The nature of the **pay-as-you-go** models will make it easy for you to get started with a cloud project, pursue a challenge, and succeed with a good price/ performance ratio. And, looking at the PaaS offerings, a project can be equipped with the required components in hours instead of weeks or even months. You can react, when the need arises, instead of purchasing new hardware and software over lengthy processes. If a project becomes obsolete or goes down the wrong path for any reason, deleting the related artifacts and eliminating the related cost can be done very easily.

Cloud technology can help you get things done more quickly, efficiently, and at a far lower cost. We will be talking about the advantages of **cloud security** later in this book.

Exploring the benefits of AI and ML

Companies start building AI projects and massively rely on those statistical functions and the math behind them to predict customer churn, recognize images, detect fraud, mine knowledge, and so much more. ML projects often begin as part of a big data or Data Warehouse project, but this can also be the other way around; that is, the start of an AI or machine learning project often leads to the development of an analytical system.

As my **data scientist** colleagues tell me, if you want to be able to really predict an event based on incoming data, a machine learning model needs to be trained on quite a large amount of data. The more data you can bring to train a model, the better and more accurate the model will be in the end.

A wonderful experiment to do with image recognition, for example, can be done at www. customvision.ai. You can start by examining one of the example projects there. I like the "*Chocolate or Dalmatian*" example.

This is a nice experiment that did not need too much of input to enable the image recognizer to distinguish between Stracciatella Chocolate ice cream and Dalmatian dogs. When you try to teach the system on different images and circumstances, you might find out that you need far more training images than six per group.

Understanding ML challenges

I have experimented with the service and uploaded images of people in an emergency versus images of people relaxing or doing Yoga or similar. I used around 50 – 60 images for each group and still didn't reach a really satisfying accuracy (74%).

With this experiment, I even created a model with a **bias** that I first didn't understand myself. There were too many "emergency" cases being interpreted incorrectly as "All good" cases. By discussing this with my data scientist colleagues and examining the training set, we found out why.

There were too many pictures in the "All good" training set that showed people on grass or in nature, with lots of green around them. This, in turn, led the system to interpret green as a signifier of "All good," no matter how big the emergency obviously was. Someone with their leg at a strange angle and a broken arm, in a meadow? The model would interpret it as "All good."

In this sandbox environment, I did no harm at all. But imagine a case where a system is used to help detect emergency situations, and hopefully kickstart an alarm those vital seconds earlier. This is only a very basic example of how the right tool in the wrong hands might cause a lot of damage.

There are so many different cases where machine learning can help increase the accuracy of processes, increase their speed, or save them money because of the right predictions – such as predicting when and why a machine will fail before it actually fails, helping to mine information from massive data, and more.

Sorting ML into the Modern Data Warehouse

How does this relate to the Modern Data Warehouse? As we stated previously, the Modern Data Warehouse does not only offer scalable, fast, and secure storage components. It also offers at least one (and, in the context of this book, six) compute component(s) that can interact with the storage services and can be used to create and run machine learning models at scale. The "run" can be implemented in batch mode, near-real time or even in real time, depending on the streaming engine used. That Modern Data Warehouse can then store the results of ML calculations into a suitable presentation layer to provide this data to the downstream consumers, who will process the data further, visualize it, draw their insights from it, and take action. The system can close the loop using the enterprise service bus and the integration services available to feed back insights as parameters for the surrounding systems.

Understanding responsible ML/AI

A responsible data scientist will need tools that support them conducting this work properly. The buzzword of the moment in that area is **machine learning operations** (**MLOps**). One of the most important steps of creating a responsible AI is having complete traceability/auditability of the source data, and the versions of the datasets used to train and retrain a certain model at a certain timestamp. With this information, the results of the model can also be audited and interpreted. This information is vital when it comes to audits and traceability regarding legal questions, for instance. The collaborative aspects of an MLOps-driven environment are another important factor.

> **Note**
> We can find a definition of Responsible AI, following the principles of Fairness, Reliability and Safety, Privacy and Security, Inclusiveness, and Transparency and Accountability, at `https://www.microsoft.com/en-us/ai/responsible-ai`.

An MLOps environment embedded in the Modern Data Warehouse will be another puzzle part of the bigger picture and helps integrate those principles into the analytical estate of any company. With the tight interconnectivity between data services, storage, compute components, streaming services, IoT technology, ML and AI, and visualization tools, the world of analytics today offers a wide range of possibilities at a far lower cost than ever before. The ease of use of these services and their corresponding productivity is constantly growing.

Answering the question

Given everything we have learned, is the Data Warehouse dead because of the availability of Data Lakes?

Not at all! What if you could combine the two philosophies and benefit from the advantages of both? The terms that are used for this new approach, which is powered by the availability, scalability, and elasticity offered by the cloud vendors, are "the Data Lakehouse" and, the term that I like more, **the Modern Data Warehouse**.

We will explore this concept in more detail in *Chapter 2, Connecting Requirements and Technology.*

Summary

In this chapter, we examined the differences between Data Warehouses and Data Lakes and the advantages of both approaches. We have the structured model of the Data Warehouse with its security, tuning possibilities, and accessibility for Self-Service BI on the one hand, and the capabilities of Data Lake systems to process vast amounts of data in high performance to support machine learning on the other.

Both concepts, when implemented in isolation, can help solve certain problems. However, even with the growing data, the disparate source systems, their various formats, and the required speed of delivery, as well as the requirements for security and usability, neither can succeed on their own: there is life in the old Data Warehouse yet!

The combination of the two concepts, together with the extended offerings of the Hyperscaler cloud vendors such as virtualization, container offerings, and serverless functions can open new opportunities in terms of the flexibility, agility, and speed of implementation. We are getting the best of both worlds.

In the next chapter, we will discuss a generic architecture sketch. You will learn about the different building blocks of a Modern Data Warehouse approach and how to ask the right questions during your requirements engineering process. In the second part of the next chapter, we will examine Azure Data Services and PaaS components. We'll explore alternative components for different sizes and map the Azure Services to the Modern Data Warehouse architecture.

2
Connecting Requirements and Technology

When you examine what cloud vendors are offering to form an analytical data estate, there is an abundance of options. Many **data architects** are investing a lot of time in carving out the right technology and the right tools to build on. This chapter will focus on the **reference architecture** of the modern data warehouse and introduce the suitable Microsoft Azure services with a short description, as well as the information you will need to select the right service to reach your goal.

In this chapter, we will cover the following topics:

- Formulating your requirements
- Understanding basic architecture patterns
- Finding the right Azure tool for the right purpose
- Understanding industry data models
- Defining T-shirt sizes
- Understanding the supporting services

Formulating your requirements

I won't try to write an exhaustive **requirements engineering** paragraph here. Almost every Data Architect is already using a comprehensive method to collect requirements and derive suitable artifacts from them. Let's try to emphasize the direction and the results that should be focused on when the requirements have been gathered and engineered.

Asking in the right direction

When you are planning for a modern data warehouse, you need to come up with questions that must be answered. Please see the *Questions* section at the end of this chapter for more examples. Let's just go through some examples here, to give you an idea:

- For the main storage component, what is your expected volume of data?
- How will you put data in your system? Are you already using an ETL/ELT tool, and can it connect to all the sources and targets that you need to maintain?
- How will data be transformed throughout the system? Do you want to perform data transformation as part of the ETL (Extract Transform Load)/ELT (Extract Load Transform) process and within the tool, or do you prefer to program your data transformation using programming languages? Which languages does your team use?
- How do you want to provide data to your users?
- Do you need to integrate **Machine Learning** (ML) capabilities into your data estate?
- What are the reporting and dashboarding needs of your data estate?
- Will you need to provide programmatic access via APIs to your data?
- What are the security features that you need to establish?
- What are your requirements regarding DevOps?

OK, these are a lot of questions to consider, we know. But please don't let them overwhelm you and don't take them as a source or a checklist to refer to when you're starting a requirements engineering process.

Answering all these questions – and your own additional ones – should be done with a clear goal in mind: to make an educated decision on the tools that you and your team will choose from the services that will be examined later in this chapter, and put them together to form your modern data warehouse.

> **Important note**
>
> When you're going through your requirements and choosing suitable services, you will always want to keep an eye on the skillset of your team. There are many great tools out there on the market, and new technologies can create big opportunities to grow. But can you train your team? Can you invest enough time? Maybe it is worth examining the new tools carefully, to leverage as much knowledge and skills as you can from the people that must work with the system. Python and Spark are great for data cleansing, for example, but to fulfill your requirements, maybe SQL is enough.

Understanding basic architecture patterns

In this section, we will examine the basic architecture pattern of the modern data warehouse (see the following diagram) and examine the components in a bit more detail:

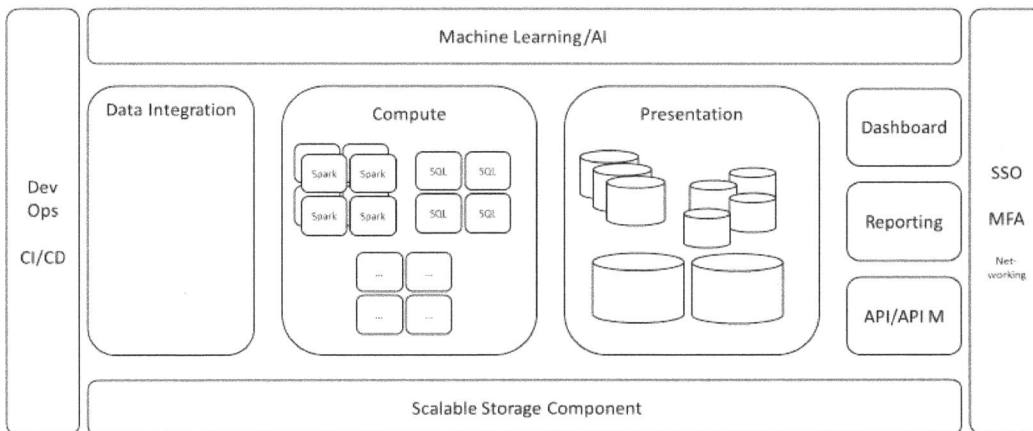

Figure 2.1 – High-level modern data warehouse architecture

We'll now dive into the scalable components in the following section.

Examining the scalable storage component

This points to a storage component that must be able to store any amount of data in any format. We need to be able to create folder structures of the necessary depth. The component needs to offer high throughput when writing, as well as reading, data. Additionally, security is of high importance, and we need at least a **Portable Operating System Interface** (**POSIX**)-like interface with the option to configure access control that's both file- and folder-based. This means we must be able to control who can read, write, or execute content that is stored here.

As we've mentioned already, we want to be able to control the cost of certain areas in the storage component. Files that are not accessed that often should be put into a cheaper tier (cool or archive storage), and files that are used constantly should be in a tier that charges less for the access (hot).

The storage should offer interfaces for all the compute components that are used in the system. This means an ETL/ELT tool should be able to use the storage and read/write files there, as well as Spark or SQL, for example, as compute components. Even better, the storage also offers an HDFS interface to allow tools from the Hadoop family to access the data.

One big advantage is to have the storage as close to the compute components as possible in the same data center. This means we need all the services of the modern data warehouse on the same network backbone to allow the highest possible throughput volume when reading and writing data, as we will see in the next section, *Looking at data integration*.

The storage component should also be able to be scaled as required by the need of arriving data. The best solution would be to scale the storage independently from the compute components that are used to process the stored data.

Looking at data integration

The data integration block shown in *Figure 2.1* represents a family of tools and methods that are used to land data in the system. ETL/ELT tools with "point-and-click" functionality count in this area, as well as programming languages for data processing such as Python. These are two sides of the same coin. It all depends on personal preference: some people prefer the "point-and-click" tools, while others want to code, just like Mathew Broderick in *Wargames*.

The main requirement for this area is having a suitable collection of connectors for the various data sources and targets, which is where you want to unload data from or load data to. Optimally, the connectors should use all the optimization features of the data sources. This ensures that extracting the data and loading it into the target can be done as quickly as possible.

Any tool or language that might be used here should offer parallelization capabilities so that you can copy data in parallel to minimize loading times.

Data must be encrypted during loading time (on the fly) to avoid security threats. Therefore, the tools must, as a minimum, implement **Transport Layer Security** (**TLS**) as a standard.

Finally, we want the tools/development environments to integrate with **Development and Operations (DevOps)** repositories for versioning and deployment features, up to automated deployments. Many tools nowadays offer transformation features. This leads us to the next point in this list.

But wait – so far, we have only talked about batch-oriented loading! In the real world, there are so many more use cases that will require more event-based streaming integration. We need to capture events *as* they pour into our system. Maybe we are talking about the **Internet of Things (IoT)**, where we need to capture and analyze telemetry as it hits the system. Maybe we need to analyze for fraud in a payment system and push every single message through a ML model.

Streaming analytics needs to be possible within our architecture. We need to be able to calculate windowing functions such as so-called **sliding** or **hopping** windows, and we need to aggregate incoming measures. Maybe we need to recognize patterns over a certain group of products over a certain timeframe; for example, *Button A was pressed twice, then Button B was pressed, then Button A again, and after that sequence the device failed*. If we can recognize this before the device fails, we can do something before the failure.

The streaming analytics component needs to be able to talk to our storage component and land the events there for later analysis.

Sorting in compute

In the modern data warehouse, you will have compute components that you will need to process and transform into many different data formats. Tabular-oriented data such as **Character-Separated Values (.csv)** and **.parquet** (a column-compressed format, optimized for analysis), but also **JavaScript Object Notation (.json)** and **Extensible Markup Language (.xml)** are the easier formats to work with. Most **DBs (DB)** technologies can read them and offer SQL extensions to break them apart for analysis. So far, we can still rely on SQL DBs and SQL as a programming language to transform the raw data into the curated zone or the core data warehouse (see *Balancing the Benefits of Data Lakes Over Data Warehouses*, the *Understanding Data Warehouse layers* and *Comparing Data Lake zones* sections).

When it comes to analyzing images, for example, or breaking apart a video and extracting information from it, SQL is not capable of doing so. Due to this, we will need a more sophisticated engine. The compute part of your modern data warehouse can consist of Spark clusters, for example. They provide a rich set of programming languages (see *Balancing the Benefits of Data Lakes Over Data Warehouses*, the *Understanding the Data Lake ecosystem* section) that can be extended by specialized libraries such as **OpenCV**, a Python library for video and image processing.

One of the biggest advantages of the compute components is their capability to scale independently from the storage component. You can add compute power right when you need it. Many services nowadays add capabilities for automated scaling (see *Balancing the Benefits of Data Lakes Over Data Warehouses*, *Figure 1.3*).

Adding a presentation layer

The presentation layer in the modern data warehouse has three main purposes:

- **Performance**: DBs still deliver superior performance over file storage. Indexes, partitioning, compression, and distribution (in the case of **Massively Parallel Processing (MPP)** DBs) will always provide the performance needed when the data is consumed for visualization.

- **Security**: Offering Row- and Column-Level Security features and similar easy-to-implement grants, revokes and data masking will give them a big advantage in comparison to file storage.

  With the ability to correlate data from several areas of the other storage components, this layer will be added as a central security point, thus eliminating the need to integrate many different systems.

- **Usability**: With the ability to abstract DBscontent into so-called DBs **Views**, the data engineer can hide DBs complexity from the consuming user. A DBs view is a SQL statement that can be stored and accessed, similar to a DBs table. Technical table and column names can be translated into understandable and usable names for the user.

  Additionally, this layer can correlate different areas of the storage component very easily and can act as a consolidation point, not only from a security point of view but also from a usability perspective.

Planning for dashboard/reporting

Self-service BI is the goal for many companies. In the area of dashboarding and reporting, we will need tools that are capable of connecting, at a minimum, to our central presentation layer (see the *Adding a presentation layer* section). This depends on your strategy regarding what freedoms you want your users to have. Tools might be discussed that allow them to correlate other sources with our data warehouse data. With all the Data Lake and Schema-on-Read features that can be used to accelerate development life cycles, you still might not be able to instantly react to user requests to add new data to the system. A tool that can access many different sources and relate them to the central data warehouse can relieve high tension between business and IT.

Often, these self-service BI tools offer an additional caching layer, where data is gathered and prepared to speed up analysis.

There are different layers of trust that you can offer to your users:

- **No self-service (totally governed)**: Users can only consume the reports and dashboards that are delivered.

- **Layout self-service with governed data models**: You deliver data models in your presentation layer and a predefined set of visualizations as reports or dashboards. Users can either just consume the reports/dashboards or create their own, based on the data models that are provided.

- **Full self-service**: You deliver data models in your presentation layer and a predefined set of reports and dashboards. Additionally, you allow the users to add their own data using the capability of the self-service BI tools. Users can correlate their own datasets with the data coming from your data models.

Important note

Take care when introducing different layers of the self-service BI, and don't forget to always ensure governance and clear rules regarding who is allowed to publish what to whom, and so on. Power users who are allowed to add their own datasets to their analysis, and also publish their reports in your collaboration system, might soon outrun the "single-version-of-the-truth" approach with their reports and information. In a more governed approach, these users can add value as scouts for your modern data warehouse and can act as the wider development team for your analytical estate. Non-governed environments may soon end up falling back into what we called **Excel-Hell** years ago, where you will find tons of similar (but not identical) reports that all pretend to show the *real* figures. Many man-years of arguments have been fought to find out who was right.

You are navigating between freedom and agility and total control. Yes, freedom and agility can add up to productivity, but remember: *"With great power comes great responsibility!"*

Adding APIs/API management

The architecture foresees a layer for automated data consumption using APIs. This layer will not only need to stick to the security rules that were formulated and implemented, but it might also need to offer different APIs to different consumers.

Versioning the interfaces is one big requirement that this layer needs to implement. If you are providing data to your customers using APIs, you might even need to manage different versions of your APIs to different customers. For example, you could provide an interface to query information about your automated product delivery to the warehouse of your customers. Here, you start with a certain setup in your interface. Soon, you will have customers asking for additional features to be added to your interface. Other customers might not be able to adjust. If you want to keep this controlled, your API management needs to support versioning, alongside security layers, to control customer access to the API and its versions.

Relying on SSO/MFA/networking

To avoid issues with authentication and different security layers, the components of the system are required to offer **Single Sign-On (SSO)** capabilities. By integrating with a centralized user and group control, all access control should focus on the users and groups that are provided by this repository. For example, this allows you to add new users to a set of groups to inherit their access rights so that none of the participating services need to be touched.

Multi-Factor Authentication (MFA) is the state-of-the-art authentication method that should be provided with the modern data warehouse.

Assuming you are planning to run your modern data warehouse in the cloud, you need to integrate networking into your architecture. Building virtual networks that hide all communication between the services that form your modern data warehouse and can be peered to your on-premises network are state-of-the-art already. Peering, in this case, means that you link the virtual network on the cloud platform to your on-premises network in such a way that your services can be accessed as if they reside in your data center. This ensures that no traffic between your on-premises world and the cloud services will traverse the public internet anymore.

Not forgetting DevOps and CI/CD

DevOps and **Continuous Integration/Continuous Deployment (CI/CD)** are vital to any software project that is developed by more than one developer and needs to uphold quality standards. A central code repository that offers versioning, branching, and merging for collaborative development and approval workflows and documentation features is the minimum requirement here.

You might want to implement CI/CD for deployment automation. In cloud environments and agile project methods, you might want to switch from monolithic waterfall project methods to more sprint-oriented developments. Waterfall projects have been around for a long time. Projects can be planned out and may contain many months of development work, extensive testing and bugfixing, and a rollout in the end. If you were to develop this in the wrong direction, a lot of money would be wasted.

This has changed with the advent of agile methods, where a software product is divided into smaller chunks that can be developed and delivered separately. This is where CI/CD comes into the picture, where you rely on the system to automatically package and deploy small chunks of the software that can stand alone.

> **Important note**
> Don't ever forget to include testing in your project plan and, please, take it seriously. The modern DevOps tools will support you in automating tests and integrating them into your CI/CD pipeline.

Finding the right Azure tool for the right purpose

Now that we have a more detailed generic picture of the components that make a modern data warehouse, let's examine the available Azure services. We will try to identify the ones that are suitable for your requirements. All the following services are classified as **Platform-as-a-Service (PaaS)** components. And, as a consultant will always tell you, it depends.

While going through your engineered requirements, you should have a picture in mind of where you want to go. You will know about the data volume, data formats and sources, your transformations, the presentation strategy, and all the questions from the first section of this chapter (see *Asking in the right direction*).

The answers to the questions about volume, for example, together with data formats, will play a vital role in you selecting your storage component. They will point you to the DBs service that you will choose to implement your presentation layer in. There is even a decision to be made here on whether you need to implement all the layers upfront or if you can skip one for the time being.

When examining Azure and the data services provided by Microsoft, I always call this *my candy shop* as there are so many useful components that can be put together. I wish I had access to such tools in earlier projects!

> **Important note**
>
> Don't take the reference architecture as a dogmatic blueprint that needs to be followed word for word. In the end, it is *your* individual system that *you* are building. If you don't need certain building blocks, just skip them! When you carefully engineer your requirements, you will end up with a clear vision of the system that you need. With cloud technology at your command, you can start small and add components to your architecture later, when you really need them.

Understanding Industry Data Models

Industry data models have been developed by different companies for a long time. Microsoft started to offer this approach with the **Common Data Model** (**CDM**), which spans all three cloud pillars of Microsoft (Azure, Office 365, and Dynamics 365). The goal is to easily interconnect data services with each other and follow a predefined structure that is known by all the components involved. If, for example, sales data is read from Dynamics 365, its structure is defined and already available for Data Factory. It can be written to Azure Data Lake Storage or Synapse SQL pools, and from there Power BI already knows the structure and can offer reports and dashboards.

The CDM can be adjusted to your needs. This means you, as the developer, can add attributes to existing entities, create additional entities, or skip the ones you don't need.

To learn more about industry data models, check out *Chapter 13, Introducing Industry Data Models*.

> **Note**
>
> The CDM service (not to be confused with the CDM model structures) can even store data if you decide not to involve a data lake or a data warehouse. Power BI and the other products of the Power Family still can directly access the service and interpret the structures provided.

Recently, Microsoft acquired ADRM Software. This company is known as a cutting-edge provider of industry data models and offers about 75 different models at the time of writing. With this merger, Microsoft can now offer industry-specific data models that are predefined and adjusted to the needs of organizations in the following industries:

- Automotive

- Education

- Financial services

- Healthcare

- Manufacturing

- Media and communications

- Nonprofit

Using these accelerators on top of the Azure data services can reduce development cycles drastically, as you don't need to reinvent complex data structures and their data types. You can benefit from the maturity of the models and the collected wisdom of several years of development.

OK, enough of the theory work – let's try and transport this knowledge into tangible Azure services by examining some use case examples.

Thinking about different sizes

When you look at your engineered requirements and go through all the Azure data services that can form a modern data warehouse, you still might find it complex to decide which services to pick. And with the generic reference architecture in mind, there is no silver bullet to provision so that everything is fine. But let's examine some considerations about sizes, performance, and cost. One of the beauties of the cloud in general, and the ADS framework on Azure in particular, is that you can always switch gears once you recognize your system is too small, you need far more punch for your calculations, or to ensure speed for your users. Don't get me wrong – some services can be cumbersome to replace. However, it is still far easier to replace something than to rebuild everything from scratch.

Let's check out the three different sizes of modern data warehouse by using S, M, and L as rough indicators of our requirements.

Planning for S size

Let's assume your data volume requirements will not exceed a few hundred megabytes, and the transformational steps and business logic that you need to implement are easy to implement and aren't too complex. In this case, you might start with a Power BI-only approach as this can already fulfill your requirements:

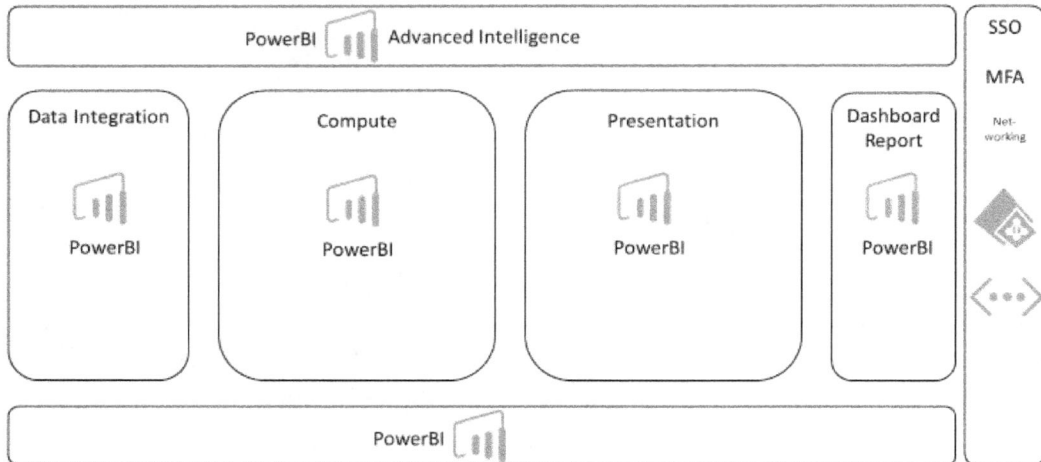

Figure 2.2 – S size Power BI-only architecture

As a self-service BI tool, Power BI enables you to access more than 120 different data sources. The connectors come for free with the tool.

This comes with an **in-memory column store DBs** that has a very high compression rate and delivers performant datasets that can boost your visualizations. Typically, we find compression rates of around 10x, which means that 1 GB of raw data will compress to 100 MB in Power BI.

For the development, Power BI uses **Power BI Desktop**, which is delivered for free and serves as the primary development tool for **Power BI reports**. You will find a data acquisition tool called **Power Query** for self-service ETL. Once you have landed your data in the in-memory column store DBs, you can create relationships between the tables there. Remember, you have 120+ different data source options and you can create a data model using all the possible sources.

Next, there's the very powerful function language known **Data Analysis Expressions (DAX)**, that which all kinds of calculations, even between hierarchical levels (and many more options). Power BI comes with a rich set of visuals that can be extended from an always-growing online gallery. If you need to add visuals that are not in the default set, you can use a Python or R library and plot your own visuals.

The Power BI ecosystem provides a portal (`PowerBI.com`) where you can share your reports, create dashboards, and create new visuals based on datasets that you upload with your reports, or that you can create directly in the portal and get going. There is a functionality here called **Q&A**. Based on datasets that are available in the portal, you can use Q&A to query the data in a natural language, without the need to learn query languages.

Of course, Power BI reports and dashboards can be consumed on mobile devices. Reports and dashboards can be adjusted to mobile usage for display on tablets and mobile phones.

Power BI will be discussed in detail in *Chapter 12, Distributing Data*:

Purpose	Tool
ETL/ELT	Power Query integrated in Power BI
Storage	Tabular Engine Power BI
Transformation Engine	DAX Power BI
Presentation Layer	Tabular Engine Power BI
Visualization	Power BI Reports/Dashboards

Figure 2.3 – Planning for S size

Planning for M size

If you have concluded that a Power BI-only approach might not exactly fit your needs – if you have found that you're ranging in the lower gigabyte volumes already, for instance – you might want to check out some of the other tools mentioned in this chapter. When your transformations get a little more complex but still can be done using SQL, we're looking at an M size modern data warehouse:

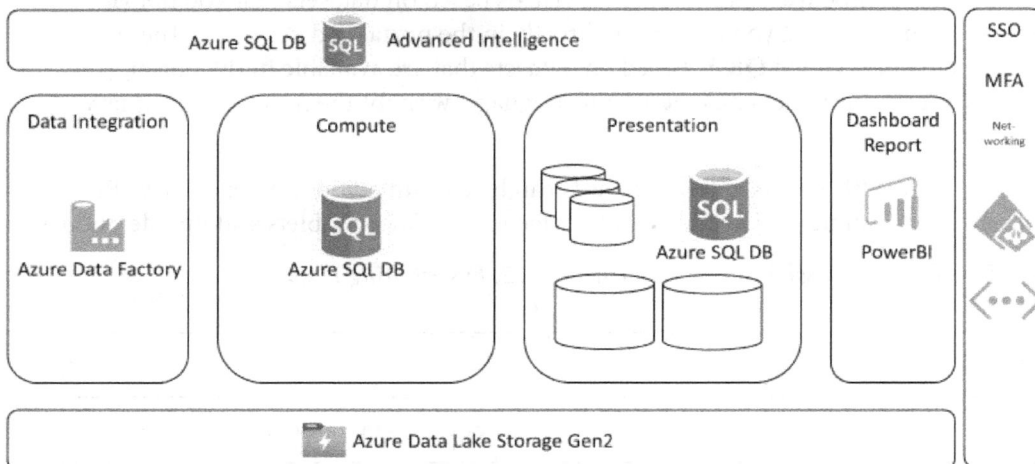

Figure 2.4 – M size architecture – adding Data Factory and Azure SQL DB

Let's consider some of the services you might want to use in a medium-sized modern data warehouse.

Adding Azure Data Factory

If you need to get data from different sources, Data Factory would be a good choice. The Azure Data Factory ETL/ELT tool offers 90+ connectors to on-premises and cloud data sources, such as relational DBs, file storage, ERP solutions such as SAP or Salesforce, and many more.

Our modern data warehouse architecture implements Azure Data Factory as the data transport component. With the available built-in connectors and a **data gateway** called **self-hosted integration runtime**, which can even be used for scale-out architectures, the Data Factory provides all the necessary components for extracting and transporting data quickly and securely to the Azure storage component of your choice.

Data Factory implements all data movement and other tasks with different types of artifacts, including **pipelines**, **activities**, **datasets**, **linked services**, **data flows**, and **integration runtimes**. It even integrates with the Microsoft CDM. This suite offers more than 75 industry data models and structures, ready to be used and filled with your data. For all these components, Data Factory can provide **data lineage** information to track down dependency information about sources, targets, and transformations.

In the portal experience of Azure Data Factory, you can develop your pipelines on a **point-and-click** basis. Another option when using Data Factory would be to create all the artifacts as JSON files. The object definitions are well documented. This gives you the option to add Data Factory to your own application.

On top of this, Azure Data Factory will act as an **orchestration tool** and will therefore provide a **monitoring environment** that integrates with **Azure Monitor,** which is the overarching monitoring service for the whole Azure platform.

For more detailed information about Azure Data Factory, please check out *Chapter 5, Integrating Data into Your modern data warehouse.*

Implementing Azure Data Lake Storage

OK, the naming is a little interesting, but if you are searching for a storage component that doesn't limit you in terms of size, amount, or the format of files that can be stored there, Azure Data Lake Store is for you! Basically, it is an Azure storage account (Blob storage) with an activated feature called the **hierarchical namespace**. This allows you to physically create any folder structure in your Azure storage account. It adds HDFS-compatible access and **POSIX** permissions to folders and files.

Data Lake Storage was built to scale with your needs. If you want to store and analyze up to a petabyte – or even exabyte –of data, this is the storage you will need on Azure.

Azure Data Lake Storage will be discussed in *Chapter 3, Understanding the Data Lake Storage Layer.*

Using Azure SQL DBs in the presentation layer

Azure SQL DBs are a different kind of RDBMS offering on Azure compared to Synapse SQL pools. SQL DBs are typical **symmetrical multiprocessing** (**SMP**) DBs (Synapse SQL Pools provide MPP technology) like you find, for example, on **Microsoft SQL Server**. SQL DBs have inherited most of the genomes of SQL Server.

On Azure, we distinguish between the following:

- **A single database**: A single DBs (PaaS) that offers you the complete T-SQL surface as a scalable service. There is no feature like a server available in this component. Typically, this service is used for web apps and similar. It can scale nicely up to 4/8 TB, depending on the provisioned tier.

- **SQL DB Hyperscale**: This is yet another MPP engine but with a different purpose and other optimization features. This engine targets DBs up to 100 TB in size for OLTP loads or even IoT datasets. This DB works very well with highly selective queries.

- **Managed Instance**: A SQL Server Instance-as-a-Service that is comparable to a SQL Server instance with a defined capacity to spin up SQL Server DBs, as well as the possibility to **lift-and-shift** SQL Server instances as-is to the cloud.

All these components can deliver the full T-SQL surface with functions such as Row-Level and Column-Level Security, data masking, stored procedures, views, and so on, similar to a SQL Server DB. There are some differences, though: you won't be able to access the storage beneath the DBs from your T-SQL, but these are functions you might not need anyway in your cloud setup. The SQL DB can even deliver AI components such as Python and R support.

We will touch on Azure SQL DB again in *Chapter 4*, *Understanding Synapse SQL Pools and SQL Options*.

Keeping Power BI

As for visualization, well, there's no change from the S size here. Power BI is, of course, the tool of choice. You can import the data from your data warehouse DBs and boost your performance again with the column store data model in Power BI. You can also implement some ML algorithms here, using either the Python- or R-script when importing the data, or the Python- or R-visual when displaying the data:

Purpose	Tool	Alternative
ETL/ELT	Azure Data Factory	N/A
Storage/Landing Zone	Azure storage accounts	Azure Data Lake Store Gen2
Transformation Engine	Azure SQL DB: SQL	Azure Data Factory
Presentation Layer	Azure SQL DB	N/A
Visualization	Power BI Reports/dashboards	N/A

Figure 2.5 – Planning for M size

Planning for L size

So, you have concluded that you'll need the Big Mac. Does your data volume go even higher, maybe three digits of gigabytes or even more? Do you need to perform complex transformations on high data volumes? Maybe you even need to analyze images, audio, or video for ML models that you need to train? OK, let's do this! Let's design the modern data warehouse so that it can handle this:

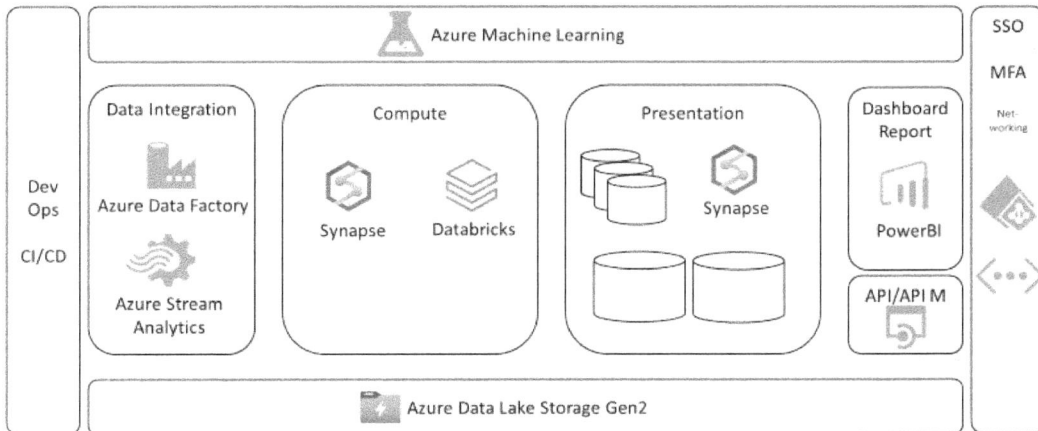

Figure 2.6 – L size Azure services in the modern data warehouse

Let's look at these component services in more detail.

Keeping Data Factory

For the ETL/ELT part, you have Azure Data Factory at your command. The scalability of the engine will give you high performance for extracting and landing the data in Azure. Do you need to transform your data during the load? You can use mapping flows in Data Factory to do so.

Adding Stream Analytics

Do you need to consume streaming data from an IoT system or want to react to any event that hits your logistics system? You can use Azure Stream Analytics to get and transport the incoming events to your data lake.

More information about streaming in a modern data warehouse will be available in *Chapter 8*, *Streaming Data into Your MDWH*.

Sticking to Azure Data Lake Storage

Looking at the amount of data that you want to store in your system, we are going beyond the standard Azure storage accounts – maybe not volume-wise, but complexity-wise, for sure. You will need to build a more complex folder structure and maybe need to archive files. Let's say you have identified different teams that need to access different folders with different access rights. This will require Azure Data Lake Storage.

You can find more details about Data Lake Storage in *Chapter 3*, *Understanding the Data Lake Storage Layer*.

Implementing complex transformations

You need to process and analyze complex files. This is referred to as **unstructured data**. This could be images, sounds, or videos, but also text files such as PDFs, and so on. With these requirements, you are forced to include other compute components in your environment as the plain SQL in your DBs won't do the trick.

You will need to check a Spark component to accomplish this task, for example. The possibility of using Python, Scala, R, SparkSQL, or even .NET, along with the capability to add any additional libraries will enable you to crack open your images and extract the required information. This can then be used to train, retrain, or run your ML models. Synapse Spark pools and Databricks could be good choices for this part of the modern data warehouse. Maybe you will even want to use Spark Streaming instead of Azure Stream Analytics. So, this points you to the two alternatives.

For details on how to implement transformational steps in your modern data warehouse architecture, check out *Chapter 4, Understanding Synapse SQL Pools and SQL Options, Chapter 5, Integrating Data in Your Modern Data Warehouse Chapter 6, Using Synapse Spark Pools, Chapter 7, Using Databricks Spark Clusters*, and *Chapter 11, Developing and Maintaining the Presentation Layer.*

Adding Azure Machine learning (ML)

This service is streamlined for the productivity of the developer/user. Using a browser-based **user interface** (**UI**), Azure ML Studio supports the data scientist in several ways.

Azure ML targets beginner data scientists, as well as the professionals. The offering here starts from **point-and-click** development with more than 40 predefined algorithms and the option to publish a trained model instantly as a REST API. The senior data scientist, on the other hand, will find everything to individually develop their data models, from the Jupyter-like notebook interface to the vast amount of different VM types and sizes with and without GPU support to form compute clusters.

Azure ML supports collaboration by implementing **ML operations** (**MLOps**), where the data scientist can collaborate and version models. The source data that is used for model training during development is also tracked and audited. ML Studio supports lineage through the whole process. This is important for transparency: there is nothing worse than not being able to track the origin of the data and the steps that were taken while developing an ML model.

Interconnectivity with other Azure services is another feature of the Azure ML service. A Synapse Spark Cluster, for example, can make use of ML features. It can spin up new ML clusters in the ML workspace, start Auto ML features for finding the best ML algorithm, and implement the results directly into a calculation.

Azure ML Studio supports Python and R and can implement frameworks such as MLflow, Kubeflow, ONNX, PyTorch, and TensorFlow.

As a new feature, ML Studio supports **responsible ML**. This means the development environment adds features to help us understand, protect, and control the source data for a certain model, as well as the model itself and the processes that are implemented to create the model. The support to find and eliminate biases in an ML model is one of its new features.

We will be talking about ML capabilities from an infrastructural point of view in *Chapter 6, Using Synapse Spark Pools, Chapter 7, Using Databricks Spark Clusters, Chapter 9, Integrating Azure Cognitive Services and Azure Machine Learning, Chapter 10, Loading the Presentation Layer,* and *Chapter 11, Developing and Maintaining the Presentation Layer.*

Adding Synapse Analytics SQL Pools

Next is the presentation layer! From my point of view, this is a job for Synapse SQL Pools. You have transformed the data throughout different layers or zones in your Data Lake Storage. Now, you want to provide a curated data model that can implement all the necessary security and still perform on the data that you need to present to your users.

Synapse Analytics SQL pools are the third evolutionary step of the Azure SQL data warehouse MPP engine. SQL pools provide petabytes in terms of volume and performance and add all the necessary DBs functionalities such as indexing, partitioning, Row- and Column-Level Security, and data masking. SQL pools typically form the presentation layer in our modern data warehouse architecture blueprint.

Synapse Analytics SQL pools use MPP techniques such as **distributed** or **replicated tables** to boost analytic query performance on the DBs side. In terms of the participating nodes, the DBs can be scaled to add more nodes when needed and fewer when less work is needed. Workload management features will help you organize the DBs and use the configured sizes as efficiently as possible.

Transact-SQL (T-SQL) features such as **Polybase** and the **COPY functionality** allow you to access your Data Lake data directly and even offer a **Native Predict Function**, which uses ML models directly in the DBs. This will enable you to efficiently work with your data from within the DBs and use SQL for many different tasks.

Synapse SQL pools integrate into the Synapse Monitor, just like all the other Synapse features, and can use the **managed virtual networks** of the Synapse environment.

Logically, Synapse SQL pools can support the compute and presentation layers in our modern data warehouse reference architecture.

You will find detailed insights about Synapse SQL pools in *Chapter 4, Understanding Synapse SQL Pools and SQL Options,* and *Chapter 11, Developing and Maintaining the Presentation Layer.*

Leveraging Power BI

Visualization will, once again, be done using Power BI. With its DirectQuery mode and composite model, which is a mix between import and DirectQuery modes, you can leverage the superior speed of the Synapse SQL pools, and then use the Power BI column store engine to boost the part of the model where speed really matters:

Purpose	Tool	Alternative
ETL/ELT/Data Integration/ Streaming	Azure Data Factory + Azure Stream Analytics	N/A
Storage/Landing Zone	Azure Data Lake Store Gen2	N/A
Transformation Engine	Synapse Spark Pools	Azure Databricks, Synapse SQL pools, Synapse SQL on-demand
Presentation Layer	Synapse SQL Pools	N/A
Visualization	Power BI Reports/Dashboards	N/A

Figure 2.7 – Planning for L size

As you may have guessed already, this is not an exhaustive collection of all the alternatives that you can build using the Azure data services. For example, you might wish to add a Spark engine such as Databricks to the M size setup. For this, you'll need a storage account to land data for your S size setup and get it from there. And, of course, there is no right or wrong approach to doing this.

In the final section of this chapter, we'll examine some of the additional services that can be employed to support your modern data warehouse in Azure.

Understanding the supporting services

Regardless of the size you are planning for, some services might catch your attention anyway. This section will give you an overview of the supporting services that you might need in your modern data warehouse approach.

Requiring data governance

When you start building your modern data warehouse, there is one thing that you need to get right: data governance! There are too many data lakes out there that have mutated into **data swamps**, and you'll find so many Data Warehouses that have lost any credibility because their users can't find the right data or are only able to find outdated information. The relationship between these data sources might not be clear to self-service BI users if they produce the wrong figures. Plus, dimension data, which describes the measures that you want to report on, might not be up to date or might rely on the wrong sources. Alternatively, your users might not be able to recognize the date and time of the figures they are looking at.

Here, you will need to implement Data Governance measures and make them available to your users.

Understanding Azure Data Catalog Gen2

At the time of writing this book, Azure Data Catalog Gen2 hasn't been announced for Public Preview but will be available within the upcoming weeks.

With the new version of Azure Data Catalog, you have the option to create automated, recurring discovery runs for your storage accounts, Data Lake Stores, and DBs. These scans can be equipped with all kinds of patterns that you might want to discover and catalog. For example, you can set up patterns for recognizing product ID patterns, credit card numbers, birthdates, and so on. This is done using **regular expressions**, so there are nearly no limits in terms of the recognition patterns that you can implement. The discovery run will collect detailed information about the scanned data and will show you where in your modern data warehouse you have data sources that contain the product ID, credit card numbers, birthdates, and everything that you can formulate as a pattern and include in the scan job.

Data Catalog will make this information available in a portal where the user can search and analyze all the information that was collected during the discovery runs.

Data Factory is another source that can be scanned for metadata by Data Catalog. This will deliver **Data Lineage** information to your Data Catalog instance that can be analyzed together with the information that was collected from your modern data warehouse.

With its upcoming releases, Microsoft plans to integrate Data Catalog Gen2 with the Azure Synapse Studio. This will add another dimension of Data Governance to the Synapse Workspace. Imagine a browser-based suite where you can develop your data lake, your Data Warehouse DBs, the necessary Data Factory jobs, Spark notebooks, and SQL scripts, and then integrate all this with your metadata layer from Data Catalog.

We will examine Azure Data Catalog in *Chapter 14*, *Establishing Data Governance.*

Sharing data with Azure Data Share

With Azure Data Share, you can deliver data from your DBs, your storage account, or your data lake to your internal systems, as well as your external customers. You can do this completely code free just by going through some easy configuration steps. Here, you select the source data to share and the email addresses of the people you want to share your data with.

You can set up **Terms of Use** that need to be accepted by the recipient to accept your invitation.

A data share can be sourced from an Azure SQL DBs from a Synapse SQL Pool. This is known as in-place sharing. Following the update cycle that you provide during the configuration, the data will be updated in the share accordingly. You can also select folders or files to share. This works for both Azure storage accounts and Azure Data Lake Storage.

The shares that the recipient then sees are snapshot-based. Depending on the sharing policy that's been selected, the share will be updated incrementally either daily or hourly.

You, as the sharing party, have full control over the share. If the recipient does not follow your Terms of Use, you can revoke the share from the recipient.

We will touch on Azure Data Share in *Chapter 14*, *Establishing Data Governance.*

Next, we'll look at how to make sure our Azure services are kept secure.

Establishing security

In a modern, cloud-based environment, security is one of the major areas that needs to be done right. When we examine the options of the cloud vendors, we can see that they invest big time in this topic: neither they nor their customers can afford leaks.

Understanding network security

Azure supports you in implementing network security by providing a vast selection of offerings. **Virtual networks** are fundamental building blocks in this case. You can implement **Network Access Control** and **Azure Firewall** here, as well as **Name Resolution**. You can also build **perimeter networks**, also known as **Demilitarized Zones (DMZs)**, which are supported by **Azure Distributed Denial of Service (DDoS) Protection**, as well as routing support with **Azure Front Door** and a **DNS**-based traffic load balancer named **Azure Traffic Manager**. All these tools and functions help ensure that you can build a secure network to embed your modern data warehouse into a safe environment.

Understanding Azure Active Directory

Another important block in your security planning should be **identity management**. **Azure Active Directory** (**AAD**) will support you in implementing SSO, as well as offering MFA When we start diving into the different services during this book, you will find that **Azure Role-Based Access Control** (**Azure RBAC**) is available for most of them.

AAD will support you in your consumer identity and access management, as well as managing devices, application registrations, and application object and service principals, all of which form a similar concept to the service users that you would create to run services in your on-premises environment.

Understanding data encryption on Azure

When we examine data encryption on Azure, we distinguish between the following:

- **Data at rest**: This is data that is stored on any storage service, such as Azure storage accounts, Azure Data Lake Storage, Azure SQL DBs, or Synapse SQL pools, for example.

    Encryption is done by default for any data that is stored anywhere on Azure. This means that Microsoft will always use 256-bit **Advanced Encryption Standard** (**AES**) when you store data on your storage of choice. Microsoft will manage the keys for this encryption for you if you don't select a different strategy.

    If you prefer to **bring your own key** (**BYOK**) to encrypt your data, you can do so and store it either in the **Azure Key Vault** or an Azure Key Vault Managed **Hardware Security Module** (**HSM**).

For Azure storage account access with Blob storage, you can provide your key when you're accessing the storage. This means you must include your key in the request.

In the DBs on Azure, you will be provided with **Transparent data encryption (TDE)** technology that has already matured to a high level in Microsoft SQL Server.

Additionally, Synapse SQL pools can implement **Column-Level Encryption (CLE)**. You can use CLE to encrypt the selected columns of a DBs table.

Azure SQL DB offers another level of encryption called **Always Encrypted**. This feature will encrypt a column at rest on the fly. It enables an application to use the encrypted column in SQL-WHERE clauses. This avoids transporting a whole data table to the client so that it can be decrypted there. This happens before the filter clause is applied. This is called deterministic mode. There is also a randomized mode that offers even higher encryption.

- **Data on the fly**: This refers to data transfer, which means that the data is actually *flying*; that is, it's being transported over a network from a source to a target.

  TLS will be used in any connection that you open to your **Azure subscription**. This will prevent your data from being sent in readable format over the network that you're using.

We will touch on these security topics throughout this book.

Establishing DevOps and CI/CD

A modern data warehouse architecture cannot skip DevOps and CI/CD. **Azure DevOps** allows you to integrate all the artifacts that you've created in your different development environments.

DevOps pipelines help you enable automated workflows so that you can approve artifacts that have been developed. This includes automated testing and the final automated rollout of your artifacts to your production environment.

Of course, DevOps will support you in versioning, branching, merging, and all the planning and documentation tasks that you need to go through in your development project. DevOps can offer Scrum support and many additional project functions that will make development easier.

You will find more details about DevOps and CI/CD in the upcoming chapters, where we will cover the development process with the different tools in more detail.

Summary

In this chapter, we talked about requirements engineering and the importance of asking the right questions. The answers, once engineered in a structured approach, will be vital to deciding on the building blocks and tools that will be used to create our modern data warehouse.

We built a generic reference architecture for a modern data warehouse and examined the building blocks that form this approach. We also talked about their major functionalities.

We then learned about industry models for acceleration purposes and how they can help you kickstart your project.

Finally, we explored three suggestions for sizes and mapped different Azure data services to them. Additionally, you learned about the assisting services that will complete your architecture. And don't get me wrong, there are even more to discover.

What I want to express here is that there are many ways to tackle challenges in a modern cloud-based environment and that, as we mentioned previously, there is no silver bullet. We'll examine the Azure data services offerings together in the upcoming chapters – and, of course, have some fun in the Azure candy shop!

In the next chapter, *Understanding the Data Lake Storage Layer,* we will start diving into Azure Data Lake Store, how to provision it, and what functionality you can expect.

Questions

The following are additional questions from the *Asking in the right direction* section:

- **General questions**: Your modern data warehouse may need to hold data for a longer period. Do you need different access tiers (hot, cool, or archive)? Is older data not accessed that often? How do you need to design the access rights to the data? Are there only automatic processes, or will users want to access data themselves? Do you need to establish replication for reliability, and to what extent?

- **Data loading**: Are you planning for a new ETL/ELT tool? Do you want to run that in the cloud or on-premises? What are your expectations for the availability of connectors? What are your expectations for usability, or do you want to code your data transport layer? What language do you prefer for this? What will the volume of the data be that is to be transported? Do you need parallel processes? Do you expect scalability?

- **Data transformation**: Do you want to perform data transformation as part of the ETL/ETL process and within the tool, or do you prefer to program your data transformation using programming languages? Which languages does your team use?

- **Data consumption**: Do you have different user groups with different requirements? What tools will your users use to access data? Data scientists still access their source data in a different way, and with different tools, from most data analysts (from controlling departments, for example). Do you need to hide certain data from certain user groups? Are there user groups that need to access the same data pool but are only allowed to see rows and attributes that are related to their user group? Do you want to centrally control access to your data? What amount of data that landed in your system will be presented? Is this going to be highly aggregated before the presentation layer will hold it? Does the presentation layer need to perform aggregations? Maybe on the fly?

- **Machine Learning**: Are you already using ML tools and need to integrate them? Do you use specialized libraries that need to be available to the processes? What programming languages do you need to support? Do you need to perform live predictions on incoming data? Do you need to track and audit your ML models?

- **Reporting**: Do you need to integrate with existing reporting solutions? Are you providing static reports to your users? Do you need to provide interactive visualizations? Are you planning for self-service BI? How open do you want to design "self-service"? Do your users need to add their own data to existing data in the modern data warehouse? Do you need to embed visuals into applications? Are you using specialized visualizations? What are the access control requirements for your visualizations?

- **Networking/security**: Do you plan to establish virtual networking as part of the solution? Do you want to run everything on PaaS without virtual networking? What are your requirements when it comes to encryption key handling? Will you use BYOK? Are you planning to use a central repository for security credentials? Do you need to establish SSO? Are you planning to integrate with AD?

- **DevOps and CI/CD**: Are you already using DevOps in your company and need to integrate with it? Are you planning to establish automated deployments? To what extent are you planning automated deployments? Do you want to create mandatory approval workflows for deployments? Do you want to establish a mandatory second set-of-eyes strategy?

- **APIs/API management**: How do you need to establish access control in your APIs? Are you planning to provide different versions for different consumers with your APIs? What will be the expected range of response times for requests?

Section 2: The Storage Layer

This section provides a brief overview of the different storage components that can be implemented on an analytical system in Azure.

This section comprises the following chapters:

3
Understanding the Data Lake Storage Layer

One of the components in our modern data warehouse Architecture that spans over all the processing steps, and is available to all the participating services, is our **Data Lake Storage Layer**.

It will serve as our landing zone, the transient storage while we're transforming and cleaning our data, and a source for queries for the analytical components. Users will establish their sandboxes, processes will store their logs here, and Data Scientists will use it as their main playground, together with their tools.

In this chapter, you will learn how to set up **Azure Data Lake Gen2 Storage**, and also find suggestions about how to organize it so that you can flexibly react to any challenge. You will learn how to access data in Data Lake Storage, and you will look at some approaches for how to monitor your Storage Account. We will discuss options for backup and **disaster recovery** (**DR**) and examine the security and networking options for your Data Lake Store. Finally, we'll cover some hints and tips.

In this chapter, we'll cover the following topics:

- Setting up your Cloud Big Data Storage
- Organizing your data lake and why
- Implementing a data model in your data lake
- Monitoring your storage account
- Talking about backup
- Implementing access control in your data lake
- Setting the networking options
- Discovering additional knowledge

Technical requirements

There are not too many technical requirements for this chapter, but there are a couple! All you will need to follow this chapter is the following:

- An Azure subscription where you have at least contributor rights, or you are the owner
- The right to provision an Azure Storage Account with Hierarchical Namespace enabled

Setting up your Cloud Big Data Storage

When you want Azure Data Lake to provision a service within your Azure subscription and you search for it, you will find a service with that exact name. We are not going to use that service, as it is the Generation 1 version of Data Lake Storage and is only there for continuity reasons.

Microsoft added the Data Lake Store Gen2 in kind of a hidden fashion. As this new version of the Data Lake is based on the standard Azure storage account, you will provision it exactly like that: an Azure storage account with an option set for Azure Data Lake Gen2 or, even better, with the **Hierarchical Namespace** enabled. We will look at this option shortly.

Provisioning a standard storage account instead

When you examine the Azure storage account limits, you might ask yourself, *why should I go for a Data Lake Store when the standard Storage Account already has these wide volume boundaries and possibilities?*

Well, the standard storage account has it limits. When you enable Hierarchical Namespace on your storage account, you transform it into Data Lake Storage. You are adding the option to create any folder structure and hierarchy that you might need. In comparison, with the standard storage account, you can only create so-called containers, in which you can place blobs.

Sure, you can mimic folders in your Blob storage account by adding a virtual path to the names of your blobs. But this is not the same as a real folder structure like the kind that you can build with the Hierarchical Namespace. Here, you can add security to each folder and establish fine-grained access control. Renaming a certain folder can be done atomically and in a single transaction on the Hierarchical Namespace, whereas with Blob storage, you would need to repeat this for each file that is stored in a certain "folder."

But let's get hands-on and create a Data Lake Store resource.

Creating an Azure Data Lake Gen2 storage account

Let's proceed and head for the Azure portal at `https://portal.azure.com/`. Let's get started:

1. Once you have logged in, you should find a similar view to the one shown in the following screenshot on your home screen in the upper area; yours might be different from what you can see in this screenshot:

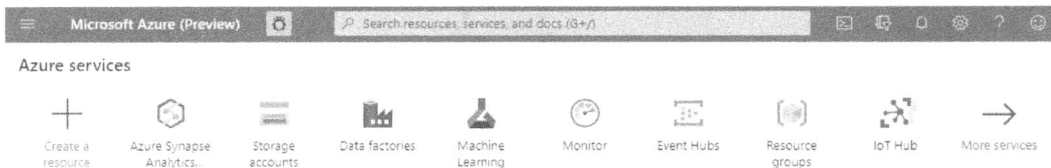

Figure 3.1 – Azure portal home screen

2. To provision a new resource, click on **Create a resource**, which is the first option on the left of all the icons you are presented with.

3. In the following **blade** (this is what all the dialogs and pages are called in Azure), you can either type `storage` into the search field and select **Storage account** directly from the result preview or kick off the search by hitting *Enter* on your keyboard and then selecting **Storage account** from the search result page. You will be taken to the start blade for storage account provisioning.

4. To kick off the provisioning sequence, hit **Create**. You will be presented with the following view:

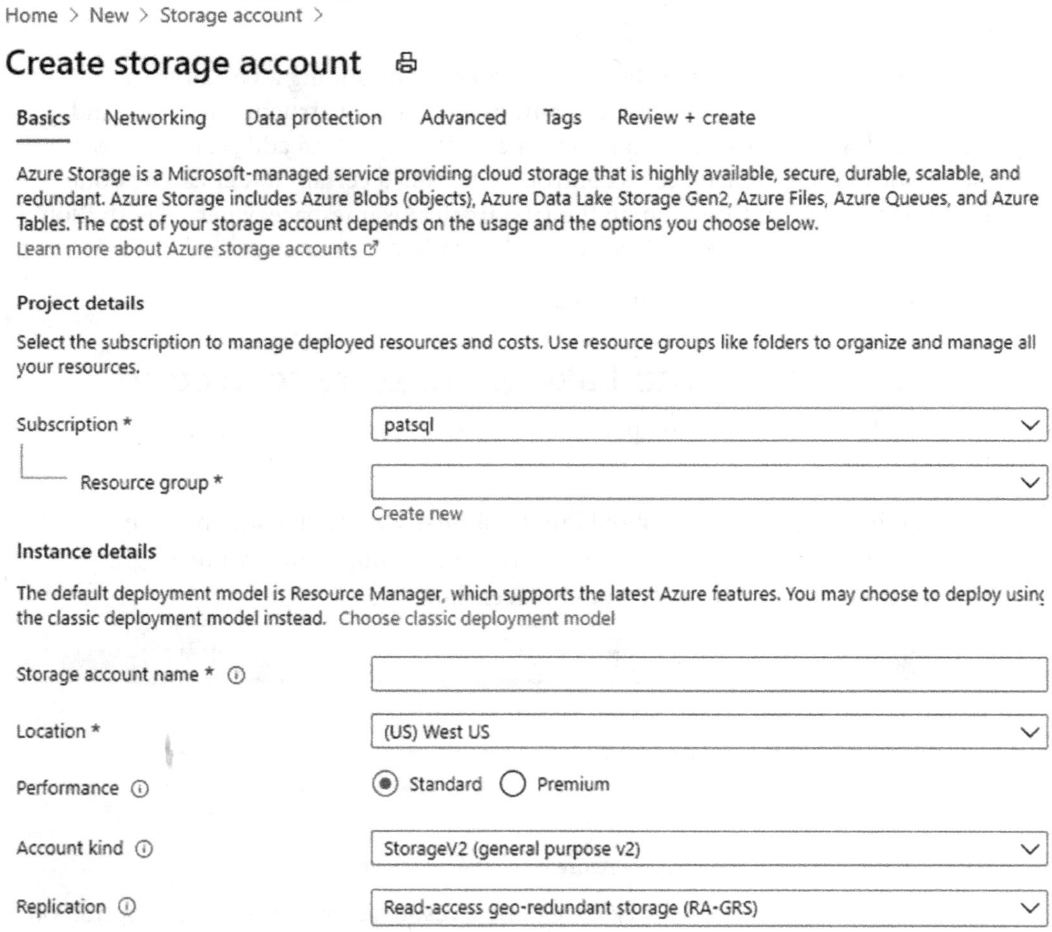

Home > New > Storage account >

Create storage account 🖶

Basics   Networking   Data protection   Advanced   Tags   Review + create

Azure Storage is a Microsoft-managed service providing cloud storage that is highly available, secure, durable, scalable, and redundant. Azure Storage includes Azure Blobs (objects), Azure Data Lake Storage Gen2, Azure Files, Azure Queues, and Azure Tables. The cost of your storage account depends on the usage and the options you choose below.
Learn more about Azure storage accounts ☐

Project details

Select the subscription to manage deployed resources and costs. Use resource groups like folders to organize and manage all your resources.

Subscription *	patsql ⌄
└─ Resource group *	⌄
	Create new

Instance details

The default deployment model is Resource Manager, which supports the latest Azure features. You may choose to deploy using the classic deployment model instead. Choose classic deployment model

Storage account name * ⓘ	
Location *	(US) West US ⌄
Performance ⓘ	⦿ Standard ◯ Premium
Account kind ⓘ	StorageV2 (general purpose v2) ⌄
Replication ⓘ	Read-access geo-redundant storage (RA-GRS) ⌄

Figure 3.2 – Starting blade – Create storage account

Let's examine the options that are available on this blade, from top to bottom:

- **Subscription**: The combo box will present all the Azure subscriptions that you have access to. Select the one that will hold your Data Lake.

- **Resource group**: Azure resource groups are containers that are used to group Azure artifacts semantically. If you are planning a modern data warehouse project, for example, you might want to put all the services into one resource group. They are a convenient way to allow access to a collection of services. And if the project has ended and you don't need the services anymore, you can delete the whole group in one go. You may already use naming conventions in your company, so maybe this one and the following will find their way into your conventions as well.

- **Storage account name**: This name must be unique in Azure as it will be used to create a URL to access this resource. This includes every storage account that anybody has created on Azure. The name for an Azure storage account needs to be at least 3 to a maximum of 24 lowercase letters or numbers.

- **Location**: Select your preferred Azure Data Center location in this combo box. You want to choose the data center that is closest to you, to ensure the lowest possible latency between you and your storage.

- **Performance**: For storage accounts, you can enable the **Premium** performance option, which will give you **Solid State Drive** (**SSD**) storage instead of magnetic drives when selecting **Standard**. With Premium, you will have even more performance for interactive workloads. But make sure that you understand the cost impact before you go down that lane.

- **Account kind**: Please select **StorageV2 (general purpose v2)** in this combo box.

- **Replication**: This option will control how your storage implements **High Availability (HA)** and **DR** features.

  a) **Locally redundant storage (LRS)** is the default for storage accounts, and means that the storage account will store your data synchronously three times on different hard drives in the Azure Data Center that you chose for the service:

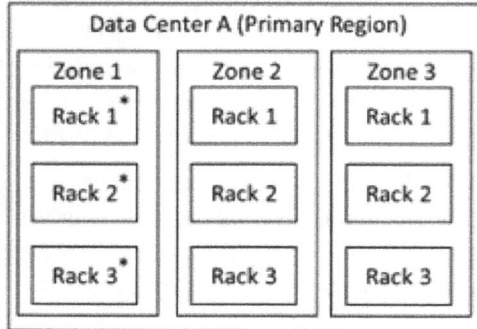

Figure 3.3 – Locally redundant storage

  b) **Zone-redundant storage (ZRS)** will store your data synchronously on three different hard drives, in three different physical locations, in the Azure Data Center that you chose for the service. This means that the data will reside on racks that are independent from each other and won't share a power supply, networking, or cooling:

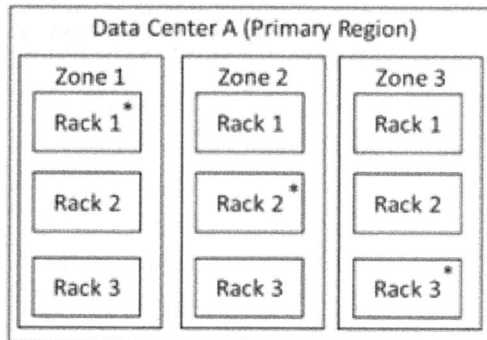

Figure 3.4 – Zone-redundant storage (ZRS)

  c) **Geo-redundant storage (GRS)** implements LRS for the primary region (Azure Data Center), and then replicates your data asynchronously to a single location in the secondary region:

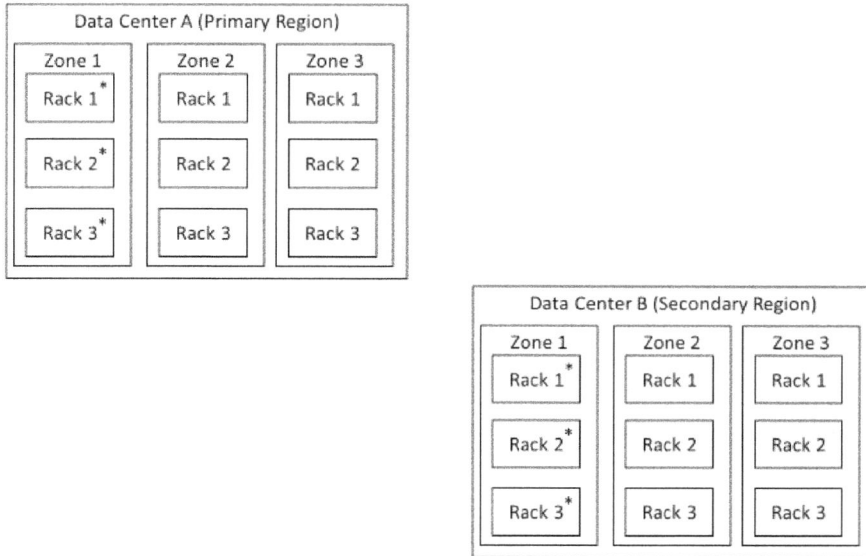

Figure 3.5 – Geo-redundant storage (GRS)

d) **Geo-zone-redundant storage (GZRS)** implements ZRS for the primary region, and then replicates your data asynchronously to a single location in the secondary region:

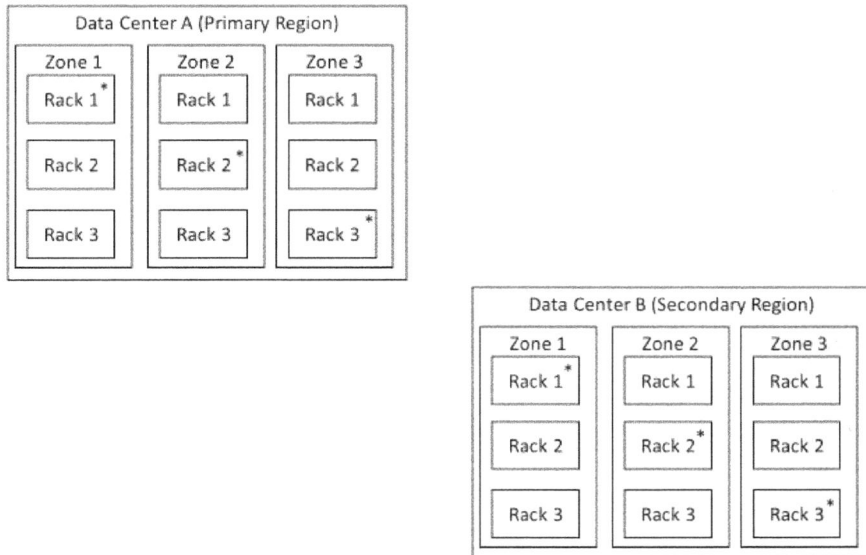

Figure 3.6 – Geo-zone-redundant storage (GZRS)

e) **Read access to Geo-redundant storage (RA-GRS)** will, additionally to the GRS functionality, provide read access to the secondary region.

f) **Read access to Geo-zone-redundant storage (RA-GZRS)** will, additionally to the GZRS functionality, provide read access to the secondary region.

5. After selecting **Next: Networking**, you will be taken to the second blade of the provisioning process:

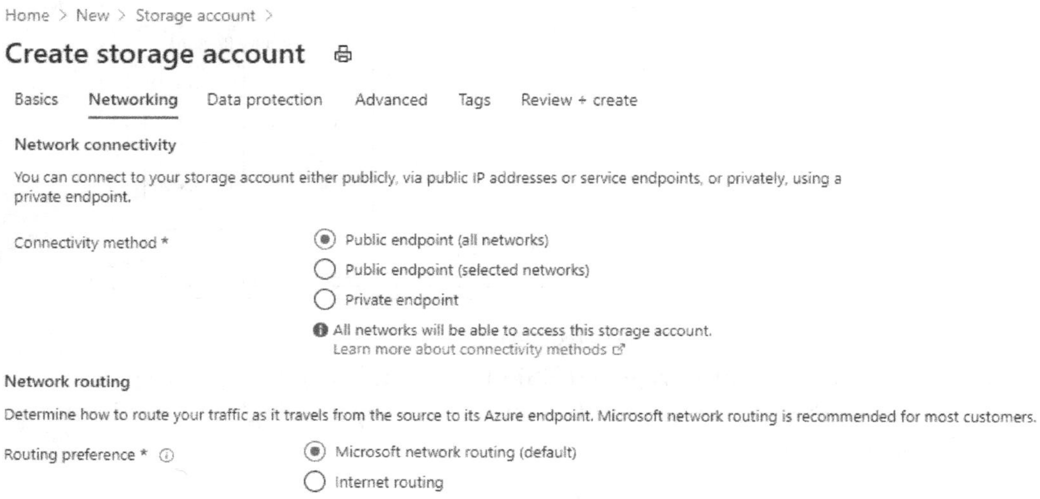

Home > New > Storage account >

Create storage account 🖶

| Basics | Networking | Data protection | Advanced | Tags | Review + create |

Network connectivity

You can connect to your storage account either publicly, via public IP addresses or service endpoints, or privately, using a private endpoint.

Connectivity method *
- (●) Public endpoint (all networks)
- (○) Public endpoint (selected networks)
- (○) Private endpoint

ⓘ All networks will be able to access this storage account.
Learn more about connectivity methods ⬈

Network routing

Determine how to route your traffic as it travels from the source to its Azure endpoint. Microsoft network routing is recommended for most customers.

Routing preference * ⓘ
- (●) Microsoft network routing (default)
- (○) Internet routing

Figure 3.7 – Networking blade – Create storage account

6. You need to select a connectivity method in the upper area:

a) **Public endpoint (all networks)**: By selecting this method, you will configure your data lake to be available for requests coming from any network.

b) **Public endpoint (selected networks)**: If you select this option, your Data Lake Storage will only be visible from within a **virtual Network (VNet)**. You will need to name the targeted VNet here. Alternatively, you can list the IP addresses or IP ranges that you want to allow in your firewall settings.

c) **Private endpoint**: Using private endpoints will narrow the availability of your Data Lake Storage down to a single IP within a VNet, and will route all traffic over the Azure backbone network to and from clients within that VNet instead of the public network.

7. From the lower area of the **Networking** blade, you can select the **Routing preference** for your Azure Data Lake Storage. This is the area where you can prioritize either network performance or cost efficiency.

   If you are connecting to Azure using your own Express Route, for example, you can use **Microsoft network routing (default)**. This will route all your traffic as quickly as possible to the Microsoft network, and will also give you better performance for communicating with Azure resources.

   The option for **internet routing**, on the other hand, will route all your traffic as closely as possible to the **point of presence (PoP)**, next to the Azure resource:

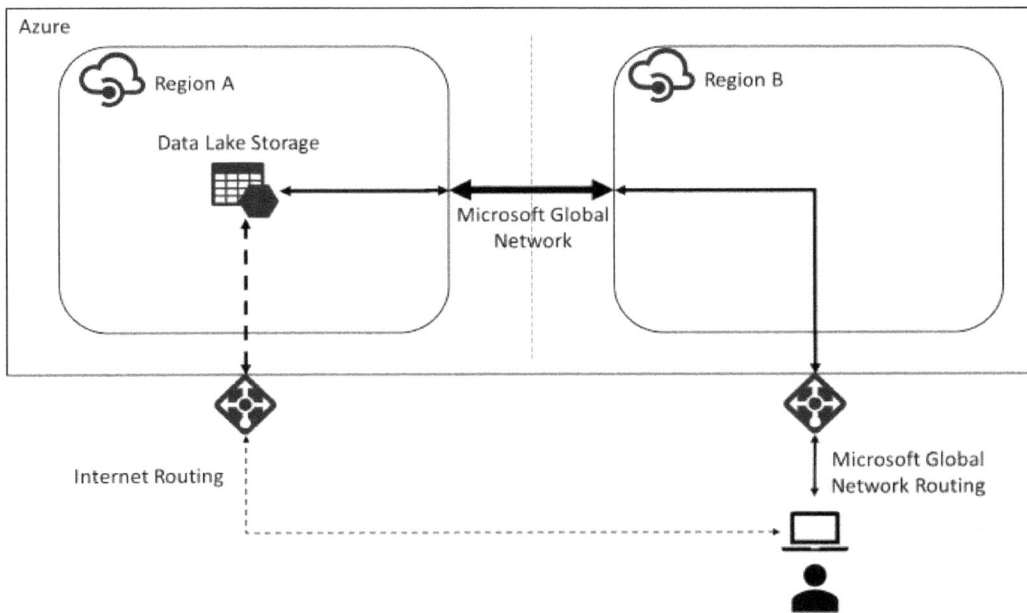

Figure 3.8 – Microsoft global networking versus internet routing

8. By selecting **Next: Data protection**, you will be taken to a blade where you will configure recovery and tracking options. At the time of writing this book, these options are not available with Data Lake Storage but are on the roadmap, right before **General Availability (GA)**. There is a high chance that some of these options will be available to you when this book gets published.

> **Important note**
>
> But for now, selecting one of these options will gray out the most important option on the next page, on the **Advanced** blade. Leave the **Data protection** blade as-is and proceed to **Next: Advanced**:

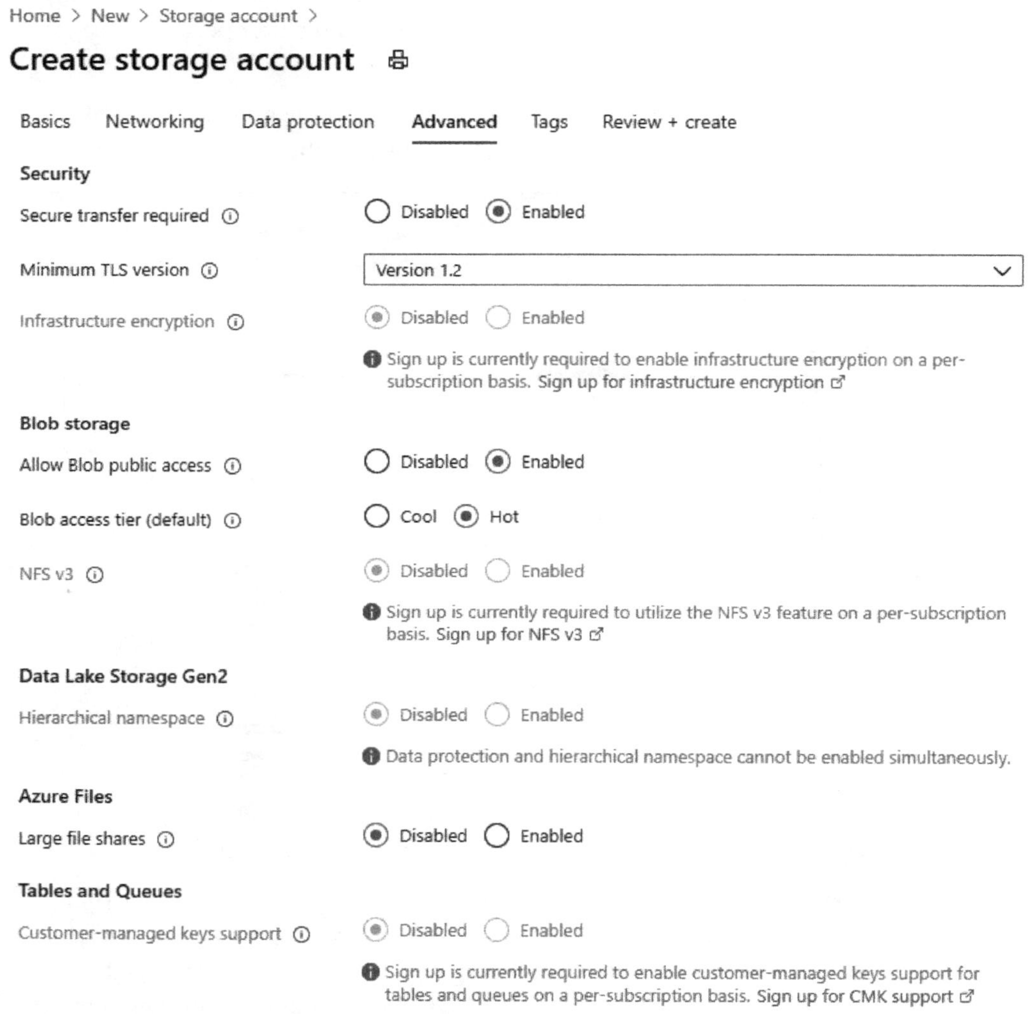

Home > New > Storage account >

Create storage account 🖶

Basics   Networking   Data protection   **Advanced**   Tags   Review + create

Security

Secure transfer required ⓘ   ◯ Disabled ◉ Enabled

Minimum TLS version ⓘ

Version 1.2   ⌄

Infrastructure encryption ⓘ   ◉ Disabled ◯ Enabled

ⓘ Sign up is currently required to enable infrastructure encryption on a per-subscription basis. Sign up for infrastructure encryption ↗

Blob storage

Allow Blob public access ⓘ   ◯ Disabled ◉ Enabled

Blob access tier (default) ⓘ   ◯ Cool ◉ Hot

NFS v3 ⓘ   ◉ Disabled ◯ Enabled

ⓘ Sign up is currently required to utilize the NFS v3 feature on a per-subscription basis. Sign up for NFS v3 ↗

Data Lake Storage Gen2

Hierarchical namespace ⓘ   ◉ Disabled ◯ Enabled

ⓘ Data protection and hierarchical namespace cannot be enabled simultaneously.

Azure Files

Large file shares ⓘ   ◉ Disabled ◯ Enabled

Tables and Queues

Customer-managed keys support ⓘ   ◉ Disabled ◯ Enabled

ⓘ Sign up is currently required to enable customer-managed keys support for tables and queues on a per-subscription basis. Sign up for CMK support ↗

Figure 3.9 – Advanced blade – Create storage account

9. The most important option on this blade is **Data Lake Storage Gen2 / Hierarchical namespace**. You should enable this if you want to use your storage account as a Data Lake Store and leverage the Hierarchical Namespace.

    But let's examine the other options:

    a) **Secure transfer required**: This will enable the TLS setting on the storage account. That means, if any connection is made to the storage account, for example, you use a REST API, you will need to use HTTPS. Using only HTTP will fail if the option is enabled. By default, this option is enabled.

    b) **Minimum TLS version**: Here, you can select the minimum TLS version that a client needs to use to interact with your storage account.

    c) **Infrastructure encryption**: At the time of writing, this feature can be activated by subscribing to the preview. It will add another layer of encryption over your data at rest in your storage account.

    d) **Allow Blob public access**: When this option is enabled, your storage account allows anonymous access to blobs using container **Access Control Lists (ACLs)**.

    e) **Blob access tier (default)**: Choose between either **Cool** or **Hot**. This determines the default access tier for all the objects that are stored with your storage account.

    f) **NFS v3**: At the time of writing, this feature can be activated by subscribing to the preview. It will enable the Network File System protocol, which will allow users to exchange and share files across networks. You will be able to mount an NFS v3 enabled storage account to your Windows or Linux client.

    g) **Hierarchical namespace**: This is the Big Mac function for our Azure Data Lake. It will enable big data analytics acceleration and allow you to control access at the folder and file level.

    h) **Large file shares**: Enabling this option will allow you to share up to 100 TB of files per account.

    i) **Customer-managed keys support**: At the time of writing, this feature can be activated by subscribing to the preview. It will enable customer-managed keys for queues and table storage. It does not directly relate to our Data Lake Storage requirements.

10. The **Next: Tags** button will bring you to the penultimate blade in the sequence.

    This blade will allow you to add key/value pairs to your resource, and will give you extended possibilities to track the usage of your resource when you are reporting and analyzing the consumption of your system.

    > Tip
    >
    > You should make use of tags whenever possible on Azure to allow for cost and usage analysis of your environment. If you plan this ahead of implementing the services, you can deliver detailed cost insights later.

11. When you select **Next: Review + create**, you will be taken to the final overview of your configuration settings, before you can create the service. This is your last opportunity to briefly check what you are going to create. At this point, you can still return to the previous blades and change your settings, if you've realized that you have misconfigured something.

12. The **Download a template for automation** link will take you to the template viewer/editor of the Azure portal. Here, you can work on the resources that you are going to provision. The **Azure Resource Manager** (**ARM**) offers an interface that can be controlled by providing JSON files that can be configured to deploy Azure resources in an automation framework.

    ARM provides PowerShell commands, a REST API, or the **Azure Command-Line Interface** (**CLI**) to automate nearly any step that can be done via the Azure portal, and it sometimes offers even more flexibility.

    > Important note
    >
    > If you have clicked the link now and you are wondering how to return to the blade where you can finally start provisioning the Data Lake, please go to the top-right corner, right beneath your username, and click the **X** button, *DON'T* click the **Back** button of your browser. It will take you somewhere else, not to the page you want.

13. To finish provisioning Azure Data Lake Storage Gen2, hit **Create** in the bottom-left corner of the summary blade. You will be transported to the provisioning view. It will update in a few seconds and provide you with a **Go to resource** button, as soon as Data Lake provisioning is finished:

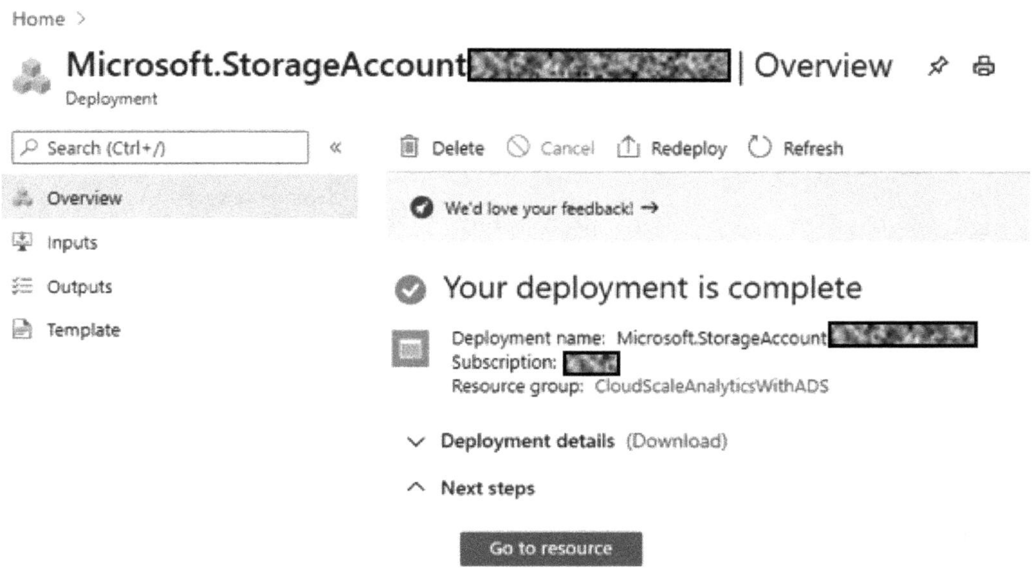

Figure 3.10 – Deployment complete – Azure Storage Account

14. Once the Data Lake has been created, you will need to perform one final step to be able to use it properly: you must create at least one so-called filesystem that acts as the root for the folder structure that you are going to implement. You can start this by navigating to the **Storage Explorer (preview)** entry in the navigation blade of your Data Lake Store. Right-click the **Containers** entry in the displayed navigator blade of the Storage Explorer, select **Create file system**, and give it a name:

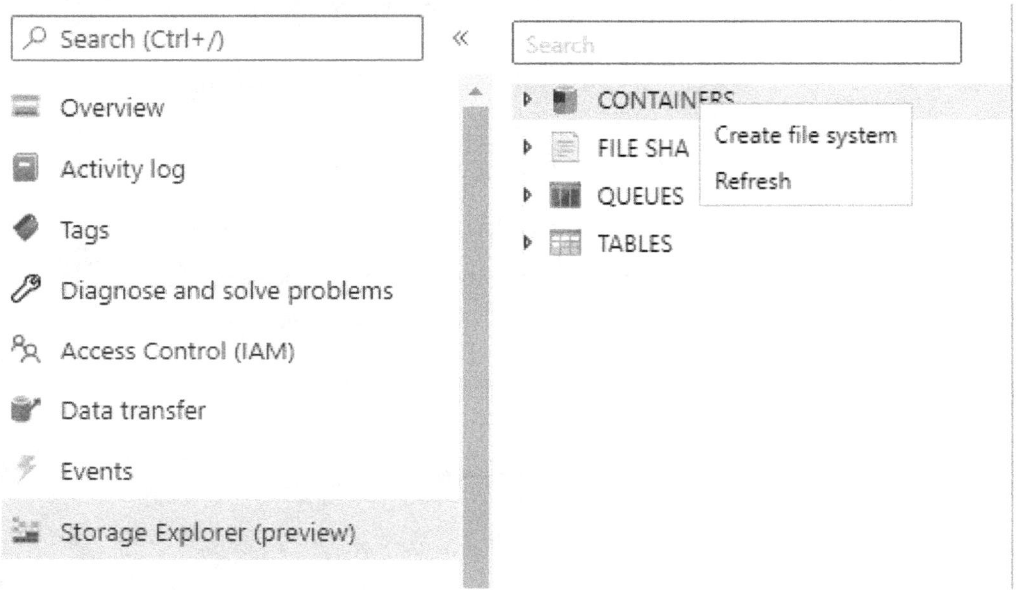

Figure 3.11 – Creating a filesystem in your new Data Lake Storage

You may wish to create several filesystems in your Data Lake Storage. They can be used to distinguish between semantical unrelated topics, for example. Other companies use only one data lake for their whole data estate. They use these filesystems to separate DEV, TEST, and PROD environments, and so on. Other companies even create different data lakes for this purpose, and divide the DEV, TEST, and PROD environments completely from each other.

Now that you've learned how to provision your storage, let's learn how to organize it.

Organizing your data lake

A well-structured system of zones/layers and folders will help you control your data lake. On the one hand, you will find a canonical approach that makes it easier to understand structures and the semantics behind these zones and folders. On the other hand, generator approaches will enable you to automate processes in your modern data warehouse.

Many Big Data projects suffer from poorly organized folder structures, and it becomes a challenge to find the right data for the right analysis at the right time. The so-called data swamp can be nearly impossible to use and will even demotivate users from leveraging the effort that must be put into it.

Talking about zones in your data lake

In *Chapter 1*, *Balancing the Benefits of Data Lakes over Data Warehouses*, we addressed the question of zones in a data lake. We compared them to the layers in a data warehouse and found that they are pretty similar, and mostly follow similar semantics:

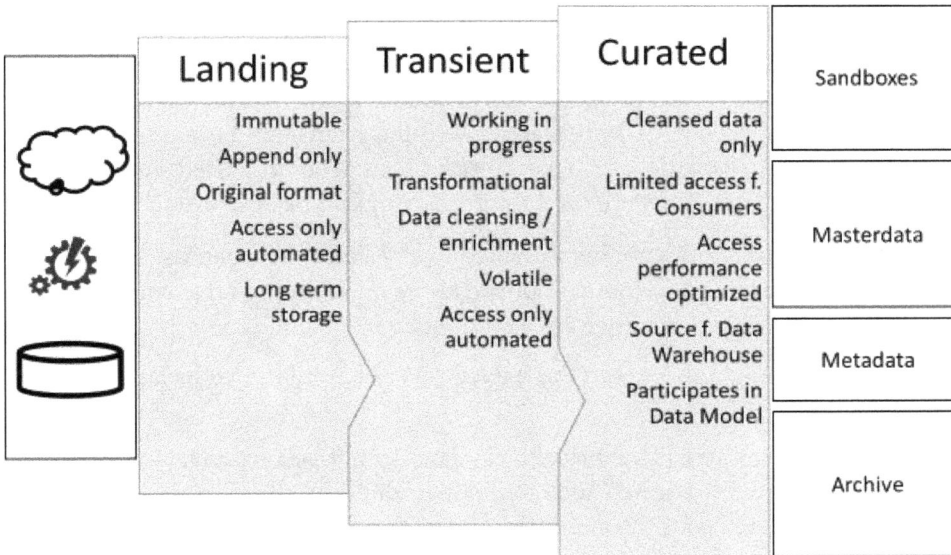

Figure 3.12 – Data lake zones

Let's examine the zones you may have in your data lake:

- **Landing zone**: This is the entry portal into your modern data warehouse. You will land every piece of data that you want to handle in your modern data warehouse in this zone. Many approaches for modern data warehouses make this zone **immutable**. This means that data that is loaded into this layer should be landed by automated processes only. Data is stored in the original format, as an exact copy of the source. The files that are stored here can't be modified by the data engineers or the Data Scientists that participate in the system. This means that this is an append-only storage option. Only administrators with a profound reason should be able to manipulate these folders.

  One big advantage of following this concept is that you will always be able to reproduce your modern data warehouse – partially or fully – based on the data that is stored in the landing zone.

  Other concepts might talk about the "Bronze" layer or similar, but most of the time, this refers to a similar concept.

- **Transient/processing zone**: This is where your Data Engineer's magic happens. Every transformational step that you have to apply to your data, every join that you need to process to add other sources to your datasets to transform raw data into information, is done using this area. We call it "transient" because the data that's stored here is volatile, and once a pipeline has been processed successfully, the intermediary files that were used for processing can be destroyed.

  The transient zone should not be presented to any staff other than the Data Engineer, and only for development purposes, or an administrative engineer for root cause analysis in productive environments.

  Other concepts might talk about the "Silver" layer or similar; this usually refers to a similar concept.

- **Curated zone**: This area is the place to get data from if you are a Data Scientist. When we compare the curated zone to the two zones discussed in the preceding paragraphs, this area concentrates more on facts and dimensions, whereas the landing zone and the transient zone will look at the data from the source angle and it will be organized to reflect that.

In this zone, you will provide data that has been cleansed from incomplete data, which means that missing values are added or at least replaced by a default that you have set. Inconsistent data will be corrected. Names such as product names, for example, will be normalized, and typos should be corrected. Data is checked, for example, for duplicate rows to be eliminated. Lookup values are either added or already represented by keys in such a way that they can easily be found when you use them to join to the related dimension. If, for example, you plan to implement a data vault in this area, you must add the necessary **hash-keys** to your data.

The Curated Zone will deliver data to consumers such as Data Scientists and will be the source for populating the presentation layer, as we discussed in *Chapter 2, Connecting Requirements and Technology*.

It is still a good idea to limit personal access to this layer to only allow users that will understand the data structures and handle bigger amounts of data with their tools.

As this is the first delivery layer, you will want to store the data in an optimal way so that accessing and querying it can be done at the most performant level. Many implementations prefer the `.parquet` file format as it represents data in columnar format. This will allow for fast scans and aggregations, as well as the concept of **column projection**. This means that a processing engine will only read the columns needed for the required query. All other columns are ignored when the file is read. In comparison, when you use a row-oriented file format, you will always need to read the whole dataset with all the columns first, regardless of whether you only want to have two of the columns in your dataset.

- **Analytical sandboxes**: These are mainly used by Data Scientists when they are following a new thesis or creating a new model. Usually, this area has a lower level of governance and acts more like a playground for development.

- **User/drop zones**: These are folders that are created for every user that actively participates in your modern data warehouse. This means, for example, the Data Scientists will need an area where they can do their personal experiments and play around. Users that need to provide data to your modern data warehouse can make use of drop zones to upload files that are then used in various processes.

- **Archive**: This is meant as a special area, where you will explicitly move your files and archive them for long-term storage. Together with storage tiering, this will enable you to save money on storage.

- **Master data**: You might want to provide master data to be used with the processes during cleansing and enriching your data on its way into the curated zone. Be aware of possible redundancy with master data that you have stored in your Data Warehouse. You will need to set up organizational measures to keep them in sync.

- **Metadata**: If you need to collect and provide metadata about your environment, you will need this area. Logging is done here, for example. You can further divide this into different substructures for security- and governance-related topics, and so on.

These are some suggestions and examples of how your top-level structures can be organized. And, of course, this is not written in stone. If you decide on other structures and find that you don't need most of these zones, leave them out. This is *your* modern data warehouse architecture. However, make sure to engineer your zones properly to avoid major redesigns later. Think twice, cut once!

Creating structures in your data lake

Once you have decided on the necessary zones that you want to implement, you can start the real work and decide on the detailed folder structures. You'll want to cover all *subject areas*, for example, that can occur in your company. And, by looking at your subject areas, you will need to define *access boundaries* so that you can control who can see what data and what that person do can with it. When we examine this further, you will find the need to reflect on the *data life cycles* in the structure for your data lake. Some files are stored temporarily, but others will need to be persisted. The *storage tier* is another topic that you might want to reflect in your structure, and *criticality* and *confidentiality* need to be displayed when you create your structures.

Talking about subject areas

Subject areas are your top-level entry points. They will represent your source systems, for example, when we are talking about raw data. When you are proceeding from left to right in your modern data warehouse structure, meaning from raw to curated, you will move more and more from source-related structures to the topics that you want to cover in your modern data warehouse.

For example, you will load data from your Enterprise resource planning system, your warehouse information system, and your sales data into your modern data warehouse. Maybe you will even need to capture and "union" data from different organizations within your company. There can be different systems that will send in customer data. In your landing zone, you will create folders for all the source systems your customer data is coming from. Throughout the loading process of your modern data warehouse, you will consolidate your data from the source system-related folders into another folder structure. This will be in the curated zone and display customer data. Here, you will either have discriminators in your data to understand which source system the information has been loaded from, or the details about your customers held in the different columns of your customer object.

Planning access boundaries

When you're designing the folder structure for your modern data warehouse, you may want to reflect access boundaries so that you can implement access control to different data areas more easily. Azure Data Lake Store Gen2 enables you to create **ACLs** based on Azure Active Directory groups. This means that when you are planning your structures carefully, you can add and remove users without manipulating your folder structures and file access rights repeatedly.

> **Note**
> We will cover ACLs and security questions in more detail later in this chapter.

You may start with the top-level folder structures, for example, when you're implementing different organizations of your company or different semantic areas that consolidate organizations, and so on.

You will come to a point where you will need to accept compromises, such as in the preceding example, when you needed to decide whether to put an organization or a topic at the top level of your structures, and in what sequence you should nest the different layers.

Here are some examples for different approaches to folder structuring:

You might look at the `Subject Area` first, then look at the `Data Source`:

```
Subject Area
++ Data Source
++++ Object
++++++ Data Loaded
...
```

But this can be seen differently:

```
Subject Area
++ Organization Unit
++++ Data Source
++++++ Object
++++++++ Data Loaded
...
```

Alternatively, you could put it like this:

```
Data Source
++ Subject Area
++++ Organization Unit
++++++ Object
++++++++ Data Loaded
...
```

Now that you have look at examples for the entry levels for your data lake structures, let's examine the leaf level and the approach you should follow there.

Planning the leaf level

You might have recognized that there were no time portions mentioned in the previous section. This is because we won't be using time slices on upper-level folders. This is because if we are applying partitioning to our data, when we read it from the data lake, we will, in almost every case, use date and time fragments for our partitions. When we put the date and time parts too high into our folder structures, we won't be able to easily read subject-oriented data in one step.

Imagine that you want to read all the customer interactions for the last 2 years in a folder structure like the following:

```
YYYY
++ MM
++++ DD
++++++ Sales
++++++++ D365
++++++++++ CustomerInteraction
++++++++++++ Callxxxxx.parquet
```

You would need to read a lot of folders to collect all the necessary data, right?

If we turn around the sequence and put date and time down to the leaf level, this could result in far fewer folders; alternatively, when you want to enable reading partitioned data, you can leave access to the CustomerInteraction folder to the reading compute engine, which can even use partition pruning, for example, when we read .parquet files. This means the compute engine can decide which folders of the date structure it is going to read based on the required dataset. If you have 10 years' worth of data but you need to process the actual year only, a Spark engine can, for example, omit all the other folders and only return the needed data:

```
Sales
++ D365
++++ CustomerInteraction
++++++ YYYY
++++++++ MM
++++++++++ DD
++++++++++++ Callxxxxx.parquet
```

We'll now check out how to understand data life cycles.

Understanding data life cycles

You want to reflect the retention time of your data when you are planning your data lake structure. Temporary files and folders should be kept apart from persistent data.

Azure Data Lake Store Gen2 implements the same data life cycle support that the storage account (Blob storage) offers. You can implement **life cycle policies** either from the Azure portal or by using PowerShell, the Azure CLI, or REST APIs.

By implementing life cycle policies, you can do the following:

- Shift files from the Hot tier to Cool, or from the Cool tier to Archive, depending on their recent usage, for example, if they have not been accessed within a certain timeframe.

- Shift files from Cool to Hot when they are accessed. This will improve performance when a file is read.

- Automatically delete files after a certain timeframe, set by you. This can be a very important function for enforcing your GDPR policies but might also be useful as a garbage collector function.

- Set rules to be run once per day at the top storage account level.

- Implement rules for folders.

You can find the **Lifecycle Management** option on the navigation blade of your Data Lake storage account:

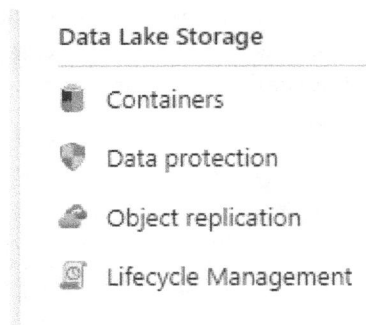

Data Lake Storage

Containers

Data protection

Object replication

Lifecycle Management

Figure 3.13 – Finding the Lifecycle Management option in the navigation blade

Let's create a policy for moving files that are older than 120 days from Hot to Cool:

1. When you navigate to **Lifecycle Management** and click **+ Add a rule**, you will be taken to the following blade:

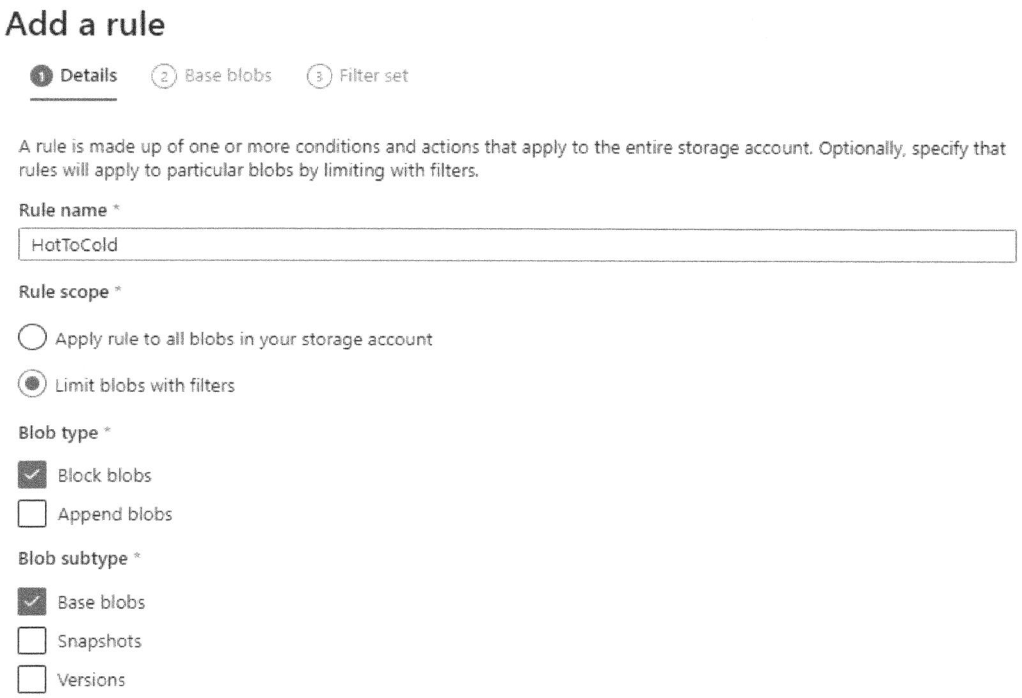

Add a rule

① Details   ② Base blobs   ③ Filter set

A rule is made up of one or more conditions and actions that apply to the entire storage account. Optionally, specify that rules will apply to particular blobs by limiting with filters.

Rule name *

> HotToCold

Rule scope *

◯ Apply rule to all blobs in your storage account

◉ Limit blobs with filters

Blob type *

☑ Block blobs

☐ Append blobs

Blob subtype *

☑ Base blobs

☐ Snapshots

☐ Versions

Figure 3.14 – Creating a life cycle rule – step 1

2.   Give it a name and select the necessary options. Let's limit the selected files using filters. When you click **Next**, you will be able to select the action that you want to perform on the files. You can enter your threshold and select the action you want to perform:

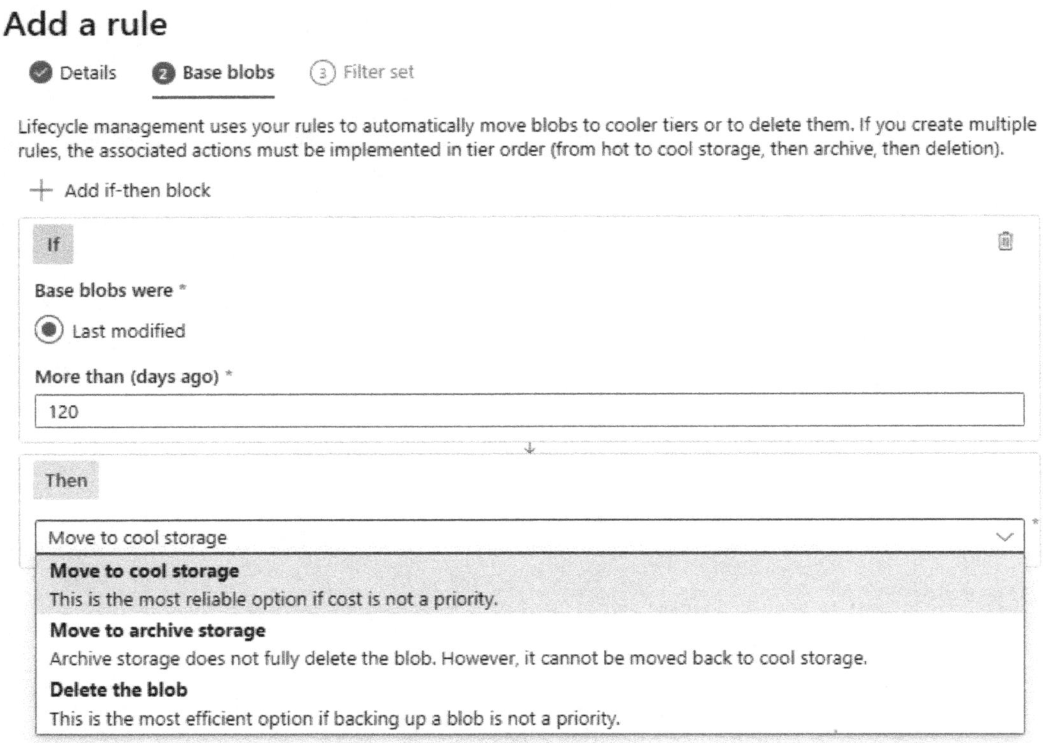

Add a rule

● Details   ② **Base blobs**   ③ Filter set

Lifecycle management uses your rules to automatically move blobs to cooler tiers or to delete them. If you create multiple rules, the associated actions must be implemented in tier order (from hot to cool storage, then archive, then deletion).

╋ Add if-then block

If	🗑

Base blobs were *

◉ Last modified

More than (days ago) *

```
120
```

Then

Move to cool storage	⌄

Move to cool storage
This is the most reliable option if cost is not a priority.

Move to archive storage
Archive storage does not fully delete the blob. However, it cannot be moved back to cool storage.

Delete the blob
This is the most efficient option if backing up a blob is not a priority.

Figure 3.15 – Creating a life cycle rule – step 2

3.   On the next blade, we will set the filters for our file selection:

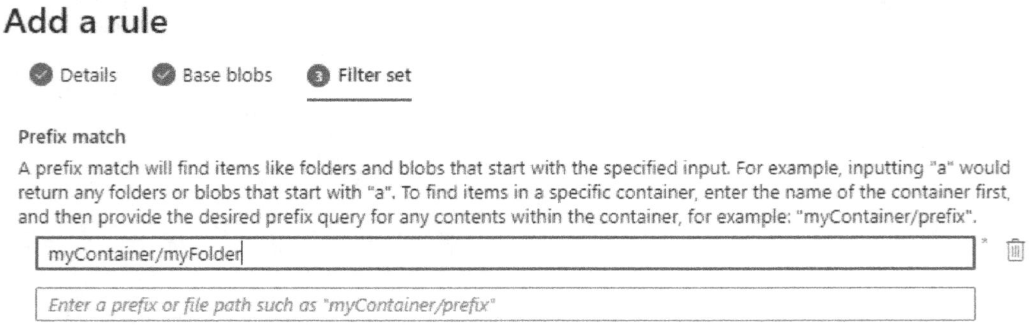

Add a rule

● Details   ● Base blobs   ③ **Filter set**

Prefix match

A prefix match will find items like folders and blobs that start with the specified input. For example, inputting "a" would return any folders or blobs that start with "a". To find items in a specific container, enter the name of the container first, and then provide the desired prefix query for any contents within the container, for example: "myContainer/prefix".

```
myContainer/myFolder
```
🗑

Enter a prefix or file path such as "myContainer/prefix"

Figure 3.16 – Creating a life cycle rule – step 3

4. Finally, you'll find all your rules in the **Overview** blade:

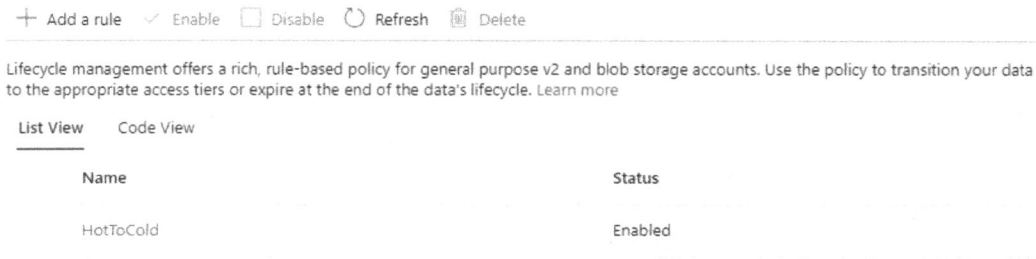

$+$ Add a rule   ✓ Enable   ☐ Disable   ⟳ Refresh   🗑 Delete

Lifecycle management offers a rich, rule-based policy for general purpose v2 and blob storage accounts. Use the policy to transition your data to the appropriate access tiers or expire at the end of the data's lifecycle. Learn more

List View   Code View

Name	Status
HotToCold	Enabled

Figure 3.17 – Life cycle rules in Data Lake Storage Gen2

This is really handy for taking care of aging data. Otherwise, you would need to either set up costly movement jobs that move data to other folders or run scripts that set the storage tier from Hot to Cool. Just leave it to your Data Lake Store.

Investigating storage tiers

You will find three different storage tiers that you can implement for the different requirements of your data lake. Deciding on the storage tier you wish to use will impact the performance of your storage when writing and reading to and from your data lake. They will also give you different price tags for the storage that you use. Let's take a look:

- **Hot**: This tier will give you the fastest access to your files for reading and writing. This can be set either at the account level for the whole Data Lake Store or at the file level. If you explicitly set the tier for a certain file, this will overwrite the default that the file may have inherited from the account level setting.

  Cost-wise, the Hot tier is optimized for accessing the files frequently. This means that reading and writing from and to a Hot tier is cheaper, but the cost of storage in the Hot tier is higher.

  Use this tier for data that you write to your Data Lake and that needs to be read frequently and repeatedly.

- **Cool**: The Cool tier will give you a better storage cost but will be more expensive when the files are accessed. This tier starts paying off if you need to store files for more than 30 days, without accessing them regularly.

  You can set the Cool tier at the account level for the whole Data Lake Store, or at the file level. If you explicitly set the tier for a certain file, this will overwrite the default that the file may have inherited from the account level setting.

Use this tier for data that you don't need for frequent analysis that is not read too often. This might be aged-out data that is only read once a year for a certain report, or once every quarter or so. Don't use this tier for regularly used files as it will drive up your costs (although storage costs are very low in the Azure offerings, we should still keep an eye on them).

- **Archive**: If you have identified data that you will need to keep for legal, organizational, or other reasons, but you don't need to actually access it, this is the right tier for you. The archive tier takes your files offline, so they won't be accessible before you hydrate them again for access. This means the process in the background will get the files for you from a permanent storage and make them available again. Files that you put in the archive tier need to stay in this mode for at least 180 days (or be deleted earlier). This tier will give you the cheapest storage mode for your data.

The metadata of your files will still be available to you, but as we mentioned previously, the data can't be read. If you need the files to be put back into a readable tier, you will need to hydrate them, which can take several hours. For smaller files, you can request a high priority rehydrate and may get them back faster.

Use this tier for long-term backups, for example, or original data from the raw layer that you don't need to access anymore but still need to keep, such as files that you need to keep for legal or compliance reasons.

Planning for criticality

When you plan and set up your structures in your Data Lake Storage, you might want to distinguish different levels of criticality to be able to configure these areas differently. If data is not very critical to your business and can be regenerated, it might not need to be part of an extended backup mechanism, along with your highly important "diamonds and pearls." This should be reflected in your structures and areas. As we mentioned previously, storage is cheap, but still, you want to keep an eye on this factor and not to back up unnecessary stuff.

Setting up confidentiality

Publicly available data will be handled differently from highly sensitive data that displays the key assets of your business, right? When you are generating your data lake structures, please keep an eye on the confidentiality of the content that you put there. As we've already discussed, the different folder structures and angles that you can use to view the data should be set up so that can control the sensitivity of the files in an easy and manageable way. Try not to mix different levels of confidentiality under the same folder. This will end up being a mess and leave your admins scratching their heads.

For example, you shouldn't place a widely usable folder into a sensitive area:

```
Root
++ NotClassifiedFolder
++++ HighlySensitiveDataFolder
++++++ ExternalWeatherDataFolder
++++++ YourSalesDataFolder
```

> **Note**
>
> In this case, try and put `ExternalWeatherDataFolder` somewhere else to avoid chaos in folder management when you're managing access rights.

It might be a good idea to not place widely accessible data beneath a sensitive classified folder, as in the previous example. You want to hide certain folders from most of your user groups. In this case, this won't be possible, as the users that need access to `ExternalWeatherDataFolder` at least need to "see" the `HighlySensitiveDataFolder` folder to be able to access the weather data. It doesn't matter if they can access the other data that is placed here. But this can cause curiosity! And you want to establish a "principle of least privilege," where everybody has the rights that they really need to do their job. This includes hiding even the existence of confidential data from users who shouldn't see it.

We will examine **Role-Based Access Control (RBAC)** and **ACLs** later in this chapter.

Using filetypes

There are a multitude of different filetypes that you will be confronted with when you are working in an agile and flexible company. You will need to process CSV data, as well as JSON files. You might be confronted with the `.avro` format, which is thought to be useful for larger datasets, as well as the `.parquet` format, which acts as the file format representation of column store technologies and is similar to the RCFile or ORC formats. So, which filetype should you use?

When you look at the source data that you store raw or in your landing zone, you might want to store this data in the format that it was provided from the source. One of the reasons for doing so is to not to lose any information about the gathered data. Even if there are missing values or only parts of the data were provided, this is information that might be of interest.

For further processing, `.parquet` is a very handy file format that is already widely used in Data Lake projects. It offers not only good compression rates via its different compression algorithms, but also the capability to partition your files when you write them to the storage. You can even ignore partitions during reads, when the reader uses filters while accessing the files. This is called **partition pruning**, and it will speed up reading.

Another advantage of the columnar storage strategy of `.parquet` files is that you can only read the columns that are needed. This is called *column projection* (see the previous section about zones; that is, *Talking about zones in your data lake*) and will improve your reading speed drastically.

> **Tip**
> Don't mix filetypes in one directory. You will add unnecessary filtering to the filetype when reading from your folders.

Next, we'll discuss how to actually use a data model in a Data Lake.

Implementing a data model in your Data Lake

You can argue that this should already be done in your Data Lake Storage. But if you examine the capabilities of the services when it comes to interconnectivity and the options to query data directly from the `Data Lake` without loading it into a database, we can set up a hybrid data model and not only use these functionalities to save a lot of money, but also reduce complexity and loading time windows in your data warehouse.

The rising data volumes that you might face can impact your loading and query strategy heavily. And even with the Azure backbone network, you will come to a point where the latency for loading your data into your data warehouse will no longer be sufficient.

Understanding interconnectivity between your data lake and the presentation layer

Later in this book, we will dive into the Synapse SQL pools (*Chapter 4* and *Chapter 11*) and Synapse SQL on-demand (*Chapter 11*). These databases – or better, SQL services – will make up a bigger part of your presentation layer. They can query the Data Lake data directly with **Transact-SQL (T-SQL)** (this the **Microsoft SQL implementation**).

We already touched on Polybase and the COPY statement for SQL Pools (the provisioned database in Synapse). SQL on-demand is provided so that you can access file-based data with SQL and deliver views, for example, that can be accessed by all kinds of reporting/dashboarding engines and other consumers that are capable of using SQL. And yes, of course, Spark is also there and can deliver a similar experience. But, in many cases, the SQL world is better integrated with visualization tools such as Power BI.

So, why not integrate the curated zone of your data lake with the presentation layer of your architecture and benefit from the technology's advantages? The modeling technique you choose to implement – whether that is a Star-Join schema or something else – to join your data warehouse data to your Data Lake Storage does not make a big difference here. But you need to consider how and where you are creating the relationships between your files and the tables in the presentation layer.

Examining key implementation and usage

Your data integration routines will need to take care of generating the keys in the right place and, if necessary, the suitable performance for your lookup routines to provide the correct keys to your data. As we are operating in a hybrid environment, leaving the key generation process to a database table identity column with a default that counts up as rows are inserted, may no longer do the trick for you. When we expect larger amounts of data, it would be necessary to have the keys already available or generated in the data lake.

You might want to think about an area where you can store the identifier from your source data, together with a surrogate key that you can create with a suitable algorithm. Of course, you can use a sequence here as well.

In terms of the Data Vault methodology, there is an interesting approach for implementing relationships with a key that won't need a lookup or an additional join when you want to add it to data that flows into your system. In a Data Vault, you can decide to use a hash key that is generated based on the identifying attributes of your source data. You might use the MD5 or SHA-1 algorithm to create your hash. But be aware: there is a statistical possibility that you will run into a hash collision. This is less probable with SHA-1, but there is no absolute guarantee that this will not happen at all:

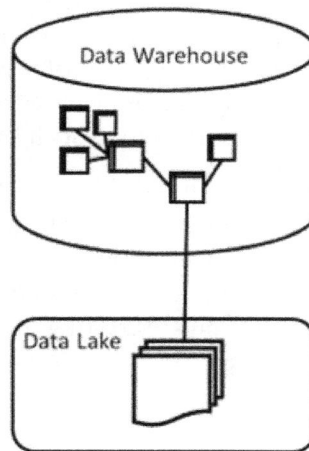

Figure 3.18 – Hybrid data estate

The next section will deal with monitoring your storage account.

Monitoring your storage account

There are different ways to keep track of your Data Lake Store on Azure. When you navigate to your storage account in the Azure portal, you will find overview charts about input and output, latency, and requests directly on the **Overview** blade. When you scroll down in the **Navigation** blade on the left, you will find more detailed views when you select **Insights** from the **Monitoring** section. You will find a subsection for **Alerts**, **Metrics**, and predefined (but configurable) **Workbooks** with predefined visuals for your storage account.

However, if you need deeper insights, or if you want to track who is doing what on your data lake, this is not displayed here. You might want to integrate your data lake with Azure Monitor and deep dive into the events.

> **Note**
> This feature is in preview at the time of writing. You can enroll in the preview on the Data Lake documentation page (`https://docs.microsoft.com/en-us/azure/storage/blobs/monitor-blob-storage`) if it is still running. At the time this book is published, this should be available.

Creating alerts for Azure storage accounts

When you want to track the events on the storage account blades of your data lake accounts, you can navigate to the **Monitoring** entry in the **Navigation** blade.

Here, you can start creating **Alert Rules** that will fire when a specific event occurs. Let's take a look:

1. After initiating the creation process, you will taken to the configuration blade for the alert. Here, you will have the option to create a detailed set of conditions that can be monitored and selected with conditions, as shown in the following screenshot:

Configure signal logic   ✕

Choose a signal below and configure the logic on the next screen to define the alert condition.

Signal type ⓘ	Monitor service ⓘ
All ⌄	All ⌄

Displaying 1 - 20 signals out of total 20 signals

🔍 Search by signal name			
Signal name ↑↓	**Signal type** ↑↓	**Monitor service** ↑↓	
Availability	〜 Metric	Platform	
Egress	〜 Metric	Platform	
Ingress	〜 Metric	Platform	
Success E2E Latency	〜 Metric	Platform	
Success Server Latency	〜 Metric	Platform	
Transactions	〜 Metric	Platform	
Used capacity	〜 Metric	Platform	
All Administrative operations	🖵 Activity Log	Administrative	
Abort Account HnsOn Migration (Microsoft.Storage/storageAccounts)	🖵 Activity Log	Administrative	
Account HnsOn Migration (Microsoft.Storage/storageAccounts)	🖵 Activity Log	Administrative	
Restore blob ranges (Microsoft.Storage/storageAccounts)	🖵 Activity Log	Administrative	
Approve Private Endpoint Connections (Microsoft.Storage/storageAccounts)	🖵 Activity Log	Administrative	
Storage Account Failover (Microsoft.Storage/storageAccounts)	🖵 Activity Log	Administrative	
List Storage Account Keys (Microsoft.Storage/storageAccounts)	🖵 Activity Log	Administrative	
Regenerate Storage Account Keys (Microsoft.Storage/storageAccounts)	🖵 Activity Log	Administrative	
Revoke Storage Account User Delegation Keys (Microsoft.Storage/storageAccou...	🖵 Activity Log	Administrative	
Delete Storage Account (Microsoft.Storage/storageAccounts)	🖵 Activity Log	Administrative	
Returns Storage Account SAS Token (Microsoft.Storage/storageAccounts)	🖵 Activity Log	Administrative	
Returns Storage Service SAS Token (Microsoft.Storage/storageAccounts)	🖵 Activity Log	Administrative	
Create/Update Storage Account (Microsoft.Storage/storageAccounts)	🖵 Activity Log	Administrative	

Figure 3.19 – Signals that can be monitored with alert rules

2. After selecting a metric or an entry from the Activity log, you can configure the logic for your signal:

Configure signal logic   ✕

Define the logic for triggering an alert. Use the chart to view trends in the data.

← Back to signal selection

Ingress (Platform)
The amount of ingress data, in bytes. This number includes ingress from an external client into Azure Storage as well as ingress within Azure.

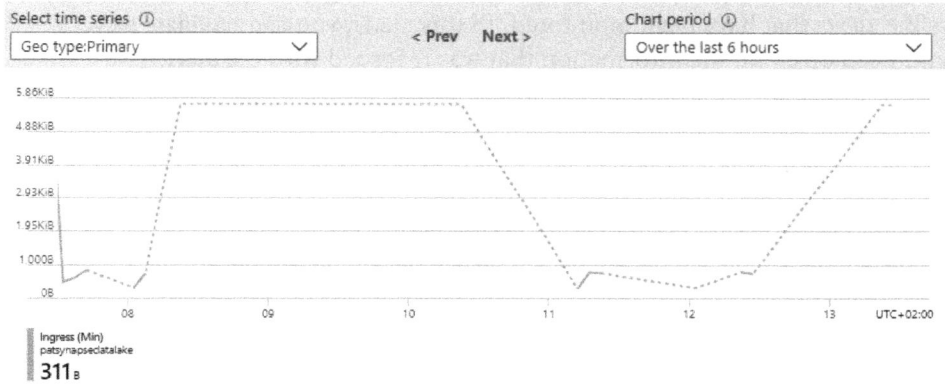

Select time series ⓘ		Chart period ⓘ
Geo type:Primary ⌄	< Prev Next >	Over the last 6 hours ⌄

```
5.86KiB
4.88KiB
3.91KiB
2.93KiB
1.95KiB
1.000B
   0B
        08          09          10          11          12          13    UTC+02:00
```

Ingress (Min)
patsynapsedatalake
311 B

Split by dimensions

Use dimensions to monitor specific time series. If you select more than one dimension value, each time series that results from the combination will trigger its own alert and will be charged separately. About monitoring multiple time series ⓘ

Dimension name	Operator	Dimension values	
Geo type ⌄	= ⌄	All current and future values ⌄	Add custom value 🗑
Select dimension ⌄	= ⌄	0 selected ⌄	Add custom value

ⓘ Monitoring 1 time series ($0.1/time series)

Alert logic

Threshold ⓘ

(**Static**   Dynamic)

Operator ⓘ	Aggregation type * ⓘ	Threshold value * ⓘ	Unit * ⓘ
Greater than ⌄	Minimum ⌄	50	KiB ⌄

Condition preview

Whenever the minimum ingress is greater than 50 Kibibyte

Evaluated based on

Aggregation granularity (Period) * ⓘ	Frequency of evaluation ⓘ
5 minutes ⌄	Every 1 Minute ⌄

Done

Figure 3.20 – Configuring signal logic

3. In this step, you will create an action group for your alert rule. You can select a **Notification type** (Email Azure Resource Manager Role or Email/SMS message/Push/ Voice) and which user to notify when this **alert rule** is satisfied. After that, select an **Action type** (Automation Runbook, Azure Function, ITSM (a ticketing solution), Logic App, Secure Webhook, or Webhook) and the related artifact with the type. Give your rule a rule and set a severity. Finally, you can tag your rule; for example, you can tag all the Azure artifacts that you can create in the framework.

In the **Alerts overview** blade, you will presented with an overview of the severity levels and possible alerts that have been monitored. In this blade, you can navigate to your details and deep dive into the information that was recorded with the alert:

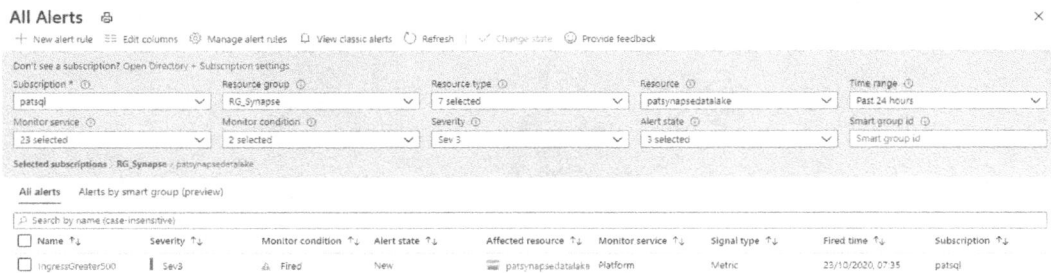

Figure 3.21 – Alerts list

That should cover the basics of account monitoring. Next, let's talk about backups.

Talking about backups

In every well-designed system, you want to have your data lake secured against data losses. As we saw at the beginning of this chapter, while creating our Data Lake Storage Gen2 account, there are different levels of redundancy that you can configure for your storage. But replication is only one facet of preventing data loss.

Configuring delete locks for the storage service

To avoid your storage service from being deleted by accident (yes, this can happen!), you must place a **delete lock** on your storage account. This will cause Azure to block any delete action and prompt the user to first get rid of the delete lock. Follow these steps to set this up:

1. Navigate to the **Locks** entry in the **Settings** section of the **Navigation** blade of your Data Lake storage account.

2. Here, you can add a lock of the **Delete** type to your account.

You can also set the whole account to **Read-only** if you need to.

Backing up your data

At the time of writing, there are many features that can help prevent data loss either in preview or on the roadmap to be delivered right around the corner. One of them is **immutable storage**, which is already in preview and can be whitelisted (at the time of writing! When this book is published, this should be available). Unfortunately, features such as **soft delete** are not available yet. Soft delete allows you to set a retention time for deleted files. If you accidentally delete a bunch of files, you can do an "Ooops-recovery" if you are within the retention time.

Until soft delete becomes available, you will need to take other measures to keep your data. One option would be to set up a copy routine that will copy the files that you need to save to another storage, where it can't manipulated and can be restored. This could be another Data Lake Storage in the Cool or even Archive tier, depending on the criticality and the **Recovery Time Objective** (**RTO**) that is intended for the files.

You could, for example, create a PowerShell script that implements a tool called **AZCOPY**. You can find download links for the tool at `https://docs.microsoft.com/en-us/azure/storage/common/storage-use-azcopy-v10`. You will also find the documentation for the tool here.

Alternatively, you might implement a copy pipeline with Azure Data Factory and create a recurring trigger there. We will cover Azure Data Factory in *Chapter 5*, *Integrating Data into Your modern data warehouse*.

Now that you have created your Data Lake, thought about its structure, and all those administrative tasks, let's dive into its security features.

Implementing access control in your Data Lake

Azure storage accounts implement different ways to control access to content that is stored there:

- **RBAC**
- **ACLs**
- Shared Key authorization
- **Shared Access Signature** (**SAS**) authorization

Understanding RBAC

To give access to a user, group, service principal, or a managed identity using RBAC, the user or the application needs to be managed by **Azure Active Directory (AAD)**. Implementing RBAC will use a so-called permission set that is put together as a role that a security principal can be assigned to.

When RBAC is assigned to Data Lake Storage, this will always be at the top level of the account or the filesystem. This means that the user or the application will have access to everything that is stored in the account or in the container that access has been granted to.

The following roles can be used to grant access to data in a data lake:

- **Storage Blob Data Owner**: This role will give you unlimited access to folders and data. It will enable the security principal (user or app) to change the owner of a file or folder and apply ACLs to all files and folders.

- **Storage Blog Data Contributor**: This role will have read and write access and can delete files and folders. It can't change the owner of files and folders, but it can apply ACLs to the files and folders owned by the user or app with the role assigned.

- **Storage Blob Data Reader**: With this role assigned, a user or app can read files and list the content of folders.

Assigning just **Owner** or **Contributor**, **Reader**, or **Storage Account Contributor** will not provide access to the data that is stored in your data lake. These roles will enable a user or app to manage a particular Data Lake storage account.

> **Important note**
>
> Owner, Contributor, Reader, and Storage Account Contributor can't access data directly, but they can access the storage keys! These can be used to access the data from client tools with using them in the URI to your Data Lake Storage Gen2. This is one example of the "principle of least privilege." So, don't assign these roles too easily.

You can assign 2,000 RBACs in one subscription. They can be set in the Azure portal using the **Access Control (IAM)** entry in the **Navigation** Blade, at the top:

1. After clicking **+ Add**, you will be presented with a combo box.

2. Select **Add role assignment**.

3. On the right-hand side of the blade, you will see another blade called **Add role assignment**. Select the applicable role from the combo box at the top.

4. Then, select the service principal type (user, group, or service principal). This combo box is called **Assign access to**.

5. Finally, you can search for a particular user or app by typing its name in the **Select** field.

6. Select what you are searching for and click **Save** at the bottom of the blade.

Understanding ACLs

ACLs will give you far more detailed control over the folders and files of your Data Lake Store. You can use them in POSIX style (this is **read**, **write**, and **execute – rwx –** remember?) on the structures.

To give access to a user, group, service principal, or a managed identity using ACLs, the user or the application needs to be managed by AAD. You can use the Azure Storage Explorer (as shown in the following screenshot) to set the needed ACL. Other possibilities would be using PowerShell or the Azure CLI. There is also a REST API that you can use, or .NET, Java, or Python.

The Azure Storage Explorer can be downloaded from `https://azure.microsoft.com/en-us/features/storage-explorer/`. There is a preview feature in the Data Lake Storage account's **Navigation** blade, in the top section, that will open a blade in the browser and display the Storage Explorer view:

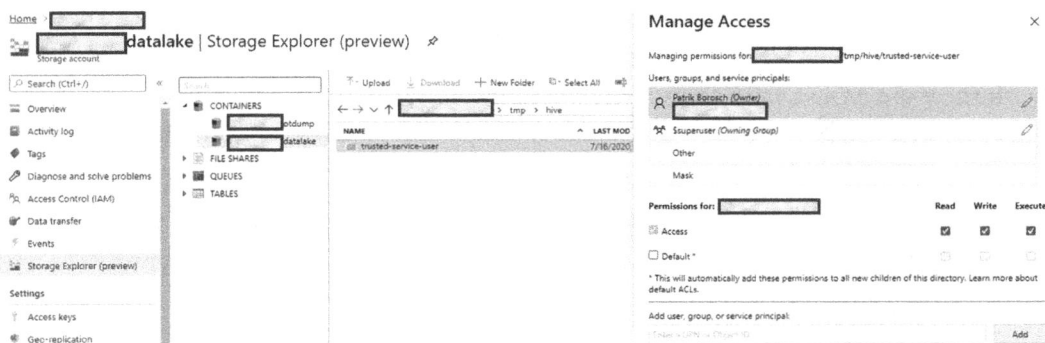

Figure 3.22 – Assigning ACL using the Storage Explorer preview feature

When you are assigning ACLs to files and folders, you will need to assign the **Execute** (X) permission from the root filesystem down to the folder. This contains the files you want to grant access to. Don't worry about the word "Execute;" this does not imply anything in the context of a Data Lake Storage account.

Let's say you want to allow a group of users to append data to a file that exists in the following location:

```
/Switzerland/Schaffhausen/Niklausen/data.txt
```

You will need to apply the following ACL for the folders and the file:

```
/--X/--X/--X/RW-
```

You can apply 28 ACLs per folder and per file in your data lake.

Understanding the evaluation sequence of RBAC and ACLs

When a user or an application attempts to access data in your data lake, authorization is evaluated in the following order:

1. RBAC assignments are checked. When the service principal (user or app) has an RBAC assigned to it, this will have the higher priority over ACLs that might be assigned.

2. When the RBAC is sufficient to access the intended data, the operation ends here, and no ACLs are evaluated any further.

3. When the RBAC is not sufficient for accessing the intended data, the ACLs are checked, if available.

4. If there are ACLs available and they are sufficient for accessing the data, the operation will be allowed. If not, the user or app will receive an "Access denied" message and the attempt will end.

If, in our preceding example, the user had the RBAC role **Storage Blob Data Contributor**, the detailed ACLs would not be needed; they would be able to append data anyway.

But if the user in the preceding example does not have an RBAC role assigned, they would still be able to append their data when the ACLs are set in the aforementioned way.

Understanding Shared Key authorization

Using Shared Key authorization will provide the incoming user or app with full authority over Data Lake Storage. This means that Shared Key authorization will always provide superuser rights to the user or app.

Shared Key authorization will not need a security principal, so an identity will be attached to the activities in that context.

You can use a Shared Key, for example, in a Databricks notebook:

```
spark.conf.set( "fs.azure.sas.<YourFileSystem>.<YourDataLake
StorageAccount>.blob.core.windows.net", "DefaultEndpoints
Protocol=https;AccountName=<YourDataLakeStorageAccount>;
AccountKey=<YourAccountKey>;EndpointSuffix=core.windows.net")
```

You will find your account keys when you navigate to the **Navigation** blade of your Data Lake Storage and use the first option in the **Settings** section: **Access Keys**.

> **Important note**
> These two keys will provide superuser access to your Data Lake Storage. This means that anyone who knows one of them will be able to change the ownership of your Data Lake Storage and manage all RBACs and ACLs.

Understanding Shared Access Signature authorization

SAS authorization is another way to allow access to your Data Lake, without the need for an Azure Active Directory identity. But compared to Shared Key authorization, you can decide on the rights that a user or application will have while using the SAS token.

When you examine the **Navigation** blade of the Data Lake Storage account, you will find the **Shared Access Signature** entry in the **Settings** section. The following screenshot the options that you have when you create a SAS token:

A shared access signature (SAS) is a URI that grants restricted access rights to Azure Storage resources. You can provide a shared access signature to clients who should not be trusted with your storage account key but whom you wish to delegate access to certain storage account resources. By distributing a shared access signature URI to these clients, you grant them access to a resource for a specified period of time.

An account-level SAS can delegate access to multiple storage services (i.e. blob, file, queue, table). Note that stored access policies are currently not supported for an account-level SAS.

Learn more

Allowed services ⓘ
☑ Blob ☑ File ☑ Queue ☑ Table

Allowed resource types ⓘ
☐ Service ☐ Container ☐ Object

Allowed permissions ⓘ
☑ Read ☑ Write ☑ Delete ☑ List ☑ Add ☑ Create ☑ Update ☑ Process

Blob versioning permissions ⓘ
☑ Enables deletion of versions

Start and expiry date/time ⓘ

| Start | 23/10/2020 | 🗓 | 21:41:13 |
| End | 24/10/2020 | 🗓 | 05:41:13 |

(UTC+01:00) Amsterdam, Berlin, Bern, Rome, Stockholm, Vienna ⌄

Allowed IP addresses ⓘ
for example, 168.1.5.65 or 168.1.5.65-168.1.5.70

Allowed protocols ⓘ
◉ HTTPS only ○ HTTPS and HTTP

Preferred routing tier ⓘ
◉ Basic (default) ○ Microsoft network routing ○ Internet routing
❶ Some routing options are disabled because the endpoints are not published.

Signing key ⓘ
key1 ⌄

Generate SAS and connection string

Figure 3.23 – Creating a SAS token

Once you have configured your options, click **Generate SAS and connection string**. Take note of the information that is displayed after clicking this button! It will disappear and you won't be able to get it back; you will need to create another one.

Important note

Microsoft recommends using RBAC and ACLs with AAD identities wherever possible. If you, for any reason, need to use key-based authentication such as SAS, you should create a user delegation SAS by using AAD credentials to ensure security: `https://docs.microsoft.com/en-us/azure/storage/common/storage-auth-aad`.

The topic in this chapter is networking.

Setting the networking options

While generating your Data Lake Storage account, you must set networking options. You have three options there and, according to your choice, you can implement Azure storage firewalls, virtual networks, and private endpoints with your Data Lake Storage.

Allowing access from all networks will cause the Data Lake Storage to be "visible" to everybody. You won't limit any network addresses, so you will need to secure the data lake with other measures, such as RBAC and ACLs. And don't forget, anybody with a Shared Key or a valid SAS will be able to reach your data lake as well.

Understanding storage account firewalls

You might want to consider setting up firewall rules to limit traffic to your data lake so that only IP ranges and addresses that you know have permission. Let's take a look:

1. When you examine the **Navigation** blade of your Data Lake Storage, you will find the entry for **Firewalls and virtual networks** in the **Settings** section:

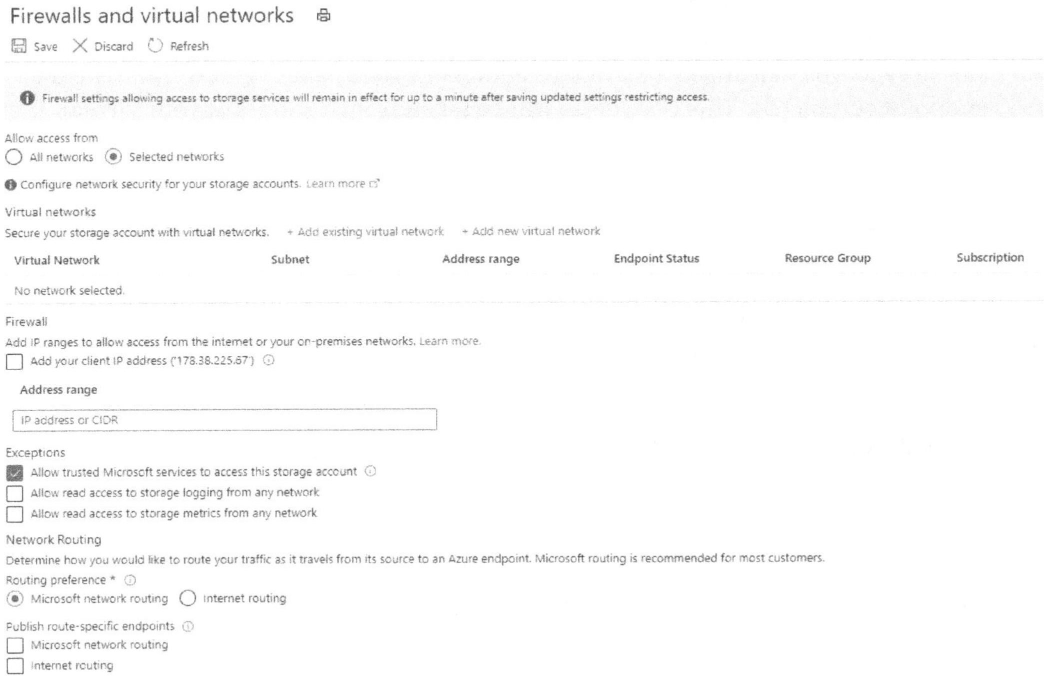

Figure 3.24 – Firewalls and virtual networks

2. By selecting **Selected networks**, you will enable the view shown in the preceding screenshot.

3. Now, in the **Firewall** section, you can start entering IP addresses or **Classless Inter-Domain Routing (CIDR)** ranges.

Now, let's learn how to add virtual networks to Azure.

Adding Azure virtual networks

Once you have implemented Azure virtual networks that hold other components of your modern data warehouse architecture, you may want to close down the access to your Data Lake Storage even more. You can limit traffic to and from your data lake to those virtual networks in the **Virtual Network** section (see *Figure 3.24*).

By doing so, you can eliminate traffic from the internet, but you will still have a public endpoint that is potentially open to visitors.

Using private endpoints with Data Lake Storage

By implementing a private endpoint within a virtual network and assigning it to your Data Lake Storage, you can attach a private IP address to your data lake and eliminate the public endpoint completely. This IP address will be taken from a subnet of your virtual network.

From this point on, you will have closed your data lake completely to your own virtual networks and their peered neighbors. Traffic will be routed over the Microsoft backbone network and will be completely hidden from the public.

You can find **Private endpoint connections** right below the **Firewalls and virtual networks** item in the **Navigation** blade of your Data Lake Store.

Discovering additional knowledge

The following is some advice that you might find useful.

Do:

* Plan for security from day one: Where are your trade-offs between security and usability?

* Enforce as much discipline as needed, but not more than is really necessary. Your data lake needs to serve your Data Scientists, as well as other communities in your company. Your modern data warehouse needs some agility.

- Structure your zones clearly and stick to the plan. If you need to redesign, don't do so in your already started structure.

- Implement a Data Catalog (we will talk about this in *Chapter 14, Establishing Data Governance*) to enable easy data discovery.

- Integrate with DevOps for a controlled and repeatable system.

Don't:

- Don't mix different formats. Always stick to one single file format per folder. You will often want to read all the files in a folder in one go.

- Don't forget naming conventions!

Summary

In this chapter, we talked about one of the main components of any modern data warehouse architecture: Data Lake Storage. You learned how to provision Data Lake Storage Gen2 on the Azure portal.

We discussed how to organize a data lake from different angles and examined the zones and folder structures that you will need to implement for efficient usage. We also learned how to implement a data model in Data Lake Storage.

After that, we looked at the administrative side of the Data Lake storage account and talked about monitoring, backups, access control, and networking.

The Data Lake Storage account is not the only component in your modern data warehouse Architecture that will hold data for analysis. Stay tuned for the next chapter, where we will shed some light on the relational database components that we can add to the architecture.

Further reading

For more information about the topics that were covered in this chapter, consult the official Azure documentation:

- Data Lake Storage scenarios: `https://docs.microsoft.com/en-us/azure/storage/blobs/data-lake-storage-data-scenarios`

- Best practices for using Data Lake Storage Gen2: `https://docs.microsoft.com/en-us/azure/storage/blobs/data-lake-storage-best-practices`

- Security recommendations for Blob storage: `https://docs.microsoft.com/en-us/azure/storage/blobs/security-recommendations`

- Using private endpoints for Azure storage: `https://docs.microsoft.com/en-us/azure/storage/common/storage-private-endpoints`

- Azure Data Lake Storage query acceleration: `https://docs.microsoft.com/en-us/azure/storage/blobs/data-lake-storage-query-acceleration`

4

Understanding Synapse SQL Pools and SQL Options

In this chapter, you will learn what **Massively Parallel Processing** (**MPP**) means in terms of a cloud PaaS database service. You will examine the concepts of distributing and replicating data in a database. Furthermore, you will see how to manage the workload in this database to your benefit by avoiding early scaling and leveraging all the performance capabilities of the service. Partitioning will extend your options when it comes to massive amounts of data when you need to grow from terabytes to petabytes. You will also learn how to load data efficiently into your database service.

Finally, we will have a look at the next evolutionary steps of the SQL pools in Azure Synapse and other SQL components, such as SQL on-demand compute.

At the end, we will compare other SQL database services and their options in Azure and how they may fit into your architecture.

You will find the following sections covered in this chapter:

- Uncovering MPP in the cloud – the power of 60
- Provisioning a Synapse dedicated SQL pool

- Talking about partitioning
- Implementing workload management
- Scaling the database
- Loading data
- Understanding other SQL options in Azure

Uncovering MPP in the cloud – the power of 60

SQL Synapse Analytics – the MPP database service formerly called Azure SQL Data Warehouse – is already well known. It has already reached its second generation and the third evolutional step is right around the corner. We will see Gen 3 in 2021. So, maybe at the time when this book is published, it will already be available, at least as a preview.

When we examine Synapse Analytics, we find some interesting concepts. Sure, the idea of MPP is not new and it has been implemented already with several appliances out there. The formerly known Microsoft **Parallel Data Warehouse** (**PDW**), now called **Analytics Platform System** (**APS**), https://docs.microsoft.com/en-us/sql/analytics-platform-system/home-analytics-platform-system-aps-pdw?view=aps-pdw-2016-au7, is one of them. There are some quite successful on-premises solutions from other vendors on the market – Teradata, Greenplum, and DB2 UDB being just a few of them.

With MPP, in comparison to **Symmetric Multiprocessing** (**SMP**) that standard databases implement, we find another concept, which is built for performance on huge amounts of data. See *Figure 4.1* for a comparison of the architectures:

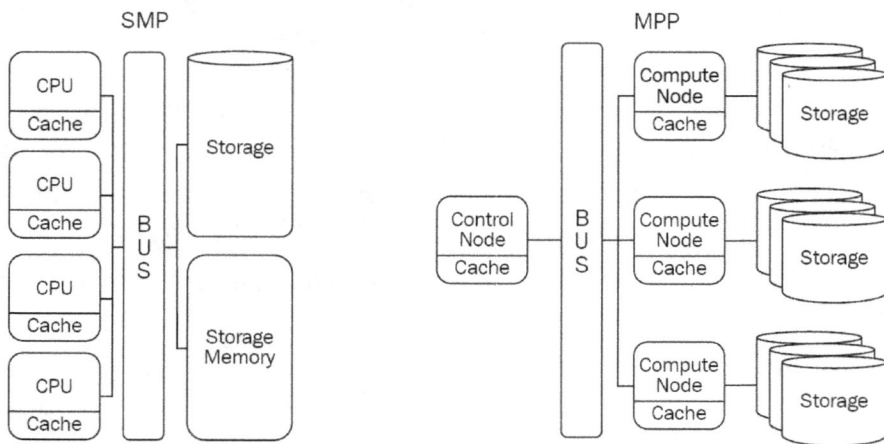

Figure 4.1 – SMP versus MPP

In the MPP database, we find a **control node**, a variable amount of **compute nodes**, and as we are on a cloud platform, Azure Storage is decoupled to allow separate scaling of the system. MPP implements a *share nothing* approach with regard to the storage that is attached to the workers.

Understanding the control node

The control node is the point of communication with the whole system. You, as a user, will only see this one and it will represent the database service to you. It is the orchestrator for all tasks and will also distribute and dispatch queries to the configured compute nodes. They will run all the queries in parallel against the distributed data.

The control node "knows" which data is stored where and will dispatch the work accordingly. This is done using the distributed query engine that is a big part of the heart of the service.

When queries are processed, the control node will always be the last instance that will catch the results of the participating workers (see the *Understanding compute nodes* section below). It is then responsible for the consolidation of the several result sets and the answer to the querying user or application.

Understanding compute nodes

Compute nodes are the processing engines in this architecture. They receive query tasks from the control node and process them in parallel (remember the first P in MPP) on their assigned storage area on their part of the data, which is called distribution.

There will always be 60 distributions and they will always be evenly assigned to the actual number of provisioned compute nodes. At this point, Synapse Analytics can scale from 1 to 60 compute nodes.

Understanding the data movement service

As you will see in the following sections about how data is distributed over the databases, there will be a need to move data between the distributions during query processing. When data is loaded to the service or is moved between compute nodes, this service will manage these tasks.

Understanding distributions

When you provision Synapse Analytics later in this chapter, you will create 60 databases in the background without seeing them or being able to control them individually. When data is loaded to Synapse Analytics, it will either be distributed to those 60 databases or be written directly to a special area of the underlying storage account. See the *Distributing, replicating, and round-robin* section for details about data distribution and what data goes to which storage option.

Provisioning a Synapse dedicated SQL pool

You have two options when it comes to creating a Synapse dedicated SQL pool: a standalone version and a version that is integrated with the Synapse Analytics workspace. As we will need the Synapse workspace several times again later in this book, for example, when you examine the Synapse Spark pools in *Chapter 6, Using Synapse Spark Pools*, we will go and provision a workspace directly:

1. Click **+ Create a resource** in your Azure portal and type Synapse into the search field. From the quick results, you can now already select **Synapse Analytics** or you hit *Enter* and select **Synapse Analytics** from the search results. On the **Overview** blade, hit **Create** and start provisioning.

2. On the following **Basics** blade, fill in the corresponding fields. You need to select the subscription where you want to create your environment. In the **Resource Group** field, either select an existing one or create a new one.

3. In the **Managed Resource Group** field, add a name for the managed resource group where Synapse will put all automatically managed resources, such as the managed virtual network resources:

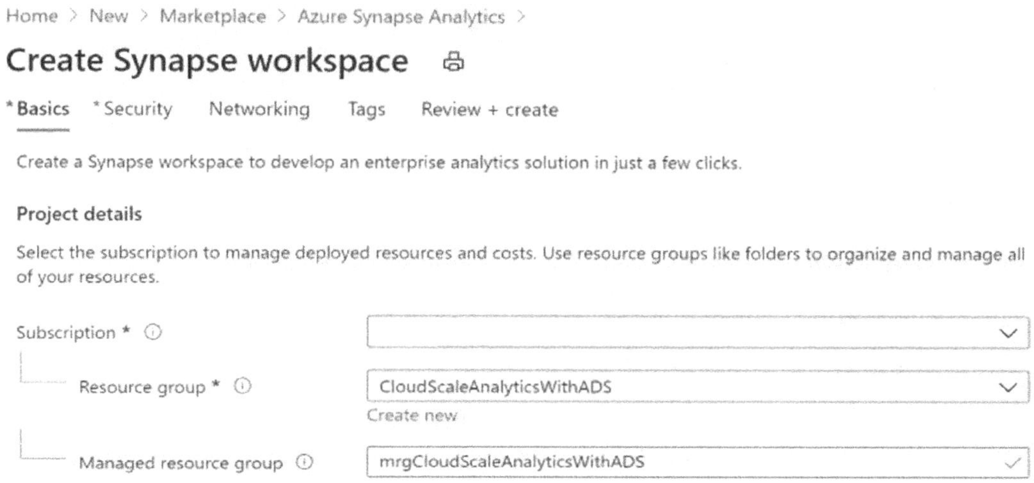

Home > New > Marketplace > Azure Synapse Analytics >

Create Synapse workspace 🖨

* Basics   * Security   Networking   Tags   Review + create

Create a Synapse workspace to develop an enterprise analytics solution in just a few clicks.

Project details

Select the subscription to manage deployed resources and costs. Use resource groups like folders to organize and manage all of your resources.

Subscription * ⓘ	⌄
Resource group * ⓘ	CloudScaleAnalyticsWithADS ⌄
	Create new
Managed resource group ⓘ	mrgCloudScaleAnalyticsWithADS ✓

Figure 4.2 – Basics blade – Provisioning a Synapse workspace

Then, name the workspace and select a region, and then select either **Data Lake Storage Account** or a filesystem that is already available in your subscription, or create new ones using the links below the fields.

Note the blue box below. During provisioning, the workspace managed identity will be added to the Storage Blob Data Contributor role in your Data Lake Storage account. Then, click **Next: Security**.

4. On the **Security** blade, add a username and a password as your SQL administrator credentials. You may select the **Enable double encryption using a customer-managed key** option. This will allow you to use your own key for the encryption of data in your workspace. Check the *Further reading, Encryption in Synapse*, section for more insights.

5. The **Allow pipelines (running as workspace's system assigned identity)** checkbox to access SQL pools will add the CONTROL permission in SQL pools to the workspace managed identity for Synapse pipelines (the Synapse version of Data Factory pipelines). Then, click **Next: Networking**.

6. On the **Networking** blade, you can decide to **Allow connections from all IP addresses** to your Synapse workspace. When you disable the checkbox, you will need to configure the firewall settings of your Synapse workspace to allow connections from IP ranges and/or virtual networks to your Synapse workspace.

The **Enable managed virtual network** setting will enable the Synapse workspace to issue managed private endpoints and use them with storage accounts and databases that are configured with virtual networks. This will facilitate your network setup effort drastically when you decide to work with virtual networks and private endpoints.

> **Note**
> You can completely hide your Synapse environment and your Data Lake Storage from other Azure services and so on by integrating them into virtual networks. This will add another security layer to your setup as access to these services can be limited to your networks only and won't be accessible from the world outside your company.

You will find further information about networking in *Chapter 11, Developing and Maintaining the Presentation Layer*:

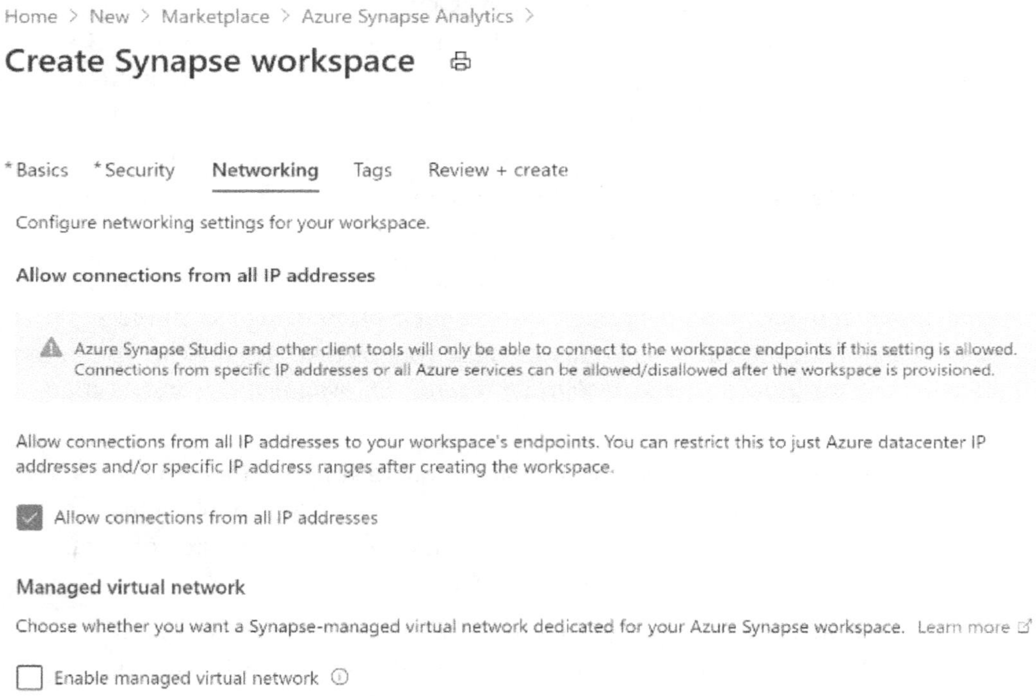

Home > New > Marketplace > Azure Synapse Analytics >

Create Synapse workspace   🖶

* Basics   * Security   **Networking**   Tags   Review + create

Configure networking settings for your workspace.

Allow connections from all IP addresses

⚠ Azure Synapse Studio and other client tools will only be able to connect to the workspace endpoints if this setting is allowed. Connections from specific IP addresses or all Azure services can be allowed/disallowed after the workspace is provisioned.

Allow connections from all IP addresses to your workspace's endpoints. You can restrict this to just Azure datacenter IP addresses and/or specific IP address ranges after creating the workspace.

☑ Allow connections from all IP addresses

Managed virtual network

Choose whether you want a Synapse-managed virtual network dedicated for your Azure Synapse workspace.   Learn more ☐

☐ Enable managed virtual network ⓘ

Figure 4.3 – Networking blade – Provisioning a Synapse workspace

With **Next: Tags**, you can then proceed to the **Tags** blade.

7. This blade will allow you to add key/value pairs to your resource and will give you extended possibilities to track the usage of your resource when you are reporting and analyzing the consumption of your system.

8. When you select **Next: Review + create**, you will be taken to the final overview of your configuration settings before you initiate creation of the service. You will have one final opportunity for a brief check of what you are going to create. At this point, you can still return to previous blades and change any settings if you realize that you have misconfigured something.

9. The **Download a template for automation** link will take you to the template viewer/editor of the Azure portal. Here, you can work on the resources that you are going to provision. **Azure Resource Manager** (**ARM**) offers an interface that can be controlled by providing JSON files that can be configured to deploy Azure resources in an automation framework.

10. ARM provides either PowerShell commands, a REST API, or the **Azure Command-Line Interface** (**CLI**) to automate nearly any step that can be done via the Azure portal, and it sometimes offers even more flexibility.

11. To finish the provisioning of your Synapse workspace, hit **Create** in the lower-left corner of the **Summary** blade. You will be transported to the provisioning view. It will update in a few seconds and provide you with a **Go to resource** button as soon as the workspace provisioning is finished.

Congratulations! You have now provisioned your Synapse workspace that you will use a lot throughout the book. However, your Synapse dedicated SQL pool still doesn't exist:

1. Go to your workspace as soon as it is ready. On the **Overview** blade, you will find a **+ New dedicated SQL pool** button in the upper area of the details. Click on that button and enter a dedicated SQL pool name.

Tip

If you aren't yet familiar with the performance measures of Synapse databases, scale down the database to the lowest possible level (DW100c) to begin with. This will avoid unnecessary costs.

---

2. On the **Additional settings** blade, you will have the opportunity to decide whether you want to create your database with or without pre-existing data:

   a) **None** will give you an empty database for your development.

   b) **Backup** will give you access to all recent versions of backups of Synapse databases in the subscription where you started the provisioning process.

   c) **Restore point** will give you a more fine-grained option to select a restore point from an existing database to be restored into this newly created pool.

   In the **Dedicated SQL pool collation** section, you can set the collation for your database. Note that you won't be able to change the collation once the database has been created. The default is **SQL_Latin1_General_CP1_CI_AS**.

   The database collation determines how the database compares data and how the database sorts data when you order it in your queries. You can exert influence, for example, in terms of whether the database orders and compares case-sensitive data, or whether "Patrik Borosch" is the same as "patrik borosch". Then, click **Next: Tags**.

3. Again, you have the option to add tags to your service. When you're finished with the tags, click **Next: Review + create**.

4. On the final **Review + Create** blade, you can again check whether everything is as you intended and switch back to earlier blades to adjust settings if needed. And, of course, you again have the option to download a template for automation. Hit **Create** when you are ready to proceed.

5. When provisioning is complete, you can return to the **Overview** blade of the Synapse workspace. Here you will find your new dedicated SQL pool in the SQL pools list:

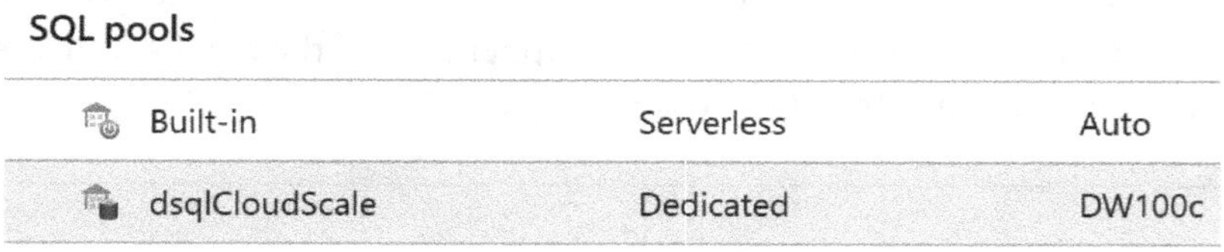

SQL pools

Built-in	Serverless	Auto
dsqlCloudScale	Dedicated	DW100c

Figure 4.4 – SQL pools in the Synapse workspace overview

Now, let's connect to the database in the next section, *Connecting to your database for the first time*.

Connecting to your database for the first time

Now that you have created your first Synapse Analytics service on Azure, you want to see that it was a success. You can use several tools to connect to your Synapse database:

- The query tool preview in the Azure portal
- SQL Server Management Studio
- Azure Data Studio
- Visual Studio using SQL Server Data Tools
- Synapse Studio

We will use the Azure portal first.

When you navigate to your Synapse Analytics workspace, you will find a button, **Open Synapse Studio**, in the **Getting started** section:

Getting started

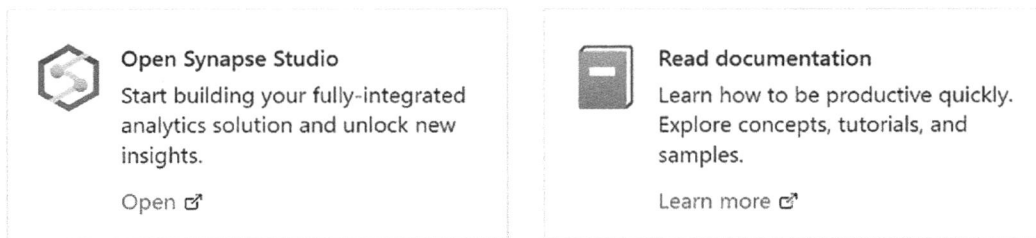

Figure 4.5 – The Open Synapse Studio button on the Synapse Overview blade

A new browser tab will open and log you in to Synapse Studio. On the left, in the **Navigator** blade, click on the **Data** hub. You will see an additional resources navigator:

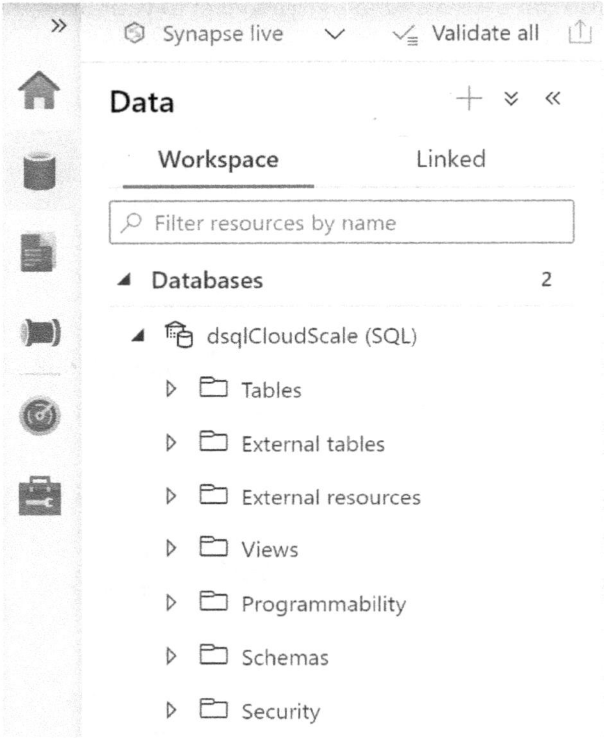

Figure 4.6 – Synapse workspace tab

Here, you can browse to your newly created database and take a look at the folders that are displayed here. You will find typical database objects. **Tables** and **External Tables** (you will learn about Polybase and external connections in the *Data loading* section) are there, as well as **Views** and **Programmability**, **Schemas**, and **Security**.

Now, when you right-click on the database name, you can initiate **New SQL Script > Empty script** and you are taken to an empty SQL interface in the details area in the middle of your Synapse Studio window.

Type in the following statement and hit **Run** once you're finished:

```
select *
from sys.schemas
;
```

Distributing, replicating, and round-robin

As mentioned previously, Synapse Analytics works on distributed datasets. You, as the developer, can decide how these datasets are distributed in the database. By selecting the right distribution strategy, you can help the database to perform at its best. There are three ways to handle distribution in Synapse Analytics.

Understanding hash-distributed tables

Hash-distributed tables are used for high volumes of data and can provide the best possible performance. We try to evenly distribute data over the 60 distributions that the database uses in the background. We do the same with round-robin distribution, but in this case, you as the architect can decide how the distribution is handled. Refer to the *Understanding round-robin distributed tables* section to understand the second method for distributing data in the database.

Hash distribution is done using a hash function that will assign each data row to one of the 60 distributions. When you create a table that will use hash distribution, you will use the column(s) to the hash function during the **CREATE TABLE** statement. See *Figure 4.7* for a visualization of hash distribution:

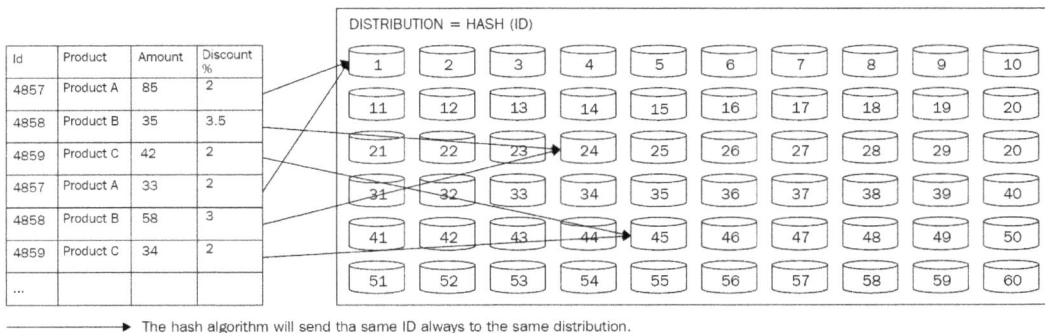

Figure 4.7 – Hash-distributed table data

The CREATE TABLE statement may look like this:

```
CREATE TABLE dbo.hashdistributedexample
(
        Tablekey          int              NOT NULL
,       Name1             varchar(20)      NOT NULL
,       Date1             datetime         NOT NULL
)
WITH
(
        CLUSTERED COLUMNSTORE INDEX
,       DISTRIBUTION = HASH ([TABLEKEY])
)
;
```

When you use hash-distributed tables, ensure that you are using a key that will provide an even spread of the data over the 60 distributions. This will be key to good performance when the data gets processed in parallel. You certainly want to avoid a situation that is known as data skew!

Data skew will happen, for example, when the column that you use for distribution has a lot of NULL values. These will all end up in the same distribution. Let's say you have NULL values in the hashed column for around half of all the rows in your table. This would mean that half of the rows will go to the same distribution.

When you now want to aggregate the rows, the compute node that works on the distribution with all those NULL values in the hash column will, for example, need to do far more work than all the other compute nodes that are configured at that moment:

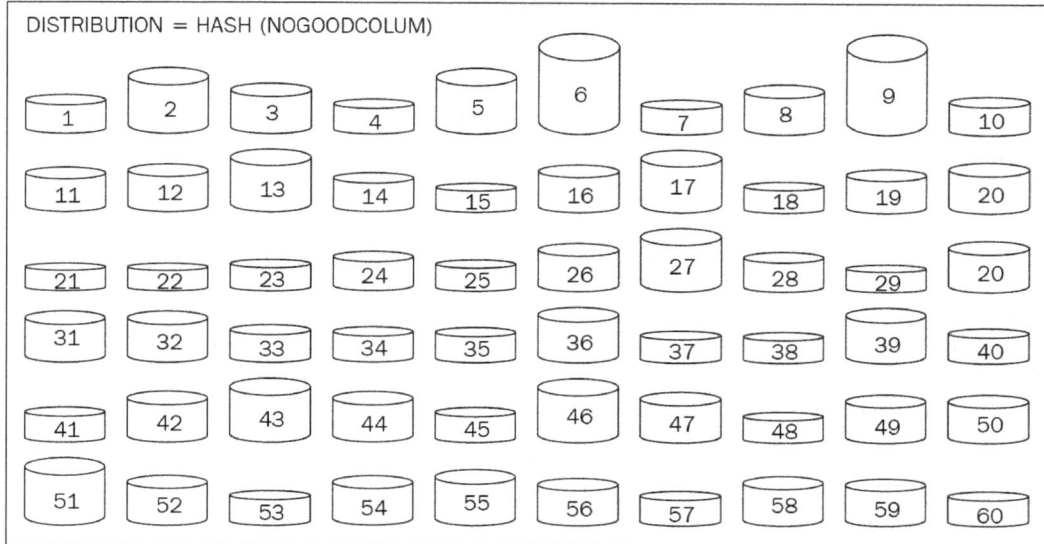

Figure 4.8 – Data skew when distributing data

When you think about good candidates for distribution columns, you might want to look for columns with high cardinality. This means that the column contains many unique values and at least 60 different values. It also should not have too many NULL values and all values should display an even distribution when distributed by 60. Date values are poor distribution columns. Users often filter over dates. This would again lead to a distribution being queried and the rest of the parallel engine would not do anything.

> **Tip**
>
> When you already know the join and aggregation criteria that your Synapse Analytics database will need to provide, the join columns and columns that are used in the GROUP BY, DISTINCT, OVER, and HAVING parts of SQL statements (except date columns) are a good option for further investigation as regards your hash distribution columns. This will help you to avoid data movement between the compute nodes when data can be processed locally and two bigger tables can be joined on the same distribution column.

The database will also help you in determining whether a table that you have created and loaded with data has data skew. The `DBCC PDW_SHOWSPACEUSED('YOURTABLENAME')` **Database Console Command (DBCC)** will give you one row per distribution with information about the number of rows and the amount of space used for the data. When you compare the figures, there should not be differences greater than 10% between the distributions. If you find higher differences, you don't have a good distribution and might suffer from skew. This will have an impact on your query performance and you will want to consider rearranging this table.

If you are wondering whether a table might be a candidate for hash distribution, you want to check the volume of data that you will store in that table. When the table is going to be greater than 2 GB of compressed data (`DBCC PDW_SHOWSPACEUSED`), you are looking at a candidate for a hash-distributed table.

Another good marker for a hash-distributed table is when the table is constantly moving as data gets inserted, updated, or deleted. These operations can then be done in parallel over all the participating compute nodes.

Where hash distribution should really pay off performance-wise is if the table is to hold more than 60 million rows. This is when performance of the **Clustered Columnstore Index (CCI)** will kick in. To understand *why*, see the *Understanding the CCI* section that follows. Of course, you can already distribute smaller tables, but then you will need to go for a different storage strategy. We will touch on that in this section.

Understanding round-robin distributed tables

Round-robin distributed tables are quite similar to hash-distributed tables, but have one big difference. They do not follow a deterministic distribution of the data. Just like the name says, the data is written to the distributions just as the data comes in.

The data movement service will take care that the data in the table will be spread evenly over the distributions. However, unlike in hash-distributed tables, a certain value in a certain column will not always be routed to the same distribution, but randomly over the system.

This seems not to make sense at all after reading the previous section, where we seek the best performance for finding a deterministic distribution for a table.

Well, there is some sense behind this. When you consider a load with billions of data rows, you want to land this as fast as possible in your database. If you are assigning a hash key based on one or more columns in that load, this will always be resolved for each row. And even in a system such as Synapse Analytics, the database cannot do this without a calculation that will take some time. Now, when you can load the data to the database without this calculation, you will definitely be faster. This, along with some other reasons, can lead to the use of round-robin distribution on a table, in the following cases:

- The table is used as a staging table (the example just mentioned).

- There are no real join criteria that you have identified.

- You are getting started and want a simple table for testing.

The most common use cases where round-robin distributed tables are used are indeed staging tables. You are loading as fast as possible to the database and take it from there:

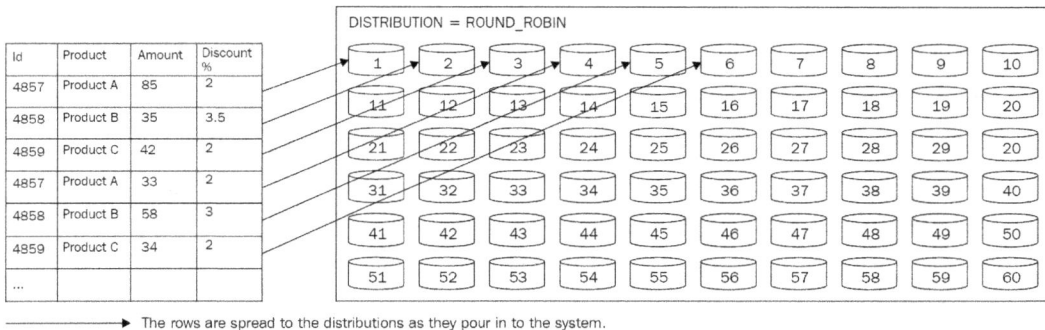

Figure 4.9 – Loading data using round-robin distributed tables

The CREATE TABLE statement for round-robin distributed tables looks quite similar to the hash-distributed one:

```
CREATE TABLE dbo.rrdistributedexample
(
        Tablekey        int             NOT NULL
,       Name1           varchar(20)     NOT NULL
,       Name2           varchar(20)     NOT NULL
,       Date1           datetime        NOT NULL
,       Attribute1      varchar(100)    NOT NULL
)
WITH
(
```

```
         CLUSTERED COLUMNSTORE INDEX
   ,     DISTRIBUTION = ROUND_ROBIN
   )
   ;
```

When transferring data from your round-robin distributed table to a hash-distributed table, you will then come across another piece in the puzzle of Synapse Analytics: the **Create Table As Select (CTAS)** statement. As we are talking about MPP technology here, we want to use that parallelism as often as possible. CTAS is an extension to the T-SQL dialect of Synapse Analytics, which will make use of all available compute nodes during a load.

When we use the CTAS statement, we will create new copies of our tables instead of loading data from one table to another. When we can use the full parallel capabilities of Synapse Analytics, this can be done far faster. Refer to the *Understanding CCI* section in this chapter for a more detailed description of how to implement this.

Understanding replicated tables

Replicated tables aim to increase join speed for small tables. By using replicated tables, you will force a full copy of the particular table on every compute node that is configured.

Under the hood, replicated tables are created as round-robin distributed tables first. When the first SQL query hits the service, which includes the replicated table, the database gets all the parts of the table from all the distributions and caches the table on every first distribution on every compute node. This will always happen when the database is scaling, when the data in the table changes, or when the table definition is changed. Stopping and resuming the database will not immediately cause a rebuild of the replicated table.

The CREATE TABLE statement for a replicated table would look like this example:

```
CREATE TABLE dbo.replicatedexample
(
         Tablekey        int              NOT NULL
   ,     Name1           varchar(20)      NOT NULL
   ,     Name2           varchar(20)      NOT NULL
   ,     Date1           datetime         NOT NULL
   ,     Attribute1      varchar(100)     NOT NULL
)
WITH
```

```
(
        CLUSTERED COLUMNSTORE INDEX
    ,   DISTRIBUTION = REPLICATE
)
;
```

As mentioned, replicated tables work best with small amounts of data. As a rule of thumb, you can decide to replicate tables that are 2 GB of compressed data or smaller. You can also replicate other, bigger data if the table is more or less static and will not change too much over time. You can find the size of your table using the DBCC PDW_ SHOWSPACEUSED('YOURTABLENAME') command:

DISTRUBUTIONS = REPLICATE

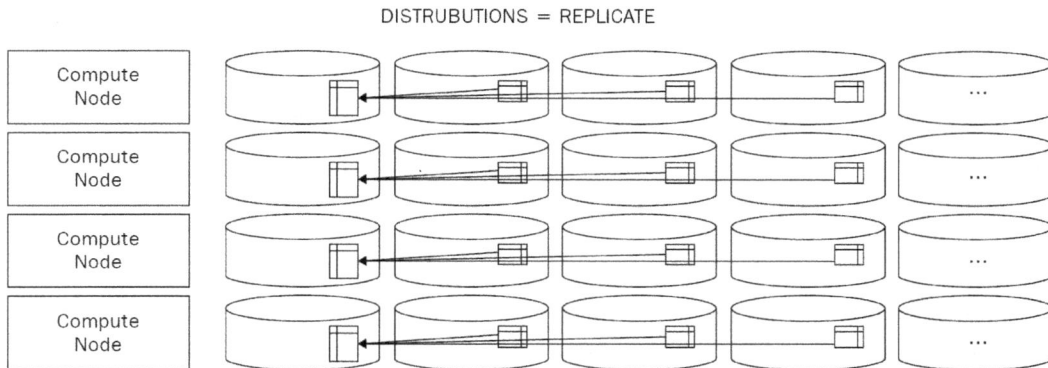

Figure 4.10 – Replicated table on the compute nodes

Dimension tables in your data warehouse are good candidates for replicated tables in Synapse Analytics.

Tables that won't be good candidates for replication would be, for example, tables that have constantly moving data in them, for example, inserts, updates, and deletes. When they constantly happen on your table, you should not replicate it due to the **cache build** mechanism mentioned above. You would constantly need to recreate your table on all the compute nodes.

If you are constantly scaling your Synapse Analytics SQL pool, you don't want to have too many replicated tables. Again, the cache build mechanism would slow you down here.

When your table holds a lot of columns, you would rather distribute your table and then create an index on the columns that are accessed frequently. When data movement happens, you will only have to move the columns needed.

Deciding when to replicate or distribute

If you don't have a clear idea yet of whether your table should be replicated or distributed, you may want to again consider the following:

- The volume of your table. If your table is bigger than 2 GB, it will be a good candidate for a hash-distributed table.

- If your table is a fact table, which might grow over time, you might already be looking at a candidate for a hash-distributed table.

- Tables that are constantly moving with data inserted, updated, or deleted are candidates for hash-distributed tables.

- If your table has many columns, you might get better performance when you hash-distribute it and create indexes for the most accessed columns.

- If your table is 2 GB or smaller, it is probably a good candidate for a replicated table.

- Dimension tables are mostly good candidates for replicated tables.

- Staging tables should be created as round-robin distributed tables as you want to load your data as fast as possible to your database and then let the database internally to the rest of the job.

- When your table does not have a clear column that is often used for joins, you might be looking at a round-robin distribution candidate. With the absence of a good column for joins, you will also not have a good column candidate for hash distribution.

Understanding CCI

The default storage strategy for tables is CCI. This will store the data of the table in column orientation, in comparison to typical database tables that store their data as rows.

CCI will not only store the data in columns instead of rows, but will also compress the data with a very sophisticated set of algorithms. When we take a Name column, for example, all the unique names will only be stored once. In a column with integer numbers, each number will also be stored only once. See *Figure 4.11* for a visualization. When you query the table, the engine will then get all the needed values from the columns and put the rows back together before returning the result set. You can imagine how much space you can save with this already. And this is only the tip of the iceberg of what is done under the hood in your column store index on your table:

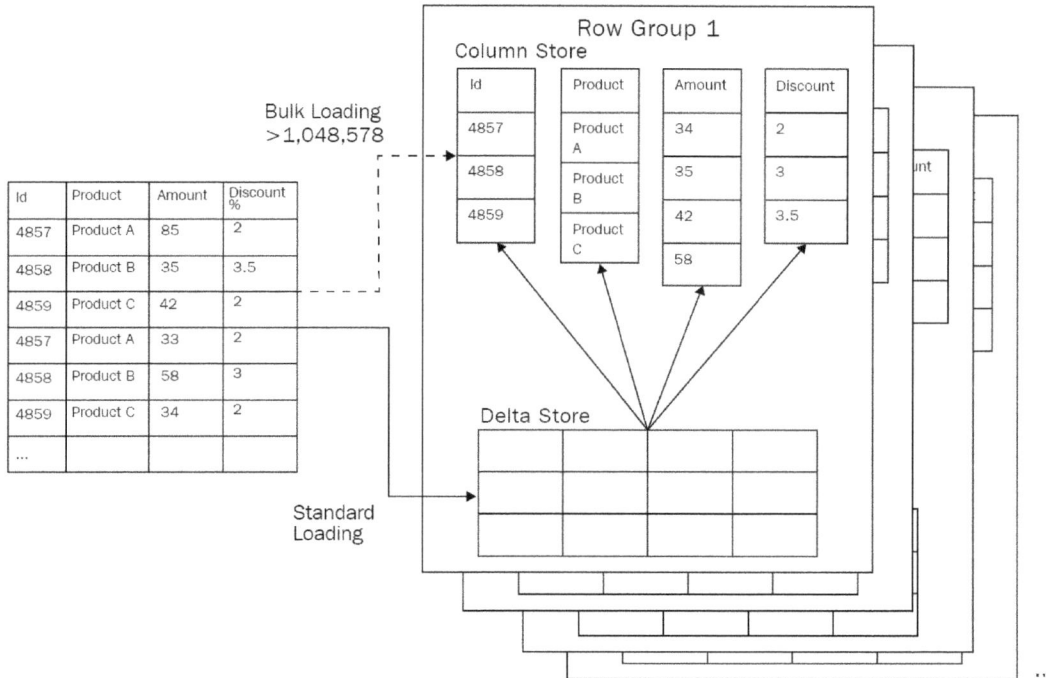

Figure 4.11 – Columnstore compression

If you examine *Figure 4.11*, you will find **row groups** as a component and there is the concept of the **delta store**. When you are loading data into your CCI table, the data first goes to the delta store, where it is stored in row format. When the threshold of 1,048,576 rows in the delta store is reached, the data gets compressed into CCI format. Before that, Synapse Analytics will always answer queries that include that table accessing the row store out of the delta store together with the data that might already be in the column store.

Now, please think for a moment about the architecture of the Synapse Analytics database. The row groups are stored in the 60 databases that build the data storage. When you now create a hash-distributed table, it will be distributed over those 60 databases, and each database will have a delta store for this table. Now, let's do the math: to have your data in your CCI on that table, you will need a minimum of 60 x 1,048,578 rows. This means that your table overall needs at least 62,914,680 rows to have a distributed column store table. As a rule of thumb, you should even go to 100 million rows for a distributed table to benefit from the speed of the CCI table.

When you have loads that already contain such a number of rows, you are kind of lucky. When you are using PolyBase or the COPY statement, the load will bypass the delta store and directly bring the data to the column store. This will boost your loading performance significantly.

> **Important info**
> When loading data to Synapse Analytics SQL pools, try and accumulate smaller files in your data lake before loading the data to the database. This will allow better performance than trying to load file by file.

But what can you do with tables that are smaller than 100 million (62,914,680) rows? You might want to create them as heap tables, which use the standard row store storage strategy. This will avoid the column store algorithms from kicking in and trying to read from the delta store and the column store when reading data from the table. This will give you better performance on smaller tables.

The CREATE TABLE statement for a heap table looks like this:

```
CREATE TABLE dbo.heapexample
(
        Tablekey        int             NOT NULL
,       Name1           varchar(20)     NOT NULL
,       Name2           varchar(20)     NOT NULL
,       Date1           datetime        NOT NULL
,       Attribute1      varchar(100)    NOT NULL
)
WITH
(
    HEAP
,       DISTRIBUTION = REPLICATE
)
;
```

Talking about partitioning

When you need to load massive amounts of data to your database, partitioning might be another optimization option. But you really should be confronted with massive amounts of data when you start considering partitioning.

Do you remember the math of the CCI and why it will only perform when you load around 63 to 100 million rows to your database (see the preceding section, *Understanding CCI*)? Now, you need to add another factor to this equation: the number of partitions that you are planning for your database.

Let's assume that you want to have one partition for every month (the most typical usage of partitions) in your table, and you plan to load data for 5 years to your database. This will add another 60 as a factor to your preceding term: 60 distributions x 1,048,578 rows per distribution x 60 months in the database. This results in 3,774,880,800 rows that your table needs to hold as a minimum in order for the CCI to be built over all the partitions and to get the best performance out of the table:

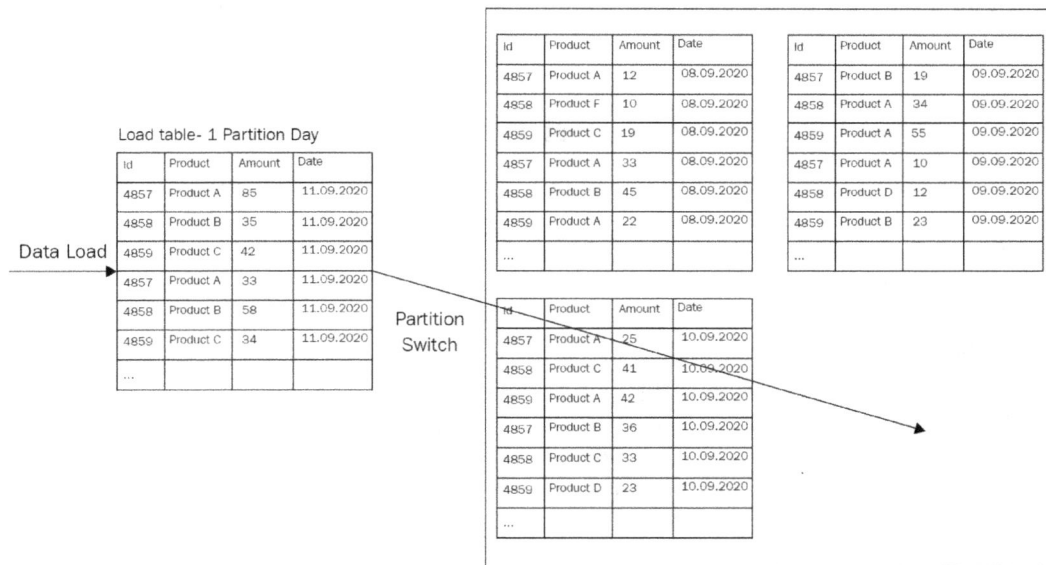

Figure 4.12 – Partition switching

One big advantage of partitions in Synapse Analytics is the **metadata operation** nature of the partition functions. As you see in *Figure 4.12*, you can load your data into an empty representation of your table that follows the same partition scheme as the original table, but only holds exactly one partition (the one you're loading to). When the load has finished, you can move the single partition of your loading table using a technique called **partition switching** to the original, the target table.

Partition switching, **partition merging**, and **partition deletion** are, as mentioned, metadata operations. This means that switching a partition into another table as well as merging two partitions or deleting one are the only operations in the metadata of the related table. This means that no data is moved at that time. Only the information of the partition is changed in the dictionary of the database.

The following statement switches partition number 2 from the source table to the target of the operation:

```
ALTER TABLE partitionswitch_source SWITCH PARTITION 2 to
partitionswitch_target PARTITON 22;
```

Refer to the GitHub repository for a full example of a load that includes partition switching.

Implementing workload management

You have read about control nodes and compute nodes in the preceding sections, and you will learn that you can add more compute nodes when your workload needs more power in the *Scaling the database* section. But keep in mind that adding compute nodes will also add cost to your bill at the end of the month.

> **Tip**
> Try and identify time slots where your database is not "hammered on" big time and either scale the database back down or, when you can afford to, hibernate it. You can save a lot of money by doing so. Development and testing environments are also good candidates for switching the database off when they aren't needed.

Fortunately, the Synapse product group has added some options to the database that will help you with optimizing the usage of a given compute environment before you need to scale the database. Workload management will enable you to set different priorities (**workload importance**) for different processes, to isolate resources (**workload isolation**) to ensure that processes can be finished, and to classify workloads (**workload classification**) to be able to assign the correct resources to queries and jobs. As a minimum, you should check **resource classes** and their usage.

Understanding concurrency and memory settings

Before we proceed with the workload management options, we need to have a basic understanding of the particular resources that are controlled with the workload management options, **Data Warehouse Units** (**DWUs**), memory, and concurrency.

DWUs are the basic measurement in Synapse Analytics. They form a combination of memory, available compute nodes, and concurrency slots for queries to be run. Check the table in *Figure 4.13*:

DWUc	Max. Concurrent Queries	Available Concurrency Slots	Compute Nodes	Available Memory (GB)	Distributions per Compute Node
DW100c	4	4	1	60	60
DW200c	8	8	1	120	60
DW300c	12	12	1	180	60
DW400c	16	16	1	240	60
DW500c	20	20	1	300	60
DW1000c	32	40	2	600	30
DW1500c	32	60	3	900	20
DW2000c	48	80	4	1,200	15
DW2500c	48	100	5	1,500	12
DW3000c	64	120	6	1,800	10
DW5000c	64	200	10	3,000	6
DW6000c	128	240	12	3,600	5
DW7500c	128	300	15	4,500	4
DW10000c	128	400	20	6,000	3
DW15000c	128	600	30	9,000	2
DW30000c	128	1,200	60	18,000	1

Figure 4.13 – DWUs, concurrency, and memory

These are the figures that you are dealing with in your data warehouse and that you can assign to certain workloads.

Using resource classes

When you examine *Figure 4.13*, the concurrency that is available in the Synapse Analytics database, the memory that your queries will be consuming, and so on. You want to be able to control which user and which query can consume how much of the available resources in a given setup.

Resource classes are the basic mechanism in Synapse Analytics for assigning users to resources. Synapse Analytics provides two types of resource classes: **static** and **dynamic** resource classes.

Static resource classes will always consume a fixed amount of memory and concurrency slots no matter what DWUc configuration you are running at. There are eight static resource classes that can be assigned: `staticrc10` to `staticrc80`. See *Figure 4.19* in the *Further reading* section of this chapter for a table regarding static resource classes and their concurrency allocation.

Static resource classes are a good decision when you have a stable, predictable workload and you know the patterns against your database. A single user in a static resource class with their queries will not gain more resources when you scale the database. But you will be able to add more users with their queries to the database, as scaling will add additional concurrency slots and memory. So, the database will be better at handling higher concurrency with static resource classes.

Dynamic resource classes work differently as they will, as the name indicates, dynamically assign resources to their members. Refer to *Figure 4.20* in the *Further reading* section of this chapter for a table that displays the resource allocation for dynamic resource classes.

Due to their nature, dynamic resource classes can be beneficial when your workload is less predictable. Single queries can benefit from scaling the database as they will get more resources allocated when the database scales up. There is no additional concurrency when scaling the database with dynamic resource classes.

There are four dynamic resource classes available: `smallrc`, `mediumrc`, `largerc`, and `xlargerc`.

> **Important note**
>
> By default, new users are assigned to `smallrc`. Don't forget this if you are wondering why loads are slow or queries don't seem to come back in an acceptable time! Many customers shoot themselves in the foot and came back yelling that the database doesn't keep its superior performance promise. Once they start diving into workload management, they change their minds.

More complex queries, and also ETL loads, will benefit from the `largerc` resource class. As more data will be held in memory, complex joins, as well as sorting operations, can be processed faster. All the query operations as well as the DML operations (insert, update, and delete) and the processes of the **Data Movement Service** (**DMS**) will be influenced by the resource classes. `ALTER INDEX – REBUILD` or `REORGANIZE`, `ALTER TABLE REBUILD`, `CREATE INDEX`, `CREATE CLUSTERED COLUMNSTORE INDEX`, and the CTAS statement are also operations that will benefit from a balanced resource class usage.

You can assign a resource class to a security account (user or group) using the `sp_addrolemember` system stored procedure. The following statement adds `loadusergrop` to the `largerc` dynamic resource class:

```
sp_addrolemember "largerc", "loadusergrp";
```

Implementing workload classification

With the new generation of Synapse Analytics, Microsoft has introduced a far more sophisticated workload management system. The concept of resource classes will vanish.

When you implement workload classification, you are creating classifier objects that can be assigned to users or applications. These classifiers will then be assigned resources from the SQL pool according to their needs and priorities. For example, you would classify an ETL load with higher priority than a user who logs in randomly, does not have high time constraints, and who queries two or three tables.

The following statement will create a workload classifier for an ETL load named `Etlload` that would be valid between 11 p.m. and 3:30 a.m. for a **workload group** named `LoaderGroup` for a security account named `LoadUser`. For the concept of workload groups, check the *Understanding workload isolation* section:

```
CREATE WORKLOAD CLASSIFIER Etlload WITH
(
WORKLOAD_GROUP = 'LoaderGroup'
, MEMBERNAME    = 'LoadUser'
, START_TIME    = '23:00'
, END_TIME      = '03:30'
)
```

Other options could also include the following:

```
,WLM_LABEL        = 'LoadSourceA'
,WLM_CONTEXT      = 'LoadDimensions'
```

As you can see, there are some options that you have to create a classifier. They will give you quite some flexibility in terms of handling resources. In your queries, you will later be able to attach to classifiers. For example, you could use the OPTION (LABEL) addition in a query to relate your query to the classifier and obtain the resources that the classifier represents:

```
SELECT
    D.CATEGORY
    ,D.SUBCATEGORY
    ,SUM(F.AMOUNT)
FROM
    DIM_PRODUCT D JOIN FACT_SALES F on D.PRODUCT_ID = F.PRODUCT_
ID
GROUP BY
    D.CATGORY
    ,D.SUBCATEGORY
OPTION (LABEL='POWERUSER')
```

Similar to the label but within a stored procedure, for example, you could set a session context:

```
EXEC SP_SET_SESSION_CONTEXT 'WLM_CONTEXT', 'LOAD_DIMENSIONS'
```

As you can have multiple classifiers that might kick in for a certain query, there is a weight for the different classifier types and how they will be precedented:

USER	64
ROLE	32
WLM_LABEL	16
WLM_CONTEXT	8
TIME	4

Figure 4.14 – Classifier weight

Adding workload importance

We can add another option to the preceding statement using IMPORTANCE. This will allow us to assign higher priority to a certain query so that it will jump ahead in the queue if needed.

The IMPORTANCE option can be one of LOW, BELOW_NORMAL, NORMAL, ABOVE_NORMAL, or HIGH:

```
CREATE WORKLOAD CLASSIFIER Etlload WITH
(
WORKLOAD_GROUP = 'LoaderGroup'
, MEMBERNAME    = 'LoadUser'
, START_TIME    = '23:00'
, END_TIME      = '03:30'
, IMPORTANCE    = 'ABOVE_NORMAL'
)
```

Workload importance can play a vital role when a query queues up waiting for the required concurrency slots.

Let's assume a workload on your database where a monthly closing is run and there are a lot of controllers from your subsidiaries working on the database. Their queries are nicely chained and consuming concurrency. As an analyst from your headquarters needs to perform an analysis with high priority, without a mechanism such as workload importance, they will need to line up and wait until a slot becomes available. However, as a VIP and with some urgency, they will need their results earlier. With workload importance, you can enable the analyst's query to bypass all the other ones and assign the next freed slot to run it.

Understanding workload isolation

For the implementation of workload isolation, you will first create a workload group. This will add resource limits to the workload classifier (see the *Implementing workload classification* section) that implements this group.

A workload group can limit the resources to a classifier in different dimensions. MIN_PERCENTAGE_RESOURCE will assign a minimum part of the resources available in the database to the classifier exclusively. This allocation will not be available to other classifiers.

If you examine the workload group settings, you will then find CAP_PERCENTAGE_ RESOURCE, which will set the upper limit for the particular workload group. It should be higher than MIN_PERCENTAGE_RESOURCE.

REQUEST_MIN_RESOURCE_GRANT_PERCENT will set the minimum resources for one request. This setting is set in increments of 0.25 and will consume resources between MIN_PERCENTAGE_RESOURCE and CAP_PERCENTAGE_RESOURCE.

REQUEST_MAX_RESOURCE_GRANT_PERCENT sets the maximum resources for one request. This one needs to be higher than REQUEST_MIN_RESOURCE_GRANT_ PERCENT.

The IMPORTANCE option can be one of LOW, BELOW_NORMAL, NORMAL, ABOVE_ NORMAL, or HIGH.

Finally, you can set QUERY_EXECUTION_TIMEOUT_SEC, which sets a timeout value in seconds. The query will be canceled once it is started. The queueing time does not depend on the timeout.

The following statement will create a workload group that guarantees a minimum resource allocation of 50%, with a maximum of 85%. The importance is set to higher than normal and queries can run for 5 minutes:

```
CREATE WORKLOAD GROUP loaderwgrp
WITH
    (
        MIN_PERCENTAGE_RESOURCE = 50
    , CAP_PERCENTAGE_RESOURCE = 85
    , IMPORTANCE = ABOVE_NORMAL
    , QUERY_EXECUTION_TIMEOUT_SEC = 300
    )
;
```

Check the *Effective values for REQUEST_MIN_RESOURCE_GRANT_PERCENT* section in the *Further reading* section for additional insights on DWUcs and the effective values for REQUEST_MIN_RESOURCE_GRANT_PERCENT.

To facilitate things for migrations and beginners, there are default workload groups that were created to mimic the former resource classes.

Scaling the database

In *Figure 4.13*, in the *Understanding concurrency and memory settings* section, you already learned about the available resource settings that you can configure within Synapse Analytics. But how do we get there?

When you browse to your Synapse Analytics service in your portal, you can scale the database from there. In the **Overview** blade, you will find the **Scale** link right next to the **Resume** or **Pause** button (displayed depending on the running state of the database):

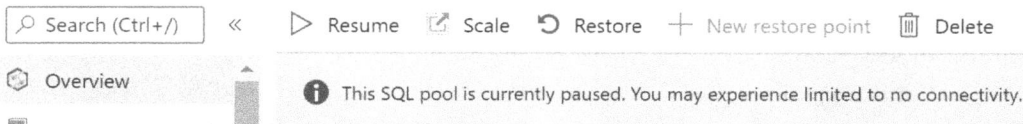

Figure 4.15 – The Resume and Scale buttons on the Overview blade

When you click on the **Scale** button, you are taken to the **Scaling** blade of the database. By dragging the slider in the **Scale your system** section, you can add or remove resources in your database:

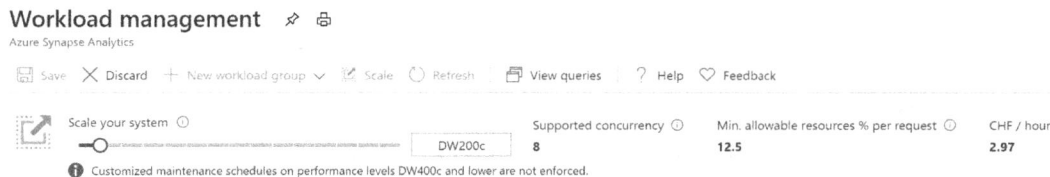

Figure 4.16 – Scaling your database

If your database contains workload groups for workload management, they will also be displayed here with an overview of their possible resource allocations according to your DWUc settings.

When your database is running, you can hibernate it from the **Overview** blade by clicking the **Pause** button:

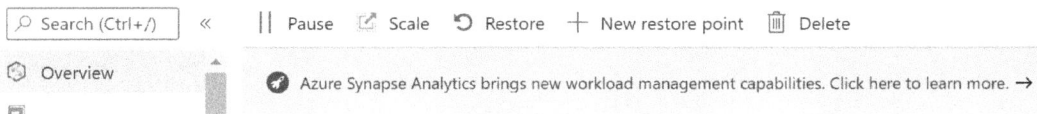

Figure 4.17 – Pausing your database

Using PowerShell to handle scaling and start/stop

Synapse Analytics can also be scaled, started, and stopped using PowerShell. The cmdlets are `Resume-AzSqlDatabase` and `Suspend-AzSqlDatabase`.

To scale your database using PowerShell, you will use `Set-AzSqlDatabase` with the `-RequestedServcieObjectiveName` option and the intended `DWxxxxc` value to set it.

To connect to your Azure subscription, you need to go through the following sequence of cmdlets:

```
Connect-AzAccount
Get-AzSubscription
Set-AzContext -SubcriptionName "YOURSUBSCRIPTIONNAME"
```

The following cmdlet will print the status of your database:

```
Get-AzSqlDatabase
```

You will get nearly the same information that you also find in the portal on the **Overview** blade (without the charts, of course).

Using T-SQL to scale your database

The third way to scale your database is from within your T-SQL code. You need to be connected to the master database of your server object. From there, you can run the following query to scale your database to the required DWUcs:

```
ALTER DATABASE YOURDATABASENAME
MODIFY (SERVICE_OBJECTIVE = 'DWxxxxc');
```

T-SQL does not have functions to stop or start your database.

> **Important information**
>
> At this point, your database will go offline during the scaling process. Synapse Analytics will roll back all running queries that have not been finished when the scaling request was issued. Then, the new DWUc setting will be applied and the database will come back online. Before the database is finally available, database recovery is performed to ensure that your database is in a consistent state once it is back and available. Note that the query cache will be lost when you scale your database. It might be a good idea to warm the cache with some queries that will bring the most important tables back into the cache. Microsoft will change this behavior in the future and switch to an online scaling functionality with the next version of Synapse Analytics. This may already be available by the time you are reading this book.

Loading data

With all the parallel options that the database can offer to you, you want to use them when you load data to your database, too. Remember the purpose of the control and the compute nodes? When loading data to your database, you want to use a technique that makes use of the compute nodes as much as possible.

Using the COPY statement

The COPY statement will support you in doing so. It will talk directly to the compute nodes and will therefore use the whole parallelism that the database can offer. It comes as part of the T-SQL dialect of the Synapse Analytics database and offers many options to influence the loading of data to the database.

When you talk to the control node, in contrast to the capability of the COPY statement, you will create a bottleneck during your load. The load would be single-threaded instead and all the rows that need to be written to the database would first flow through the control node and would then be spread to the distributions using the available compute nodes (see *Figure 4.18*):

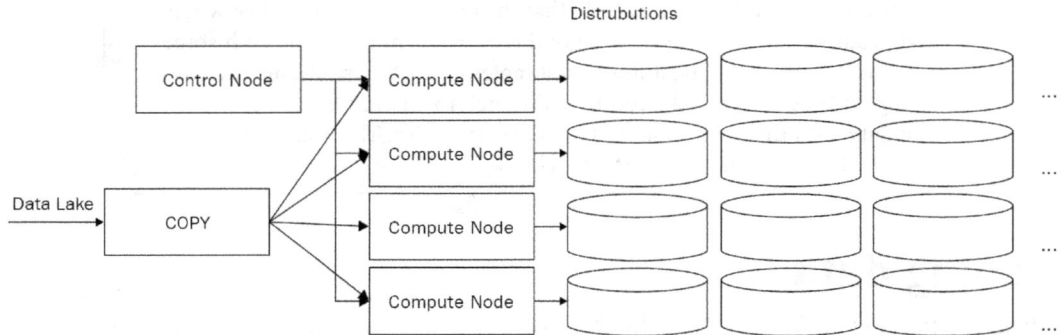

Figure 4.18 – Loading data using the COPY statement

At the time of writing this book, the COPY statement can process CSV, Parquet, and ORC files. You can configure one or more external locations from which you will read your data.

Options such as a configurable error file, the compression type of the files to process, field quotes, the field terminator, the row terminator, the number of the first row in the file, a configurable date format, the encoding of the file, and an option to allow an identity insert, complete the feature list of the COPY statement.

Check out *Chapter 5, Integrating Data in Your Modern Data Warehouse*, for a deep dive into loading data into your Synapse Analytics database.

Maintaining statistics

One thing that you should never forget when you have manipulated data in your database is to keep your statistics up to date. Your Synapse Analytics database offers the option to switch on the automatic generation and maintenance of statistics. If this is not enabled in the database, you can check it with the following statement:

```
SELECT is_auto_create_stats_on
FROM sys.databases
where [name] = 'YOURDATABASENAME';
```

It's not enabled? You can enable it by running the following statement:

```
ALTER DATABASE 'YOURDATABASENAME'
SET AUTO_CREATE_STATISTICS ON;
```

Statistics are auto-created when the SELECT statement or one of the DML statements (INSERT-SELECT, CTAS, UPDATE, DELETE) are fired. Also, the EXPLAIN statement will cause an auto-update of the statistics, too, if it contains a join or a predicate (a where or a having clause).

Understanding other SQL options in Azure

In *Chapter 2*, *Connecting Requirements and Technology*, we talked about different size options for your modern data warehouse. When we look at the M size, we have added an Azure SQL database for the presentation layer. When we compare the Azure SQL database to Synapse Analytics databases, the main difference is the SMP character of the Azure SQL database. See also *Figure 4.1* for a comparison of SMP versus MPP.

In a SQL database, data is, by default, stored in a row orientation as it is done in SQL Server. In general, you can think of the Azure SQL database as a single database that you would spin up and use. Almost all of the functionality of a SQL Server database is available with an Azure SQL database as well.

You will also have the option to create CCIs on your tables. This will give you high analytical performance on your data stored there. In comparison to Synapse Analytics, you won't get the same scale-out architecture with a flexible amount of compute nodes for the parallel execution of jobs. But for smaller amounts of data, this might not even be necessary when you can use the CCI in the SQL database.

When you scale your Azure SQL database to the serverless version, you will get a scale-out architecture that will be able to automatically add additional reading nodes to the database, which will scale with your needs.

Another difference between the services is the absence of the COPY statement. There is also no PolyBase engine available, as in SQL Server, since version 2016. If you need to integrate data from your data lake or other sources, you will need to implement other ways to do so. We will see options for this in the upcoming *Chapter 5*, *Integrating Data in Your Modern Data Warehouse*.

When it comes to a comparison of other features, such as row-level or column-level security or data masking, the Azure SQL database can also shine with these functions. In the end, the Azure SQL database is a relational database engine that will meet your requirements for smaller data warehouse workloads together with other components in the Azure framework.

Summary

In this chapter, you examined the Synapse Analytics database relational storage option. You learned about the MPP architecture and the control and compute nodes, as well as how tables can be distributed or replicated in a database and how partitioning influences the data in the database.

You read about the CCI and how you can benefit from its performance.

Furthermore, you learned about resource allocation, concurrency, and the limits of the DWUc configurations. We touched on workload management and how you can optimize your database workload before you need to scale the database and pay more money for it. But when you need to scale for more concurrency and memory, and therefore more performance, you now know where to search for this functionality.

At the end, you covered the basics of loading data and why it is a good idea to use the COPY statement. You then learned how to maintain your statistics and how to rebuild your CCI to optimize it.

Finally, we compared the Azure SQL database with Synapse Analytics for the case where you don't need the massive scale of Synapse and you want to save some money.

In the upcoming *Chapter 5, Integrating Data in Your Modern Data Warehouse*, you will learn about the options to extract data from all kinds of sources and load it into your modern data warehouse.

Further reading

To add to the content covered in this chapter, please find additional links in the Microsoft documentation and some tables that display the availability of resources for workload management.

Additional links

- https://docs.microsoft.com/en-us/azure/synapse-analytics/ sql-data-warehouse/massively-parallel-processing-mpp- architecture?view=azure-sqldw-latest

- https://docs.microsoft.com/en-us/azure/synapse-analytics/ sql-data-warehouse/what-is-a-data-warehouse-unit-dwu- cdwu?view=azure-sqldw-latest

- https://docs.microsoft.com/en-us/azure/synapse-analytics/ sql-data-warehouse/cheat-sheet?view=azure-sqldw-latest

- https://docs.microsoft.com/en-us/azure/synapse- analytics/sql-data-warehouse/sql-data-warehouse-best- practices?view=azure-sqldw-latest

- https://docs.microsoft.com/en-us/azure/synapse-analytics/ sql-data-warehouse/sql-data-warehouse-service-capacity- limits?view=azure-sqldw-latest

- https://docs.microsoft.com/en-us/azure/synapse-analytics/ sql-data-warehouse/sql-data-warehouse-workload- management?view=azure-sqldw-latest

- https://docs.microsoft.com/en-us/sql/t-sql/statements/ create-workload-group-transact-sql

- https://docs.microsoft.com/en-us/sql/t-sql/statements/ copy-into-transact-sql?view=azure-sqldw-latest

Static resource classes and concurrency slots

These are the static resource classes and their available concurrency slots. Use them for known, recurring workloads:

DWUc	Max Conc. Queries	Conc. Slots Avail.	staticrc10	staticrc20	staticrc30	staticrc40	staticrc50	staticrc60	staticrc70	staticrc80
DW100c	4	4	1	2	4	4	4	4	4	4
DW200c	8	8	1	2	4	8	8	8	8	8
DW300c	12	12	1	2	4	8	8	8	8	8
DW400c	16	16	1	2	4	8	16	16	16	16
DW500c	20	20	1	2	4	8	16	16	16	16
DW1000c	32	40	1	2	4	8	16	32	32	32
DW1500c	32	60	1	2	4	8	16	32	32	32
DW2000c	48	80	1	2	4	8	16	32	64	64
DW2500c	48	100	1	2	4	8	16	32	64	64
DW3000c	64	120	1	2	4	8	16	32	64	64
DW5000c	64	200	1	2	4	8	16	32	64	128
DW6000c	128	240	1	2	4	8	16	32	64	128
DW7500c	128	300	1	2	4	8	16	32	64	128
DW10000c	128	400	1	2	4	8	16	32	64	128
DW15000c	128	600	1	2	4	8	16	32	64	128
DW30000c	128	1,200	1	2	4	8	16	32	64	128

Figure 4.19 – Static resource classes and concurrency slots

Dynamic resource classes, memory allocation, and concurrency slots

These are the dynamic resource classes with their memory allocation and concurrency slots. Use them for ad hoc and less deterministic workloads:

DWUc	Max. Conc. Queries	Conc. slots avail.	Smallrc Conc. Slots	Smallrc % memory	Mediumrc Conc. Slots	Mediumrc % memory	Largerc Conc. Slots	Largerc % memory	Xlargerc Conc. Slots	Xlargerc % memory
DW100c	4	4	1	25	1	25	1	25	2	70
DW200c	8	8	1	12.5	1	12.5	1	22	5	70
DW300c	12	12	1	8	1	10	2	22	8	70
DW400c	16	16	1	6.25	1	10	3	22	11	70
DW500c	20	20	1	5	2	10	4	22	14	70
DW1000c	32	40	1	3	4	10	8	22	28	70
DW1500c	32	60	1	3	6	10	13	22	42	70
DW2000c	32	80	2	3	8	10	17	22	56	70
DW2500c	32	100	3	3	10	10	22	22	70	70
DW3000c	32	120	3	3	12	10	26	22	84	70
DW5000c	32	200	6	3	20	10	44	22	140	70
DW6000c	32	240	7	3	24	10	52	22	168	70
DW7500c	32	300	9	3	30	10	66	22	210	70
DW10000c	32	400	12	3	40	10	88	22	280	70
DW15000c	32	600	18	3	60	10	132	22	420	70
DW30000c	32	1,200	36	3	120	10	264	22	840	70

Figure 4.20 – Dynamic resource classes memory and concurrency slot allocation

Effective values for REQUEST_MIN_RESOURCE_GRANT_ PERCENT

When you implement workload management and control the database workload on a more fine-grained level, please use this table for an overview of the resources available:

DWUc	Lowest effective REQUEST_MIN_ RESOURCE_ GRANT_PERCENT	Max. Conc. Queries
DW100c	25%	4
DW200c	12.5%	8
DW300c	8%	12
DW400c	6.25%	16
DW500c	5%	20
DW1000c	3%	32
DW1500c	3%	32
DW2000c	2%	48
DW2500c	2%	48
DW3000c	1.5%	64
DW5000c	1.5%	64
DW6000c	0.75%	128
DW7500c	0.75%	128
DW10000c	0.75%	128
DW15000c	0.75%	128
DW30000c	0.75%	128

Figure 4.21 – Effective values for REQUEST_MIN_RESOURCE_GRANT_PERCENT

Section 3: Cloud-Scale Data Integration and Data Transformation

This section describes the Azure data services that are used to create modern data warehouses in Azure. From the provisioning of the service via the possible individual structures and best practices, through to security settings, backup, and Disaster Recovery settings, the possible interfaces, and APIs, and also the monitoring of the service in Azure and known issues, we're going to describe the use of certain services in the modern data warehouse architecture pattern.

This section comprises the following chapters:

- *Chapter 5, Integrating Data into Your Modern Data Warehouse*
- *Chapter 6, Using Synapse Spark Pools*
- *Chapter 7, Using Databricks Spark Clusters*
- *Chapter 8, Streaming Data into Your MDWH*
- *Chapter 9, Integrating Azure Cognitive Services and Machine Learning*
- *Chapter 10, Loading the Presentation Layer*

5
Integrating Data into Your Modern Data Warehouse

Extracting data from source systems and incorporating it into your modern data warehouse is one of the most important pieces of the foundation that you will create in relation to your analytical data estate.

The sheer number of possible different source formats can leave you scratching your head when you are confronted with the requirements of your users and how quickly they require their analyses and reports.

In this chapter, you will learn how to implement **ETL/ELT (Extract, Transform, Load/ Extract, Load, Transform)** pipelines with Synapse pipelines or alternatively using Azure Data Factory. This depends on your choice of services and whether you are using the integrated Synapse experience or you only need a standalone integration service.

You will examine the different source connectors and learn how to orchestrate your integration jobs. The data flows in Synapse pipelines/Azure Data Factory will extend your toolset with a flexible and scalable point-and-click transformation environment. Finally, we will investigate how to monitor your data integration environment.

In this chapter, we will cover the following topics:

- Setting up Azure Data Factory
- Examining the authoring environment
- Using wizards
- Adding data transformation logic
- Understanding integration runtimes
- Integrating with DevOps

Technical requirements

There are not too many technical requirements! All that you will require in order to follow this chapter is the following:

- An Azure subscription, where you have at least contributor rights or you are the owner.
- Your Synapse workspace or the right to provision an Azure Data Factory service.
- An Azure DevOps Git or GitHub account. This is optional, required only if you want to integrate your data factory with a DevOps repository.
- **Azure Storage Explorer**: You will need this tool to interact with your data lake, for example. Please download it from here: `https://azure.microsoft.com/en-us/features/storage-explorer/`.
- Azure Data Lake Storage and two folders, one for the source and one for a target. Refer to *Chapter 3*, *Understanding the Data Lake Storage Layer*, for a refresher.
- An arbitrary file that you will upload to your source folder and use in your first copy job.
- Access to an Azure DevOps or GitHub repository.

Setting up Azure Data Factory

If you only need a standalone data integration service, then this is the service of choice for you (see the following *Note* box). Setting up an Azure Data Factory service is as straightforward as you would expect. Like the other services that we set up in *Chapter 3*, *Understanding the Data Lake Storage Layer*, and the Synapse Analytics database in *Chapter 4*, *Understanding Synapse Pools and SQL Options*, there are some more or less easy-to-fill-in dialog blades in your Azure portal to go through in order to create a new instance of the service.

> **Note**
>
> If you are using Synapse Analytics, you can skip this section completely as Synapse pipelines will already be available for you in the Synapse workspace. Please see the *Further reading* section, *Synapse pipelines and Data Factory differences*, for a link to a feature comparison of the two.

Creating the Data Factory service

Please proceed to your Azure portal (`https://portal.azure.com`) and click on **Create a resource**. In the search box, type in `Data Factory` or navigate to **Integration** in the navigation blade and choose **Data Factory**. The provisioning sequence will commence with the **Basics** blade (see *Figure 5.1*):

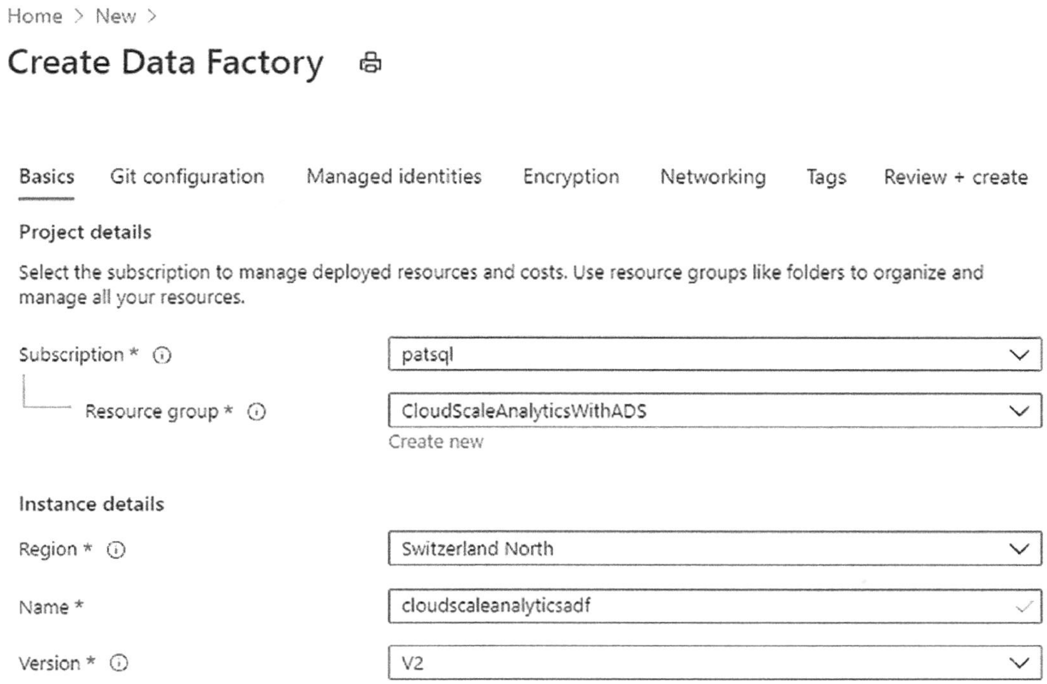

Home > New >

Create Data Factory 🖶

| Basics | Git configuration | Managed identities | Encryption | Networking | Tags | Review + create |

Project details

Select the subscription to manage deployed resources and costs. Use resource groups like folders to organize and manage all your resources.

Subscription * ⓘ
> patsql ⌄

└── Resource group * ⓘ
> CloudScaleAnalyticsWithADS ⌄
> Create new

Instance details

Region * ⓘ
> Switzerland North ⌄

Name *
> cloudscaleanalyticsadf ✓

Version * ⓘ
> V2 ⌄

Figure 5.1 – The Create Data Factory Basics blade

You will need to select the subscription to build your Azure Data Factory and then select either an existing resource group or create a new one. Refer to *Chapter 3*, *Understanding the Data Lake Storage Layer*, for a description regarding resource groups.

In the **Instance details** section, you will need to select the region where your service instance will be created. You'll need to complete the **Name** field and set **Version** to **V2**. Then, please proceed to the second blade, **Git configuration**, by clicking **Next : Git configuration** in the footer area of your blade:

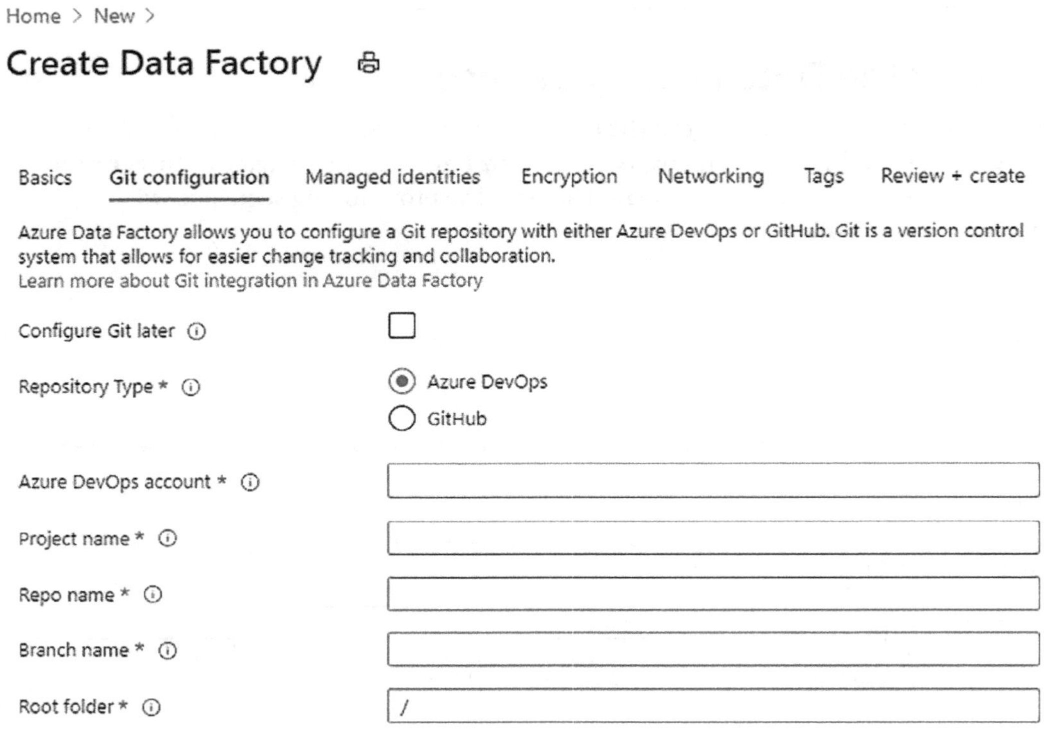

Home > New >

Create Data Factory 🖨

| Basics | **Git configuration** | Managed identities | Encryption | Networking | Tags | Review + create |

Azure Data Factory allows you to configure a Git repository with either Azure DevOps or GitHub. Git is a version control system that allows for easier change tracking and collaboration.
Learn more about Git integration in Azure Data Factory

Configure Git later ⓘ   ☐

Repository Type * ⓘ   ◉ Azure DevOps
                                          ○ GitHub

Azure DevOps account * ⓘ   [   ]

Project name * ⓘ   [   ]

Repo name * ⓘ   [   ]

Branch name * ⓘ   [   ]

Root folder * ⓘ   [/   ]

Figure 5.2 – The Create Data Factory Git configuration blade

In the **Git configuration** blade, you will be able to connect your data factory to an available **Azure DevOps** or **GitHub** repository. You will require an existing account in either of these repositories as you can't create a repository from here.

If you wish to proceed without an integration at this point, you can check the **Configure Git later** checkbox. As the option says, you can configure your Git integration later on if you so desire.

When you have finished on this blade, you can proceed to the next one by clicking **Next : Managed Identities**, followed by **Next : Encryption**, and then **Next : Networking**. The **managed identity** will be created automatically and encryption is done by default with **Microsoft managed keys**. You won't have to do anything in relation to this.

On the **Networking** blade, you can decide whether potential self-hosted **integration runtimes (IRs)** (see the *Understanding integration runtimes* section) will connect to your data factory on a public or private endpoint:

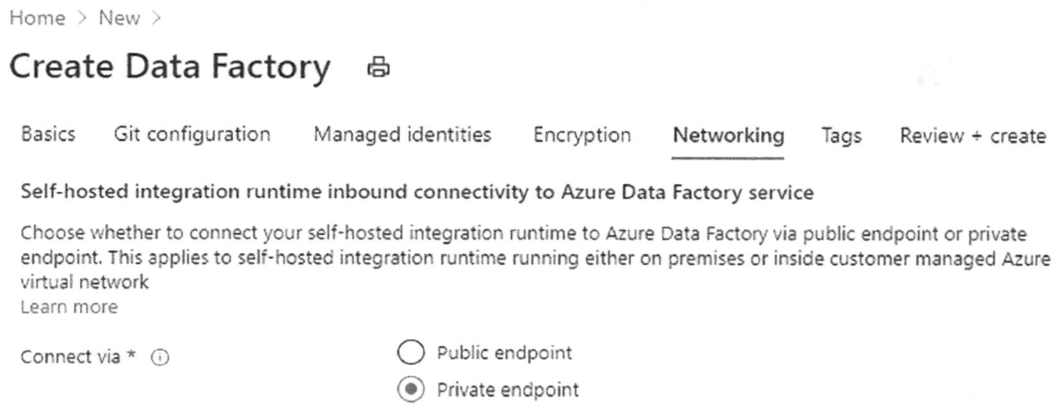

Home > New >

Create Data Factory 🖶

Basics   Git configuration   Managed identities   Encryption   **Networking**   Tags   Review + create

Self-hosted integration runtime inbound connectivity to Azure Data Factory service

Choose whether to connect your self-hosted integration runtime to Azure Data Factory via public endpoint or private endpoint. This applies to self-hosted integration runtime running either on premises or inside customer managed Azure virtual network
Learn more

Connect via * ⓘ   ◯ Public endpoint
                       ⦿ Private endpoint

Figure 5.3 – The Create Data Factory Networking blade

Finally, you will once again pass the **Tags** blade, where you can add your own tags for better reporting. You have already seen tags during the provisioning of your first Data Lake account in *Chapter 3, Understanding the Data Lake Storage Layer.*

The **Review + create** blade will again give you a final overview of the configurations for your Azure Data Factory. Clicking **Create** will finally kick off the provisioning step.

Once again, you can select **Download a template for automation** to be taken to the ARM template of the database that you are about to create.

Again, you are provided with the deployment overview until the deployment is complete and you can click **Go to resource**.

The following is a summary of the minimum steps required to create an Azure Data Factory instance:

1. On the **Home** blade of your Azure portal, select **Create a resource**.

2. Select **Azure Data Factory** and click **Create**.

3. On the **Basics** blade, create or reuse a resource group, select a region, and then select a Data Factory name. Select **V2** for the actual version (this should be the default setting).

4. On the **Git configuration** blade, select **Configure Git later**.

5. Proceed directly to the **Networking** blade and select **Public endpoint**.

6. On the **Tags** blade, enter any tag you need to report on your services.

7. After checking the **Summary** blade, click **Create and provision your Azure Data Factory**.

Examining the authoring environment

After you have provisioned your service, you can navigate to the resource and you will find the following view:

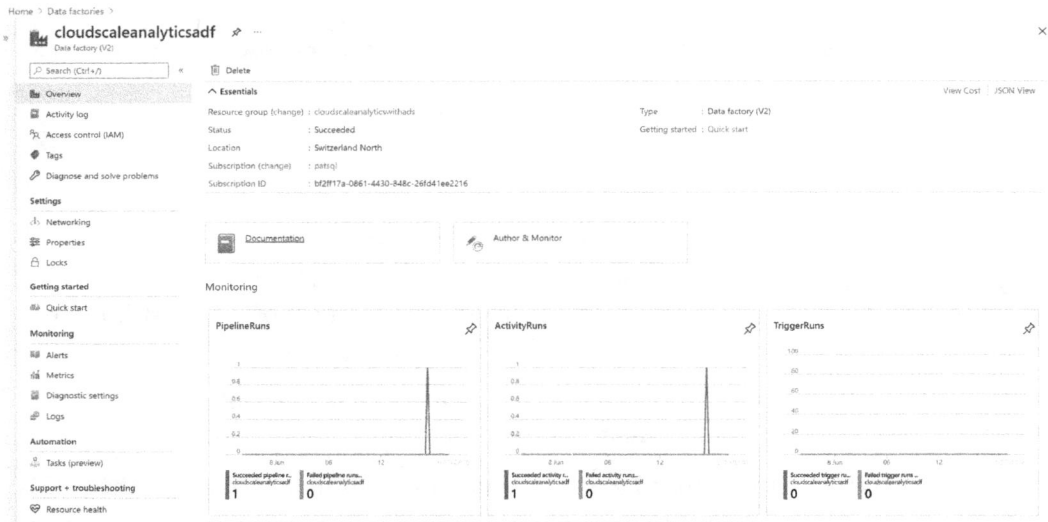

Figure 5.4 – Azure Data Factory Overview blade

Right below the **Essentials** overview section, you will find two buttons, **Documentation** and **Author & Monitor**. To access the web authoring experience of Azure Data Factory, you need to click on **Author & Monitor**. This will take you to the authoring environment, which looks like this:

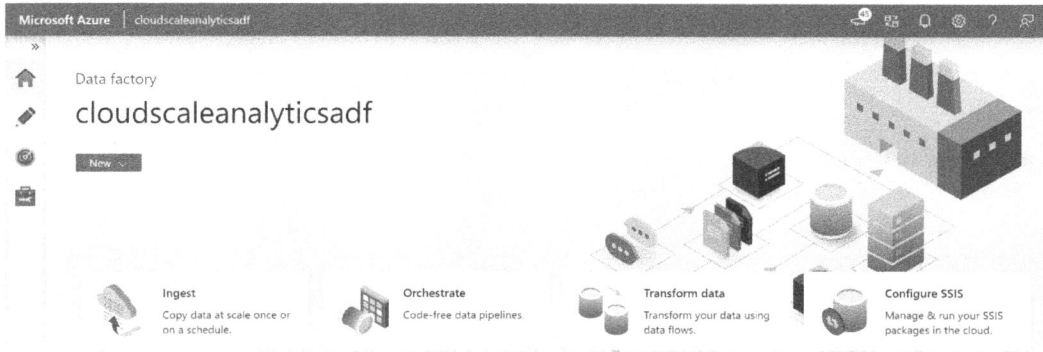

Figure 5.5 – Azure Data Factory home screen

On the home screen, you will already find some options to help you make a start. You can examine a number of videos. There is a nice overview video that is worth watching, for example. Below the **Videos** section, there is a **Tutorials** section with an introduction and other information pertaining to Azure Data Factory, and in the footer of the page, you will find additional links regarding Azure Data Factory.

> **Note**
>
> There is a similar home screen in the Synapse Analytics workspace where you will find some information about Synapse pipelines and additional information about Synapse in general.

If you examine the screen, you will find a navigator pane on the left. With the help of the >> button, you can expand the navigator or leave it minimized. There are three sections available: **Author**, **Monitor**, and **Manage**.

These sections are pretty much the same in both environments, the Synapse workspace and Azure Data Factory. In the Synapse workspace, you will find the **Authoring** section for Synapse pipelines in the **Integrate** hub. The **Monitoring** hub of the Synapse workspace will give you access to the pipeline monitor as well, and in the **Management** hub of the Synapse workspace, you will find all the pipeline settings and artifacts such as **Linked Services**, **Integration Runtimes**, and **Triggers** next to the other Synapse artifact groups.

Understanding the Author section

From here on, the actions and interfaces are the same for Synapse pipelines and Data Factory.

This is where development happens, at least if you prefer "point-and-click" development. Development happens using the following controls:

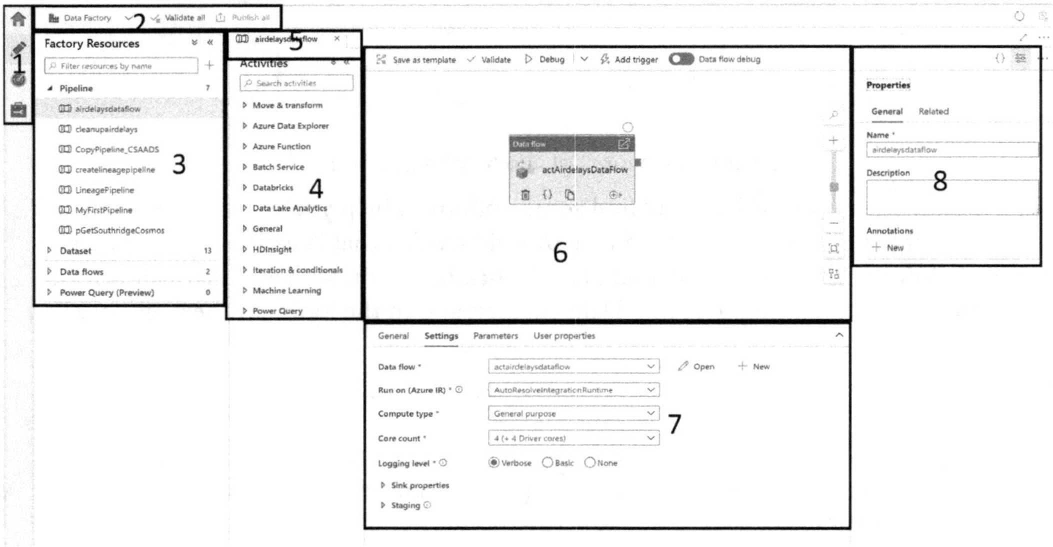

Figure 5.6 – The Synapse pipelines/Data Factory Author screen

We'll check them out in more depth now:

1. This is the aforementioned navigation pane.

2. With the options presented in the menu ribbon, you will save, validate, and refresh your environment. Here you will find the option to select a Git repository, if you have set one up. The combo box next to the Git repository allows you to branch your work from the repository and create pull requests once you have finished your work. On the right-hand area of the ribbon, you will find a switch that will toggle the debug mode for data flows that you create here.

3. In the resources navigator, you will find all the artifacts that you create within Synapse pipelines/Data Factory.

4. The **Activities** pane provides access to the different activity types that you can use to create your pipelines. You just expand the particular folder and drag the requisite activity to the development canvas, where you can start to configure it and relate it to others.

5. Every artifact you develop will show as a tab.

6. This is the development canvas. This is where you create your voodoo. You will add activities and put them into the right sequence to fulfill your needs based on their visual representation.

7. In the settings pane, you will configure the activities and the pipelines displayed on the canvas. Don't forget to set the focus to the correct object in order to display its options.

8. In the **Properties** pane, you will finally name the pipeline, enter a description, and make a decision regarding concurrency. This controls the number of parallel pipelines that will be allowed to run at the same time. The **Related** tab will give you an overview of all the connected pipelines.

> **Note**
>
> The difference in the Author section between the Synapse workspace and the Azure Data Factory studio is the resources navigator. The Synapse one will only show pipelines whereas the Data Factory one will show datasets, data flows, power queries, and templates.
>
> Synapse displays these object types at different places to smoothly integrate the pipeline artifacts into the Synapse philosophy. You will find the datasets, for example, in the data hub in the **Linked** section and the data flows and power queries in the development hub.

Understanding the Monitor section

The Monitor section will give you access to information about pipeline runs and the activities within the pipelines. You can get an overview of successes and errors and cancel jobs if needed from here:

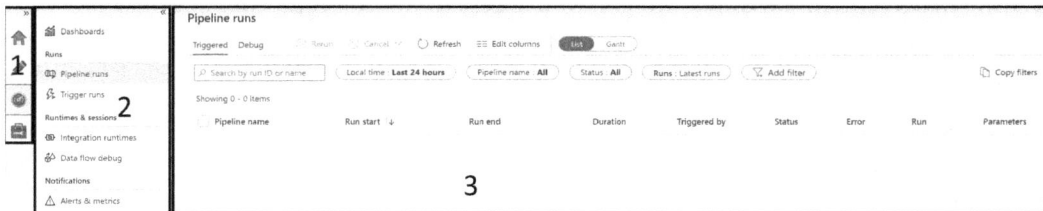

Figure 5.7 – The Synapse pipelines/Data Factory Monitor screen

1. This is the aforementioned navigation pane.

2. The monitoring navigator will give you access to the different options of the monitor, including pipeline or trigger runs, IRs, or data flow debug information.

3. The details section will show the details according to the selection made in the monitoring navigator. Most of the time, you want to visualize the pipeline runs and drill down into their details in this area.

Understanding the Manage section

In the Management section, you have access to all the artifacts that your Synapse pipelines/Data Factory needs to implement connectivity, triggers, and the DevOps settings:

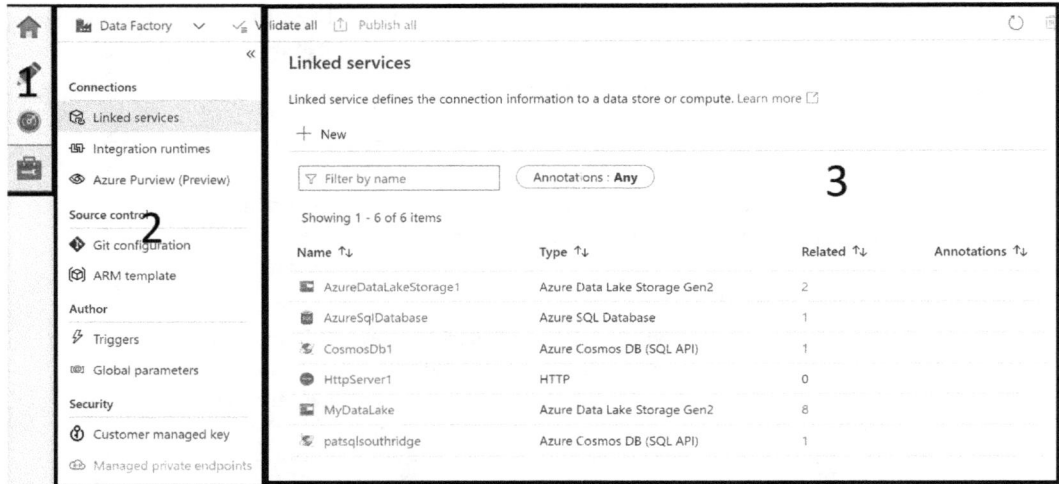

Figure 5.8 – The Manage screen

We will check this out in detail using the following steps:

1. This is the aforementioned navigation pane.

2. The management navigator will give you access to the configuration options. You will configure your connectivity, your DevOps options, and more in this area.

3. The details section is where you will find the artifacts according to your selection in the management navigator.

Understanding the object types

After you have made yourself familiar with your development environment, let's examine the typical objects that you will be creating with your Synapse pipelines or in your data factory. The best way to do this is to create a first pipeline and do something.

Creating your first copy job

We will create a pipeline with one activity that will do a binary copy between two folders in your data lake.

The binary copy will just copy the file 1:1 to the target and, for example, won't open the file or examine the file structure or the file content. This is the quickest way to get data from a to b if you just want to land it into your data lake.

Getting started

After downloading and installing Azure Storage Explorer, you need to connect to your data lake and upload a file to your destination source folder:

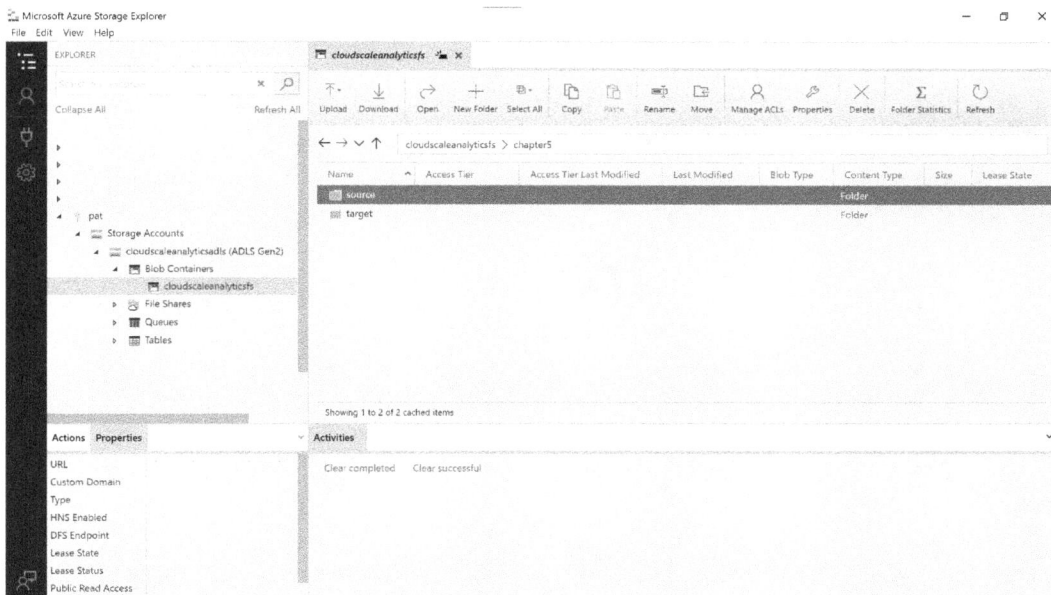

Figure 5.9 – Azure Data Explorer browsing the data lake

You will find the **Upload** button very prominent in the upper-left corner of the details section of the explorer.

Building your copy pipeline

Please proceed to your newly created Azure Data Factory or go to your Synapse Studio. In the resources navigator (see the *Understanding the Author section*), click the + button to create a new resource and select **Pipeline** from the list. You will be presented with an empty development canvas and the **Properties** pane is also shown, giving you the opportunity to name your new pipeline. Let's do so and name it `MyFirstPipeline`:

1. Now open the **Move & transform** entry into the **Activities** pane by clicking on it. Drag the **Copy data** activity to the canvas of your pipeline. You should have a view that's pretty similar to this one in *Figure 5.10*:

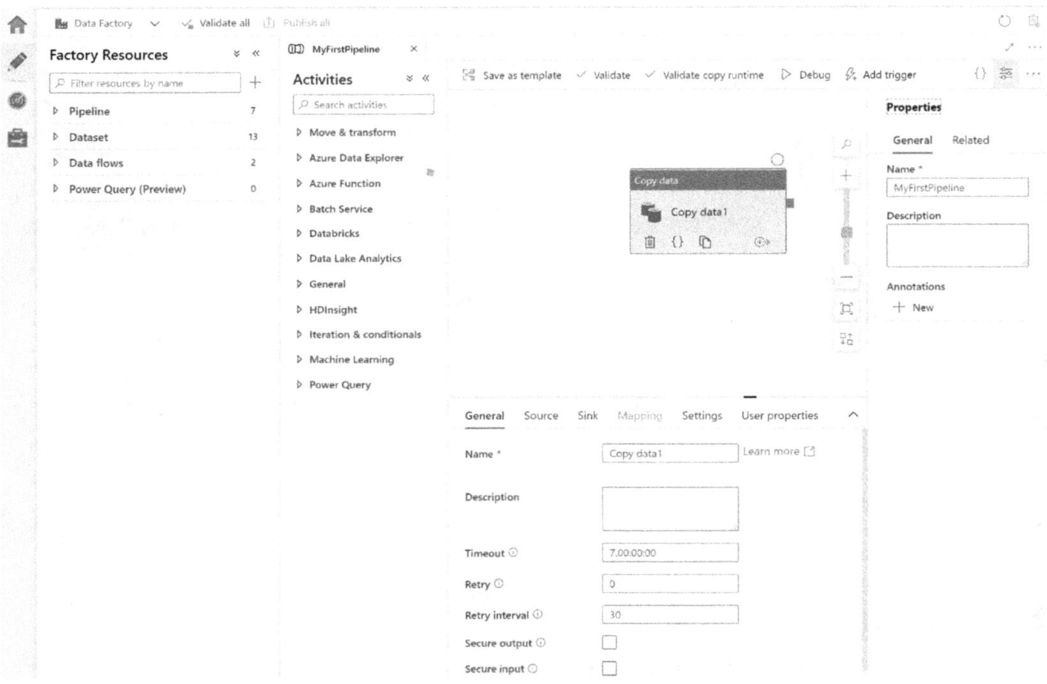

Figure 5.10 – Development canvas with a new copy activity

2. Please check the options in the settings pane. **Source** and **Sink** are marked with a red **1** to their upper right. This means that they have mandatory configurations that need to be done.

3. Please name your activity `CopyArbitratryFile` or similar and proceed to the **Source** settings. Click **+ New** to create your first source dataset. The **New dataset** dialog will open from the right:

4. Select **Azure Data Lake Storage Gen2** and then click **Continue**

   In the next step of the dialog, you will be able to pick a file format. See *Figure 5.12*:

5. As we want to binary copy the file from our source folder to our target folder, select **Binary** and then click **Continue**.

6. In the **Set properties** dialog, you can now name your dataset something such as `BinaryArbitraryFile`, and then, from the **Linked Service** combo box, select **+ New**.

7. In the **New linked service (Azure Data Lake Storage Gen2)** dialog, name your newly created linked service something such as `MyDataLake`.

8. In the **Azure subscription** drop-down box, select the subscription where you have created your data lake. Below that, in the **Storage account name** drop-down box, select the data lake service. Please leave the other options as their default settings. In the lower-right corner, you can now click on **Test connection** and hopefully, you will get a **Connection successful** message, as in *Figure 5.13*:

9. Then, click **Create** and select your file.

10. In the **Set properties** dialog, please click on the small folder icon next to the **File** box of the **File path** option. You will be taken to your data lake, where you now can select the file that you uploaded at the beginning of this section:

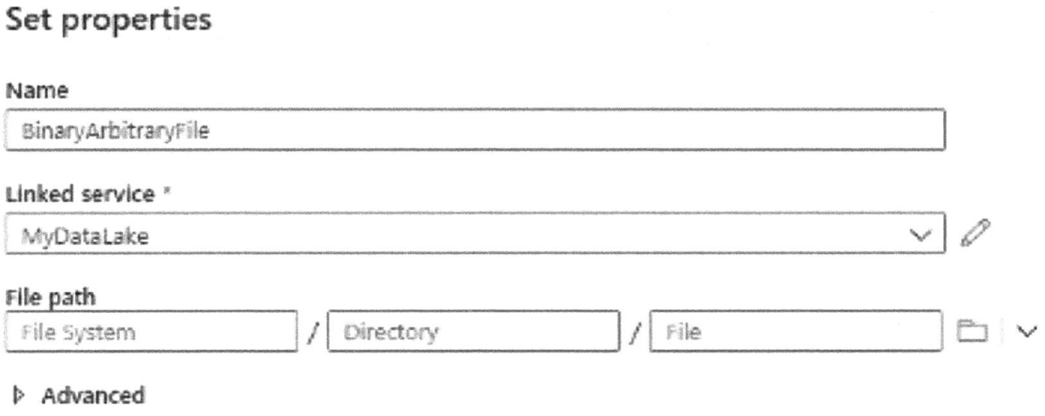

Set properties

Name

| BinaryArbitraryFile |

Linked service *

| MyDataLake | ⌄ | 🖉 |

File path

| File System | / | Directory | / | File | 🗀 ⌄ |

▷ Advanced

Figure 5.11 – Setting the properties dialog to select the source file

11. After clicking **OK**, you will return to your Copy data activity. Your **Source** configuration is now complete. Good job!

12. When you now proceed to the **Sink** configuration, you will find a very similar sequence.

> **Important note**
>
> Don't select the `BinaryArbitraryFile` dataset that you just created. This would mean trying to write to your source and that would not work.

13. Create a new dataset using the aforementioned sequence using the **MyDataLake** linked server that you have already created. The only difference compared with your target dataset (which may have a name such as `BinaryArbitraryTargetFile`) will be that when you configure the folder path in the **Set properties** dialog, you won't need to set a filename. The file will be named as per the source.

14. Once you click **OK**, you will again return to the development canvas. You're doing great so far and are at the point of firing your first Azure Data Factory pipeline. At this stage, however, you won't yet be able to start it. You will first need to save it. If you now check your menu ribbon, you will find the **Publish all** button in blue and with an indicator that shows the number of new or changed artifacts. In your case, this should be **3** (see *Figure 5.15*):

Figure 5.12 – Menu ribbon with the Publish all button highlighted

15. Why 3? Well, you have created two new datasets and a new pipeline with a copy activity. Please click the button and save your work. The **Publish all** dialog will appear and provide you with an overview of the pending changes that are saved (or checked in, if you are connected to Git).

16. Now, publish your changes and finalize your pipeline.

> **Important information**
>
> At any point during the creation of linked services, for example, you may have registered the **Storage account key / Azure Key Vault** setting. We won't touch Azure Key Vault in this sequence. However, it is a good idea to make yourself familiar with the concept. Azure Key Vault will help you to modularize your security settings and even enable developers in this case to use credentials, connection strings, and more that are stored in Key Vault without knowing them at all. This enables a whole new dimension of security settings.

Triggering and monitoring your first pipeline

Now, finally, let's initiate our pipeline and get the job done! Please click **Add trigger** above the development canvas. Select **Trigger now** and, in the following dialog, just click **OK**.

If you now change the view to the **Monitor** section, you can monitor the running of your first pipeline:

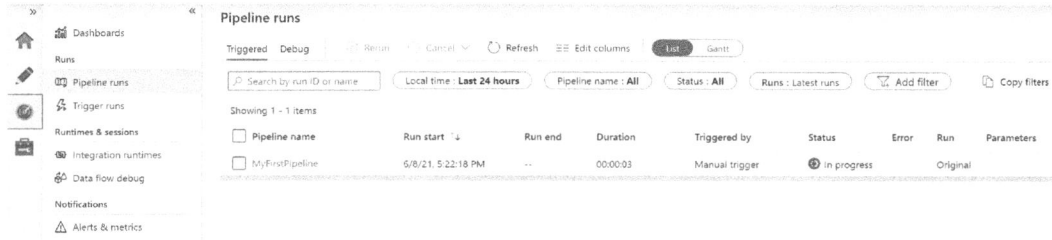

Figure 5.13 – Monitoring pipeline runs

If you are as impatient as me, you can, of course, click **Refresh** in the menu ribbon above the detail section. There should not be too much of a job to do and your pipeline should finish very quickly.

When you click the pipeline name in the **Pipeline runs** detail section, you will be taken to the details of your pipeline run. If you hover the mouse pointer over the particular entry in the list of activity runs, you will be presented with a number of buttons that will provide more insights into the particular activity run of your Copy data activity:

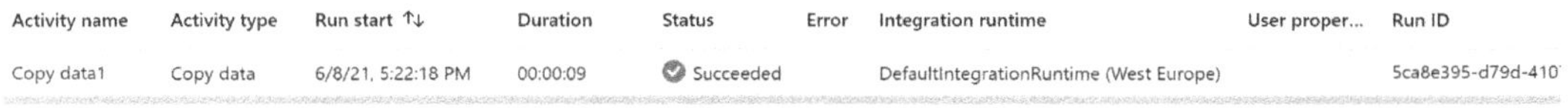

Figure 5.14 – Activity run details

By clicking on the glasses button, you will finally be taken to a detailed overview of what happened during your first ETL job with Synapse pipelines or Azure Data Factory:

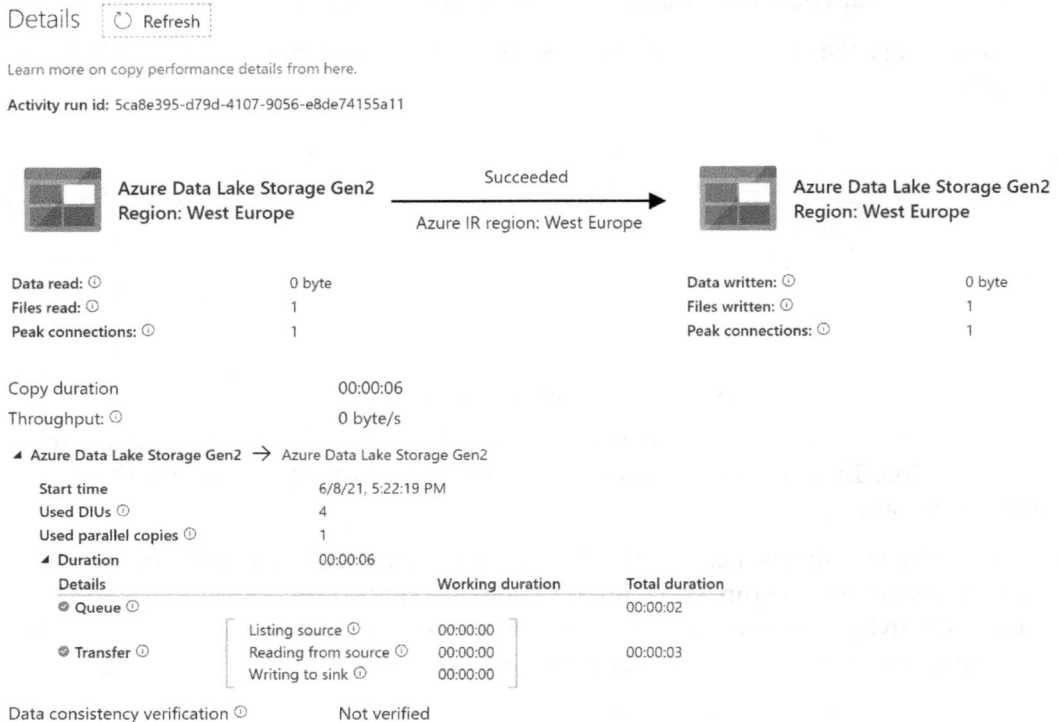

Details   ○ Refresh

Learn more on copy performance details from here.

Activity run id: 5ca8e395-d79d-4107-9056-e8de74155a11

		Succeeded		
	Azure Data Lake Storage Gen2 Region: West Europe	Azure IR region: West Europe		Azure Data Lake Storage Gen2 Region: West Europe

Data read: ○	0 byte		Data written: ○	0 byte
Files read: ○	1		Files written: ○	1
Peak connections: ○	1		Peak connections: ○	1

Copy duration   00:00:06
Throughput: ○   0 byte/s

▲ Azure Data Lake Storage Gen2 → Azure Data Lake Storage Gen2

Start time	6/8/21, 5:22:19 PM		
Used DIUs ○	4		
Used parallel copies ○	1		
▲ Duration	00:00:06		
Details		Working duration	Total duration
⊘ Queue ○			00:00:02
⊘ Transfer ○	Listing source ○	00:00:00	
	Reading from source ○	00:00:00	00:00:03
	Writing to sink ○	00:00:00	

Data consistency verification ○   Not verified

Figure 5.15 – Activity details view

You might now check the target folder with Azure Storage Explorer, if the file was transported.

What have you done? You have created a pipeline and added an activity. In this case, it was a copy activity. To get access to your source file, you needed to set up a linked service to your data lake and you needed a linked service again (in this case, it was the same as it points to the same data lake) for your target. Finally, you triggered your pipeline to run it. The source and target in particular were then created as datasets. And this is roughly how artifacts are tied together (see *Figure 5.19*):

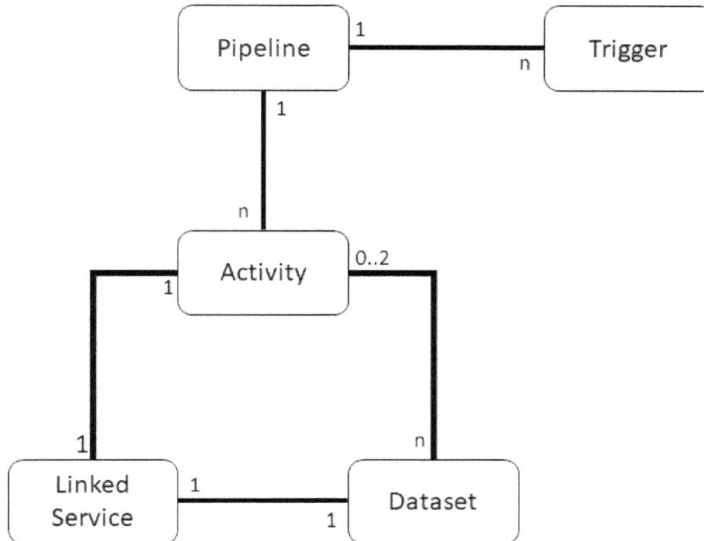

Figure 5.16 – Object hierarchy for Synapse pipeline/Data Factory artifacts

We will dive deeper into data loading with Azure Data Factory and Synapse SQL pools, for example, in *Chapter 11*, *Developing and Maintaining the Presentation Layer*.

Now that you have a basic understanding of the nuts and bolts of Synapse pipelines/Azure Data Factory, you may want to examine the following section about wizards.

Using wizards

Hey, it's Microsoft. They love wizards! And Synapse pipelines and Azure Data Factory are no exception here. If you click the **Home** button in the navigator pane of your UI (see *Figure 5.5 – Azure Data Factory home screen*), you will find a link that says **Copy data**. In Synapse Studio, you will find a link with the name **Ingest**. I would like to encourage you to examine the **Copy Data Wizard**. It will give you very good examples of how to use the copy activity.

You may perhaps want to try to copy a bunch of database tables in a single step. As soon as you have created a linked service to your database, the Copy Data Wizard will prompt you for the table selection. You can select one or more tables in this window:

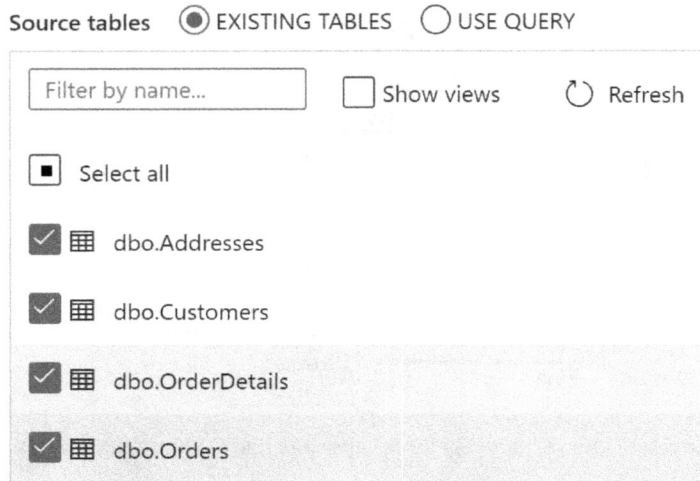

Figure 5.17 – Multi-table selection in the Copy Data Wizard

Following additional settings for the single tables (you can name each file explicitly or leave this to the wizard), you can finish your pipeline and even run it from the wizard. Alternatively, you can connect to your Synapse SQL pool from *Chapter 4, Understanding Synapse Pools and SQL Options*.

Perhaps you created the SQL pool with the sample. This might come in handy now.

> **Note**
> This wizard comes in quite handy when you need to perform a one-shot action to provide data quickly to any of your internal customers or when you want to move a whole bunch of tables or files from one place to another when you, for example, re-organize your data lake or similar.

But let's examine what happens after you have completed the wizard. If you check the pipeline that was created, you will find something similar to *Figure 5.21*:

Figure 5.18 – Result of the Copy Data Wizard for multiple database tables

If you examine the pipeline now, first, click somewhere in the white space next to the ForEach activity and you will see the configuration of the pipeline itself:

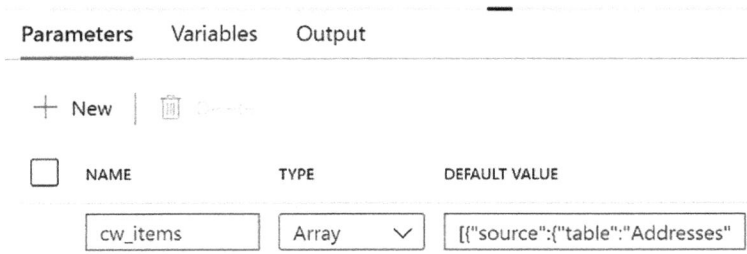

Figure 5.19 – Pipeline configuration pane below the development canvas

Now, we'll check out how to work with various parameters.

Working with parameters

You can use parameters in every field that shows the text **Add dynamic content [Alt+P]** as a link (see *Figure 5.23*). You just simply click on that link and select the parameter from the list of functions and values that appears:

cw_table   @item().source.table

Add dynamic content [Alt+P]

Figure 5.20 – The Add dynamic content link below a field

But let's move on with our example from the wizard.

In the settings pane of the pipeline, you can now see a parameter on the **Parameters** tab. This one is created through the wizard when you select the tables of the source database. The value is a JSON array that contains the names of the tables that you have selected. This is a parameter that is now available to all activities in the pipeline:

```
[
  {
    "source": {
      "table": "Addresses"
    },
    "destination": {
      "fileName": "dboAddresses.parquet"
    }
  },
  {
    "source": {
      "table": "Customers"
    },
    "destination": {
      "fileName": "dboCustomers.parquet"
    }
  }, ...
```

It contains the database tables that we have selected always as a source, but also always as a destination. Keep an eye on those two throughout the following section.

If you now click once on the **ForEach** activity and go to the **Settings** tab in the settings pane, you will see how the **ForEach** activity implements the pipeline parameter in the **Items** field. The **Items** field, as you could derive from its name, represents the list of items that the **ForEach** activity will iterate through:

General	**Settings**	Activities (1)	User properties

Sequential	☐
Batch count	[ ] ⓘ
Items	[@pipeline().parameters.cw_items]

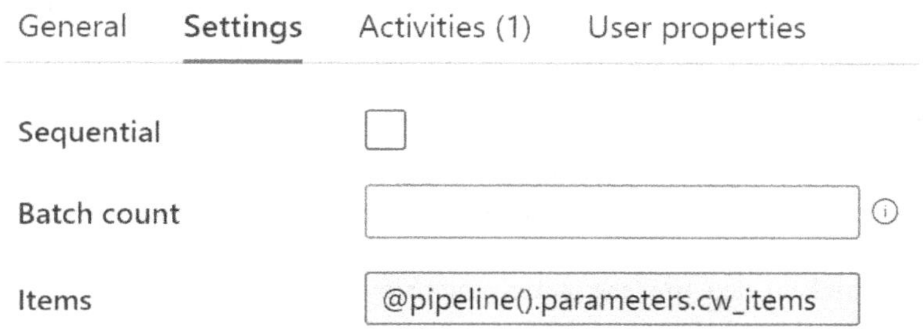

Figure 5.21 – The Settings tab on the settings pane of the ForEach activity

A single item will always then hold the following:

```
{
  "source": {
    "table": "Addresses"
  },
  "destination": {
    "fileName": "dboAddresses.parquet"
  }
}
```

However, the use of this parameter goes further. When the ForEach activity iterates through the array of items, it will provide every item to the activities that it contains. If you go to the **Activities** tab in the settings pane, you will see the number of activities there, and if you click on the pencil icon next to the list, you are taken into the activities that are run for each item that the ForEach activity will iterate through.

Now, when you arrive there (you can also click into the **ForEach** activity and click on the pencil icon there), you will see all the activities that are run per item in the array of table names from previously. When you now click on the **Source** tab in the settings pane of the **Copy data** activity, you will see another parameter that is used there (see *Figure 5.24*).

Note

The ForEach activity comes in handy when you need to iterate through collections. But keep an eye on the performance and the cost that your pipeline will cause. If you iterate through many files and create a `foreach file in folder` construct, remember that each activity run will be charged. This can add up if you process hundreds of thousands of files in a load.

A better solution would be to load all the files in one go to your analytical environment and operate based on sets of data. `DateTime` and `Window` functions and even stored procedures in your target DWH database or Spark notebooks can later take the task of iterating through sets.

An example: if you need to load 200,000 files per day and need to process, let's say, eight different steps, this will already cause 200,000 times 8 equals 1,600,000 activities to be issued per day, if you run them in a ForEach loop. This will definitely slow down your process.

If you manage to create set-based loads, you will end up with eight activities to load your data and maybe one or two more for a stored procedure or a Spark job to iterate through your sets.

Just like the pipeline itself, the source dataset that was created using the wizard contains a parameter, **cw_table**. You can examine the source dataset by clicking on the pencil icon next to its name. There you will again find a **Parameters** tab in the settings pane where this parameter was configured:

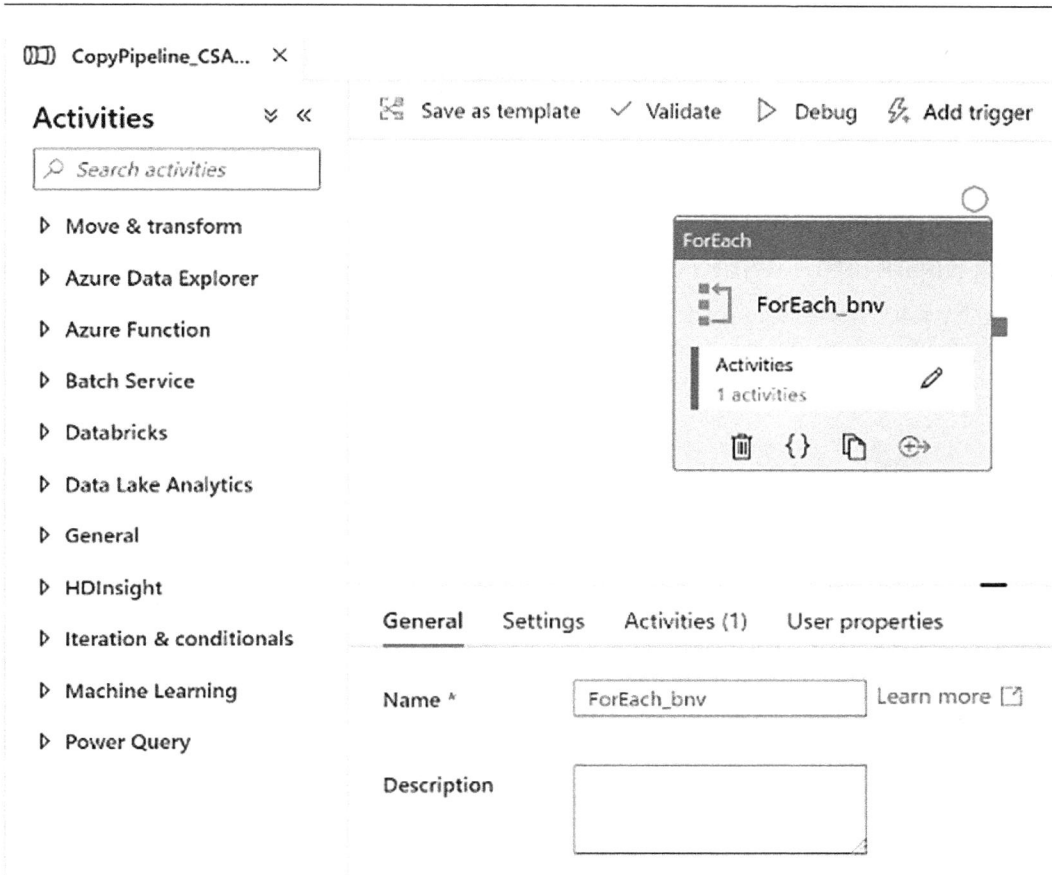

Figure 5.22 – The Source tab of the copy activity with a parameter

The parameter of the source dataset receives the value that was pushed from the ForEach activity:

```
@item().source.table
```

This is now the detail level where the single table name from the JSON array of the pipeline is used.

If you click on the **Sink** tab of the **Copy data** activity, you will see another parameter, **cw_fileName** (this time from the sink dataset), and the usage of the second part of the JSON array of the pipeline:

```
@item().destination.fileName
```

Setting parameters when a pipeline is triggered

If you go back to your main pipeline and click on **Trigger**, again, as in a previous section, *Triggering and monitoring your first pipeline*, this time the dialog that will open before you finally kick the job will show the parameter and give you the option to change it before the job is run:

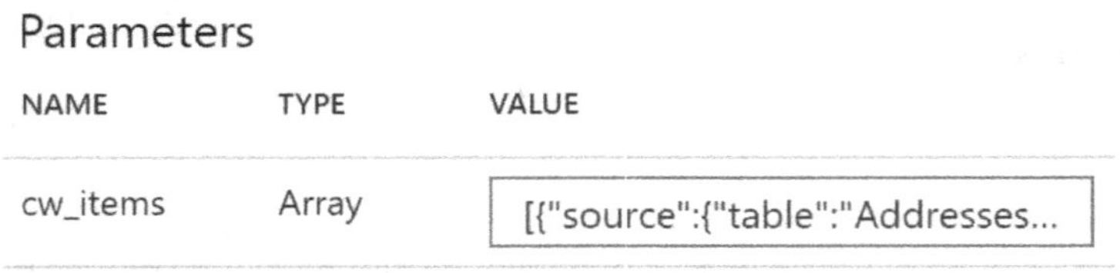

Parameters

NAME	TYPE	VALUE
cw_items	Array	[{"source":{"table":"Addresses...

Figure 5.23 – Parameters dialog when triggering a pipeline

> **Note**
>
> If you carefully read the upcoming *Chapter 6, Using Synapse Spark Pools*, you will find the concept of parameter cells in Spark notebooks (the *Handling cells* section). When you toggle a parameter cell in your Spark notebook, the parameters that you implement there will be exposed to your Synapse pipeline when you use a Synapse Notebook activity.
>
> Stored procedure activities will give you the same options for input parameters, when the used stored procedure exposes them.
>
> This is quite a powerful integration. Just as described previously, you can use your pipeline parameters in these activities and keep your integration routines very flexible.

Using variables

Variables are used in a similar manner to parameters with one big difference: you have an activity in your toolset that can be used to manipulate variables during a pipeline run.

> **Note**
>
> Variables in comparison to parameters have only a limited list of data types possible: string, Boolean, and array.

If you browse through the **Activities** pane to the **General** section, you will find the **Append variable** and **Set variable** activities. You can use the two to influence the values that a pipeline variable can have during the pipeline run. Again, the **Value** field in the **Set variable** activity will show **Add dynamic content**:

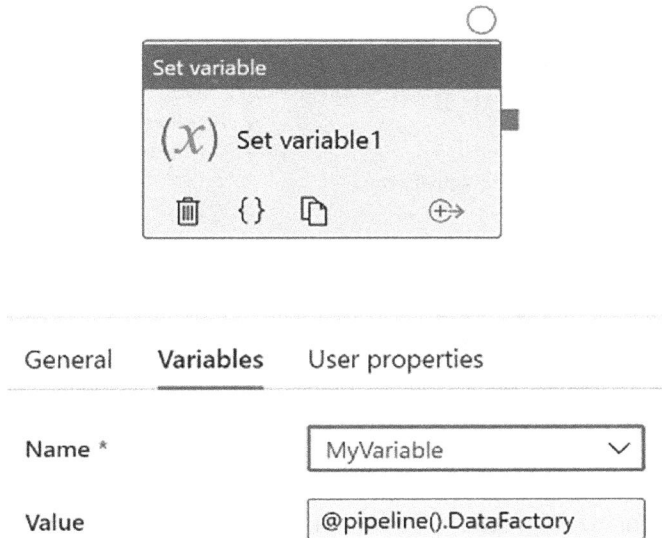

Figure 5.24 – Setting a pipeline variable to hold the name of the data factory

[**Alt+P**] brings you to the formula editor, where you can set up the formula for your variable.

The **Set variable** activity will do exactly what the name says: it will set a variable to a specific value. The **Append variable** activity, in contrast, can be used if you are collecting values into an array for subsequent use.

Adding data transformation logic

Up to now, we have examined the ELT mode of Synapse pipelines/Azure Data Factory. But like other data integration tools, Synapse pipelines and Data Factory offer the option to create logical dependencies in your pipeline run. You can put together different activities in a single pipeline and connect them to form your loading logic.

Figure 5.28 shows the possibilities for adding dependencies between two activities. In this way, you can set up complex loading logic. The displayed dependency would run the **ForEach** activity only when the **Copy data** activity has been run successfully:

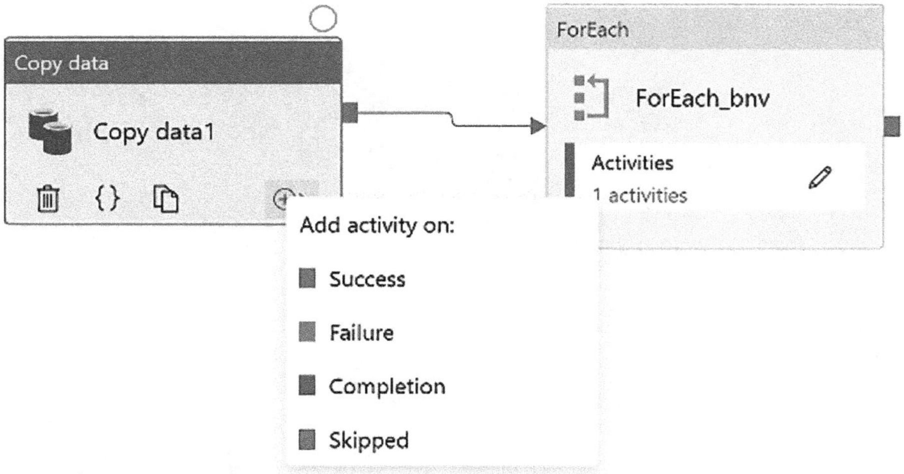

Figure 5.25 – Adding dependencies to your pipeline

But what about transformation logic? What about the need to calculate, filter, or aggregate data and add other information from elsewhere to your data?

You can use **data flows** to do so. Data flows will give you a graphical interface to "program" your transformation logic and you will only need to code calculate, aggregate, or filter formulas using the built-in expression language.

Synapse pipelines and Azure Data Factory offer two options when it comes to creating data flows in the UI: **mapping flows** and **wrangling flows**, each with a different toolset targeting different purposes.

Understanding mapping flows

Mapping flows target use cases where both the source and sink are known and structures and data types are available. The graphical interface will give you access to source and target transformations and offers a set of transformational tools, including filters, aggregators, derived column transformations, and many more.

The graphical representation of your mapping flow will cause the generation of Scala code in the background. This code fragment will be sent for processing to a Spark cluster that is controlled by Synapse pipelines/Azure Data Factory.

Before we can move on to your first mapping flow, please get some example data to work with. The following example uses a dataset about airline on-time statistics and delay causes that are provided by the Bureau of Transportation and Statistics of the United States. Please go to the GitHub repository that belongs to this book and get the `airdelays.csv` sample file from the folder for *Chapter 3, Understanding the Data Lake Storage Layer*:

`https://github.com/patsql/CloudScaleAnalytics`

Creating a mapping flow

Mapping flows are represented as activities in pipelines, just as you would use a copy activity like the one you created in a previous section, *Building your copy pipeline*:

1. Please navigate to the authoring view, create a new pipeline, and name it `airdelaysdataflow`. In the **Activities** pane, navigate to **Move & transform** and drag the **Data flow** activity to the authoring canvas.

2. Rename the activity `actAirdelaysDataFlow` on the **General** tab in the settings pane (below the authoring canvas; see *Figure 5.29*):

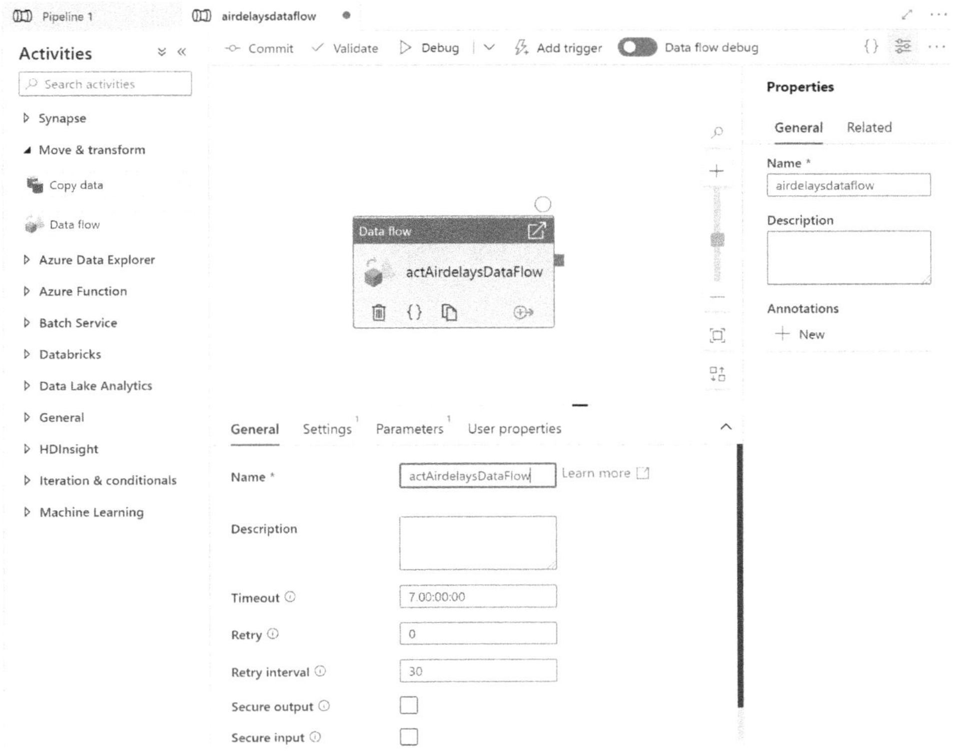

Figure 5.26 – New data flow activity

3. On the **Settings** tab, create a new mapping flow by clicking on **+ New**. You will now enter the authoring mode for mapping flows. Please click **Finish** to end the guided tour.

4. In the upper-left corner of the authoring canvas, you will now see a dotted rectangle with the text **Add Source**. Please click it and name the new data source something such as `airdelayscsv`.

5. In the **Dataset** field, click **+ New** and create a new delimited file dataset from the source that you have just uploaded:

a) Select **Azure Data Lake Storage Gen2**.

b) Select **DelimitedText**.

c) Name the dataset `airdelayscsv` and select the Data Lake linked service that you created during the creation of your first pipeline (`MyDataLake`). Then, browse to the file after clicking on the folder symbol next to the **File path** fields. Please check the box for **First row as header**. You can then finally choose the **From connection/store** option in the **Import schema** section (see *Figure 5.30*):

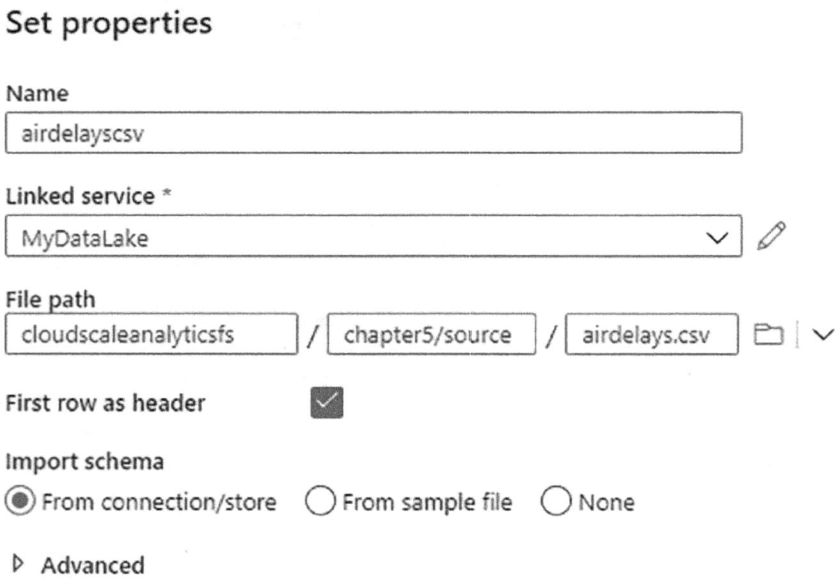

Set properties

Name

| airdelayscsv |

Linked service *

| MyDataLake | ∨ | ✎ |

File path

| cloudscaleanalyticsfs | / | chapter5/source | / | airdelays.csv | 🗁 | ∨ |

First row as header   ☑

Import schema
◉ From connection/store   ◯ From sample file   ◯ None

▷ Advanced

Figure 5.27 – Properties for the data flow dataset

6. Now you can start to develop your mapping flow visually. Click the + icon right next to the data source representation in the mapping flow. A selection of possible transformations will be displayed:

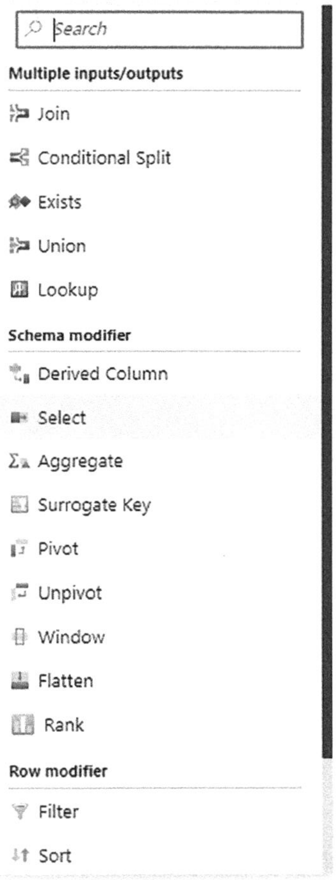

Figure 5.28 – Selection of transformation options in the mapping flow

7. We will first narrow down the number of columns used in the mapping flow. Please use **Select**. At the moment, we have 44 columns coming from the source, but we don't need them all.

> **Note**
> It is a good habit to always name your objects. You have a visual development environment and you should use all options to create understandable and maintainable pipelines.

Please select the following columns: YEAR, MONTH, FL_DATE, UNIQUE_CARRIER, TAIL_NUM, FL_NUM, ORIGIN_AIRPORT_ID, ORIGIN, ORIGIN_STATE_ABR, DEST_AIRPORT_ID, DEST, DEST_STATE_ABR, DEP_TIME, DEP_DELAY, ARR_TIME, ARR_DELAY, and DISTANCE:

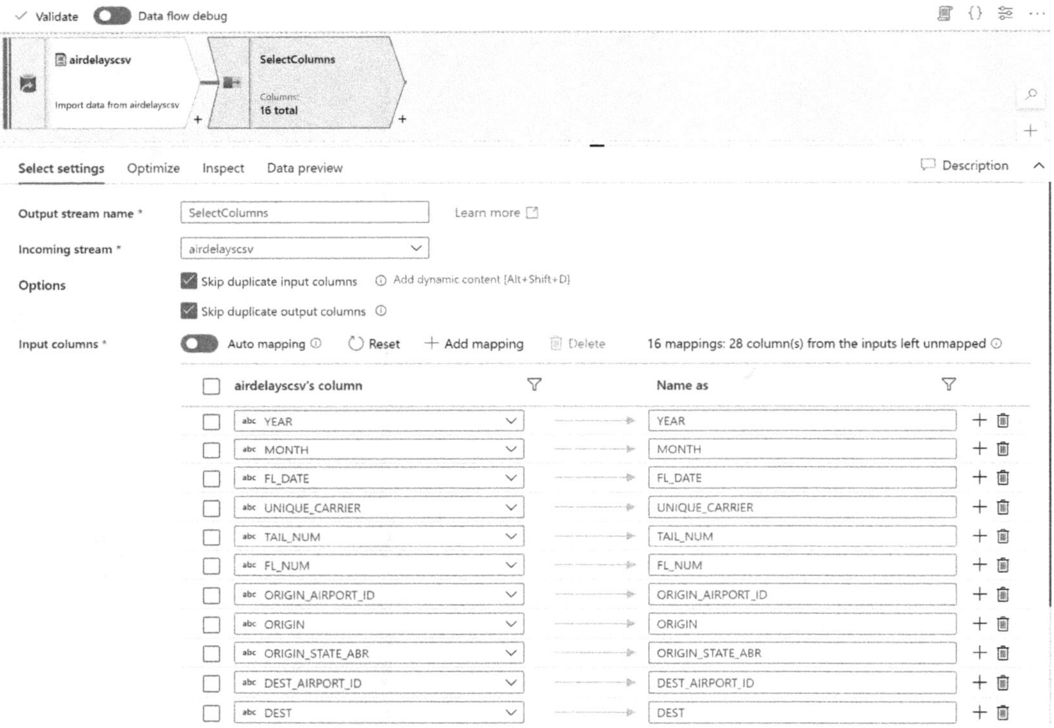

Figure 5.29 – Mapping flow with the Column selector

8. Now, let's calculate whether a delayed departure always led to a delayed arrival. Please click the **+** icon and add a **Derived Column** transformation. Name it CalcDepDelayCauseArrDelay or something similar, something that you will be able to recognize again when you return later. In the **Settings** tab of the **Derived Column** transformation, create a new column called DepDelayCauseArrDelayFlag and enter the following formula in the **Expression** field:

```
iif(toInteger(DEP_DELAY) > 0 && toInteger(ARR_
DELAY) > 0, 1, 0)
```

With **OK**, you can leave the expression editor and move on:

9. Now, let's finally write the data to a sink. Once again, click + next to **Derived Column** and select **Sink**. Name it something such as `airdelaysDepDelayCauseArrDelay` and create a new dataset in your Data Lake store. Once again, select **Azure Data Lake Storage Gen2** and then **Parquet**. In the following **Set properties** dialog, browse to the "target" folder. Don't provide a filename once you have selected the folder. However, please do add an addition to the **Folder** field. This folder will then be created and the Parquet files will be placed there. Below the **File path** fields, please select **None** and then click **Create**:

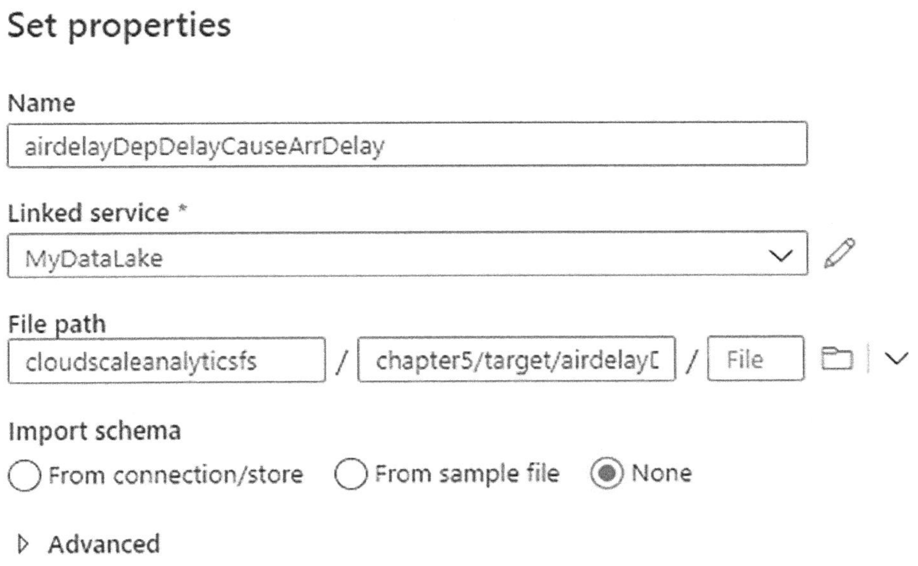

Set properties

Name

| airdelayDepDelayCauseArrDelay |

Linked service *

| MyDataLake   ∨ | 🖉 |

File path

| cloudscaleanalyticsfs | / | chapter5/target/airdelay[| / | File | 🗀 | ∨ |

Import schema

○ From connection/store   ○ From sample file   ⦿ None

▷ Advanced

Figure 5.30 – File properties for the sink in your first mapping flow

10. When you now publish your mapping flow, it will be validated and if there are no errors, it will be checked in. Finally, trigger the pipeline and go to the monitor area to follow what happens.

Important note

When the pipeline with a mapping flow gets triggered, please check the monitor area and then check the upper-right corner of the pipeline details. A Spark cluster gets created on the fly. If you are using Synapse pipelines, a Synapse Spark cluster will be used. If you are working with Data Factory, a Databricks cluster will be created instead. This will take between 1 and 4 minutes and the pipeline will not kick in immediately. This can be mitigated by provisioning a cluster and attaching your pipeline to it (refer to the following section entitled *Understanding integration runtimes*).

You will find more information about mapping flows in the *Further reading* section of this chapter. You will read about the **debug mode** of mapping flows, for example, and how to manage performance. Additionally, you will find a reference to the **expression language** used in mapping flows.

Understanding wrangling flows

In comparison to mapping flows, wrangling flows target use cases where the sink is not known, for example, when a data engineer needs to quickly provide a new dataset based on new requirements from a data scientist or a data analyst. Wrangling flows implement the **Power Query** experience that was introduced with Power BI and was then later used in newer Excel versions as the **Get data** experience.

Regarding the engine used in wrangling flows, you will see them also running on the Spark engine in the background.

Power Query will enable the data engineer to quickly "massage" source data into the required format and structure. Just like the mapping flows, there is no need to code more than some expressions for new fields, for example. Even that can be done without coding if preferred. This is a functionality that is called **Column from examples**.

This feature enables the data engineer to point the engine to one or more source columns and literally write the expected value into a new column. The AI component in the background will derive a suitable formula that leads to the given value using the source column(s), and if the function isn't sharp enough for subsequent rows, the engine will ask for additional input from the data engineer. Clever, isn't it? But let's see how to implement a wrangling flow.

Implementing wrangling flows

1. Please go to your authoring UI and, in the **Activities** pane, search for Power Query. Alternatively, in the Synapse pipeline UI, you would create a new data flow and can then select between a mapping or a wrangling flow. On the **New power query/New wrangling flow** dialog, please name it first (perhaps something such as airdelaysparquetpq).

2. Then, as a source dataset, select the dataset that points to our airdelays.csv file from above airdelayscsv.

3. As a sink dataset, please create a new .parquet dataset in your airdelaysparquetpq data lake:

   a) Click + **Add**.

   b) In the drop-down box, select + **New**.

c) Select **Azure Data Lake Storage Gen2**.

d) Then, select **Parquet**.

e) Name the dataset `airdelaysparquetpq` and select the Data Lake Storage: `cloudscaleanalyticsadls`.

f) Select the target folder by clicking on the folder symbol next to the **File path** section.

g) Add the target name to the end of the string in the **Directory** field (the middle one) of **File path**. Don't provide a filename yet.

h) In the **Import schema** section, select **None** to prevent the dialog from searching for an output schema as this isn't known yet and will be derived from the final output that our wrangling flow will create.

i) Click **OK**.

4. Back on the **New power query** dialog, you can extend **Sink properties** and click **Output to single file**. In comparison to the preceding mapping flow, we will try to only create a single file this time. A field for the filename will appear. Please name it there (`airdelayspg.parquet`).

5. Finally, move on to the authoring area of your wrangling flow by clicking **OK**.

Now that you have your source dataset open in the authoring pane, you will see something like this:

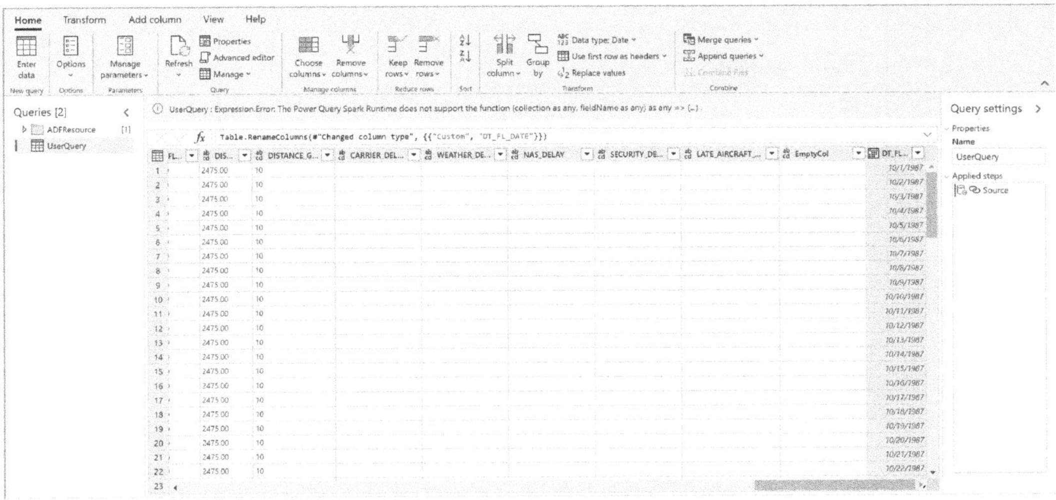

Figure 5.31 – Power Query/wrangling flow authoring

Let's explore the authoring experience here and meet the **Column from examples** functionality.

What instantly grabs your attention in this dataset is the FL_DATE column in the middle of the display. It is shown with the / character to separate the day from the month from the year. Perhaps you want to change this. Let's examine two ways of doing this.

Using the more conservative way, we replace the / with a . using a **replace** function. We can nicely replace the character as the column is in string format at the moment. You can see this from the **ab** next to the column name in the display. **123** would point you to a numeric column type and a schedule icon would tell you that you're looking at a date.

But let's go on.

6. Please select the column header of FL_DATE; this will select the whole column.

7. Either right-click and select **Replace values** on the column header or go to the **Transform** menu entry and search for **Replace values** in the **Menu** ribbon. The following dialog will ask you for the **Value to find** and **Replace with** values:

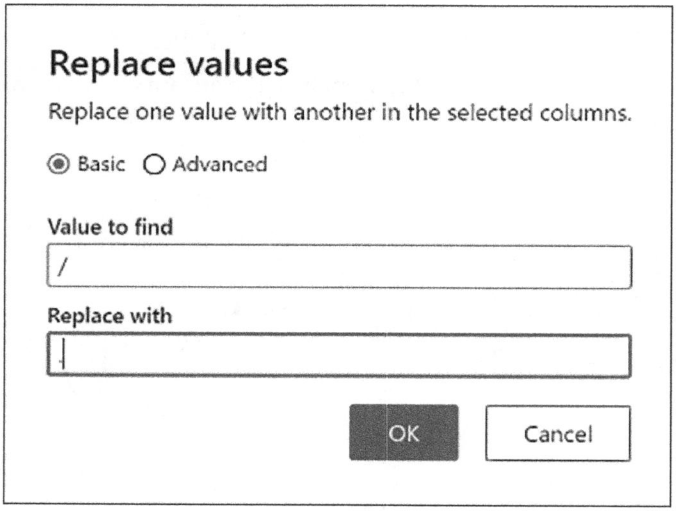

Figure 5.32 – The Replace values dialog

8. Type in / and . as shown in the screenshot and click **OK**.

9. The Power Query M formula will be displayed in the entry field above the table view and the values in the column are now showing a . instead of a /.

10. Now, let's try and change the data type of the column to **Date**. Please click on **ab** next to the column title. The data type selector appears, where you can now select **Date** as the type.

11. Ouch! There are now errors in the formula. Let's check this. All the days displayed that are higher than 12 are causing this error. The engine thinks that the first two digits of the date display the month and not the day, as in the American date format.

12. Let's revert the steps: if you look to the far-right side of your authoring window, you will find the **Query settings** pane. In the **Applied steps** area, you will find all the steps that you implement as a separate row:

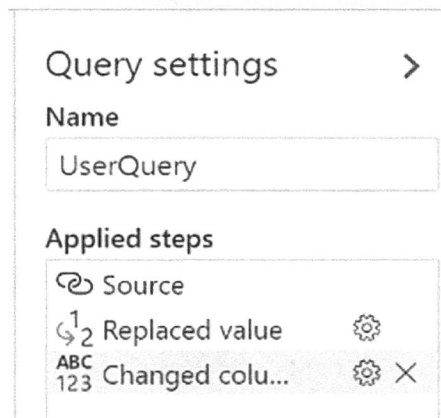

Figure 5.33 – Applied steps in Power Query

13. Please remove the last two, **Changed column type** and **Replaced value**, by clicking on the **X** sign right next to them.

14. But how can we now deal with that value? Select the column again by clicking on the title.

15. In the menu bar, select **Add column** and, from there, **Column from examples/ From selection**.

16. A new and empty column will be displayed. In the first row of the new column, please type your date in the suitable sequence with a / or a ., as you wish. In my dataset, the date in the first row is 01/10/1987. Therefore, I will type 10/01/1987 in my example column and then check what happens. The algorithm initially does not really understand what to do. So, let's go ahead and type in the example for the second row: 10/02/1987. Now the first 12 rows have been set, but still the algorithm is not yet convinced that it has it right. Please go to row 13 and type in another example, 10/13/1987. Now, all the rows are set with a value according to your examples. Pretty nice, isn't it?

17. Now, let's try and change the format to **Date** for the newly created column. This time it worked, and without you writing a single line of code. Finally, you might name the new column something such as DT_FL_DATE. If you examine the formula that was created on your behalf in the **Added Custom Column** step, you can confirm that writing the examples was far easier instead, right?

18. Finally, you will again need to publish your work to save it and then run the wrangling flow on your data.

19. The only thing you now need to do to integrate this power query into a pipeline is to drag a **Data flow** activity into a pipeline and, in the settings pane, select the wrangling flow. As for the rest, you know the drill from before.

20. Wrangling flows rely on the same implementation language as Power Query: **M**. So, if you are already a proficient M developer, you might prefer to write your query into the **Advanced editor** section. You will find the button for that on the **Home** menu in the **Query** section:

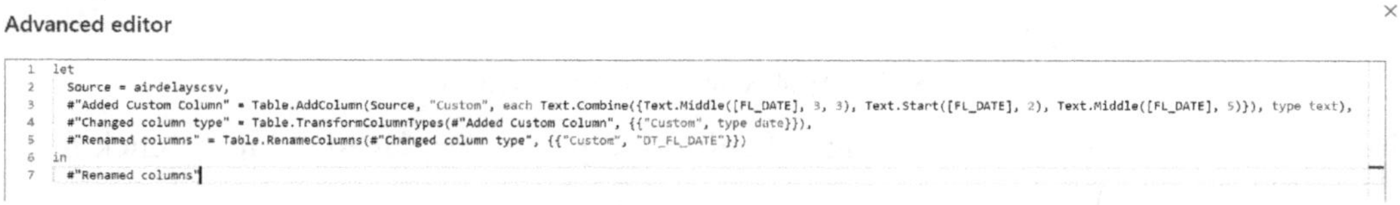

Figure 5.34 – Advanced M editor for wrangling flows

And yes, you can port M queries from Power BI to wrangling flows. There are some functions that are not yet supported, but for most of the features, this will work exactly as in Power BI. So, imagine working on a small sample dataset and then moving the query to a really scalable environment, where you then can do the heavy lifting!

Please find additional material and references to the M language in the *Further reading* section.

Understanding integration runtimes

IRs are responsible for all the compute jobs in Synapse pipelines or Azure Data Factory. When it comes to data movement as well as running the data flow activities (refer to the preceding *Understanding mapping flows* and *Understanding wrangling flows* sections), additionally, the IRs invoke activities and take care that they are run on their respective compute environments. Running **SQL Server Integration Services** (**SSIS**) packages is another responsibility that IRs handle.

There are different types of IRs available in Synapse pipelines/Azure Data Factory:

- **Azure IR**: This will take care of data movement activities such as data copy, control all activity runs, and run data flows (mapping and wrangling flows).

- **Self-hosted IR:** This must be installed in any on-premises network to "see" data sources that are loaded from on-premises to the cloud. They act as a data gateway and are responsible for the data movement from on-premises to the cloud. In this role, they control all activity runs with regard to their location.

- **Azure SSIS IR:** This will run SSIS packages. This is important when you have invested heavily in SSIS on your premises and don't want to write everything again in Azure Data Factory. SSIS packages can be deployed with only minor changes in the connection managers of the packages to reflect the new location that they need to connect to.

> **Note**
> The Azure SSIS IR is only available in Azure Data Factory.

Examining the Azure IR

This IR will participate in all kinds of workloads that are running on Azure sources, implementing data flows and invoking all kinds of activities. The Azure IR will be the compute provider for the activities that are involved with data movement and data flows.

If you examine the copy activity carefully, you will find the setting to control the **data integration units**. This setting directly relates to your IR where your copy activity is run. When there are several copy activities running in parallel, for example, and each one has its data integration unit settings adjusted, the IR will scale accordingly to fulfill the requirements of your workload without you taking care of this.

Please refer to the *Further reading* section for more detailed information about the Azure IR, how to set it up, and how to manage it.

Examining the self-hosted IR

The self-hosted IR will act as a gateway between your on-premises data sources and your modern data warehouse. The Azure part of Synapse pipelines/Data Factory will never access your on-premises data directly without you being able to control precisely what data can be retrieved from your data sources.

The self-hosted IR will be installed in your network on a VM or a physical machine that is next to your data as close as possible. By doing so, you can control the self-hosted IR and the actions possible for this gateway.

> **Note**
>
> To install a self-hosted IR, different approaches are possible. The shortcut method would be to log in to the Synapse workspace/Azure Data Factory from the exact machine where the self-hosted IR should be installed. You can go to the **Administration/Management** pane and start the creation of a new self-hosted IR from there. Once you have initiated creation, you will be offered two options: **Express setup** and **Manual setup**. If you select **Express setup**, you will install everything and configure the environment using a wizard. Following the setup, your self-hosted IR will instantly be available and can run jobs. Of course, you can execute the manual setup or even do a scripted and automated setup for your enterprise-grade installation.

Integrating with virtual networks and private endpoints

Up to now, we haven't cared about network boundaries and hiding our data traffic from the public internet. But you will need to prepare your services to route the data in a more private manner. Synapse and Azure Data Factory support you in doing so.

An Azure IR can be created within the so-called Azure Data Factory Managed **Virtual Network (VNET)/Synapse managed VNET**. This means that the service creates a VNET for you where your IR will be placed. You can trigger this setting when creating a new Azure IR. In **Integration runtime setup**, search for **Virtual network configuration** and enable it. This Azure IR will then run within a newly created virtual network on Azure that will only be reachable by peered networks (network peering is the method whereby two networks are "introduced" to one another so that the components within those networks can "see" each other).

Additionally, Synapse and Data Factory support **managed private endpoints**. A private endpoint in general can be seen as a resource interface that uses a private IP address from the virtual network the resource is associated with. The managed private endpoints, just like the managed virtual networks, relieve the admin from the management tasks to set up those network resources and IP addresses.

By setting up managed private endpoints in the managed virtual network, the admin creates such a private endpoint and assigns it to the target resource. This could be your Data Lake Storage, for example (see *Chapter 3*, *Understanding the Data Lake Storage Layer*), or a database (*Chapter 4*, *Understanding Synapse Pools and SQL Options*) that holds data that needs to be extracted. Following the assignment, the administrator of the referred resource finds a private endpoint pending and can approve or reject the creation on the targeted resource.

Using managed virtual networks and managed private endpoints, you will be able to route your data traffic completely over the Azure backbone network. Your data will no longer travel over public resources.

You will find some more information in the *Further reading* section.

Integrating with DevOps

Finally, you want to take care that work that was developed not only gets versioned and saved in a reliable manner; you also want to be able to deploy it automatically to a test and later to your production environment. This can be achieved by integrating your Data Factory or Synapse workspace with Azure DevOps, for example, or with GitHub.

Let's examine the Azure DevOps integration. When we created our data factory in the first section of this chapter (*Setting up Azure Data Factory*), we skipped the DevOps integration at that moment. Now we are going to finish that task as well:

> **Note**
>
> The Synapse workspace can also be integrated with either Azure DevOps or GitHub. We will further examine this in *Chapter 11, Developing and Maintaining the Presentation Layer.*

1. First of all, create a new project in your Azure DevOps environment. You can create it as a private project, and please choose **Git** for the **Version control** setting:

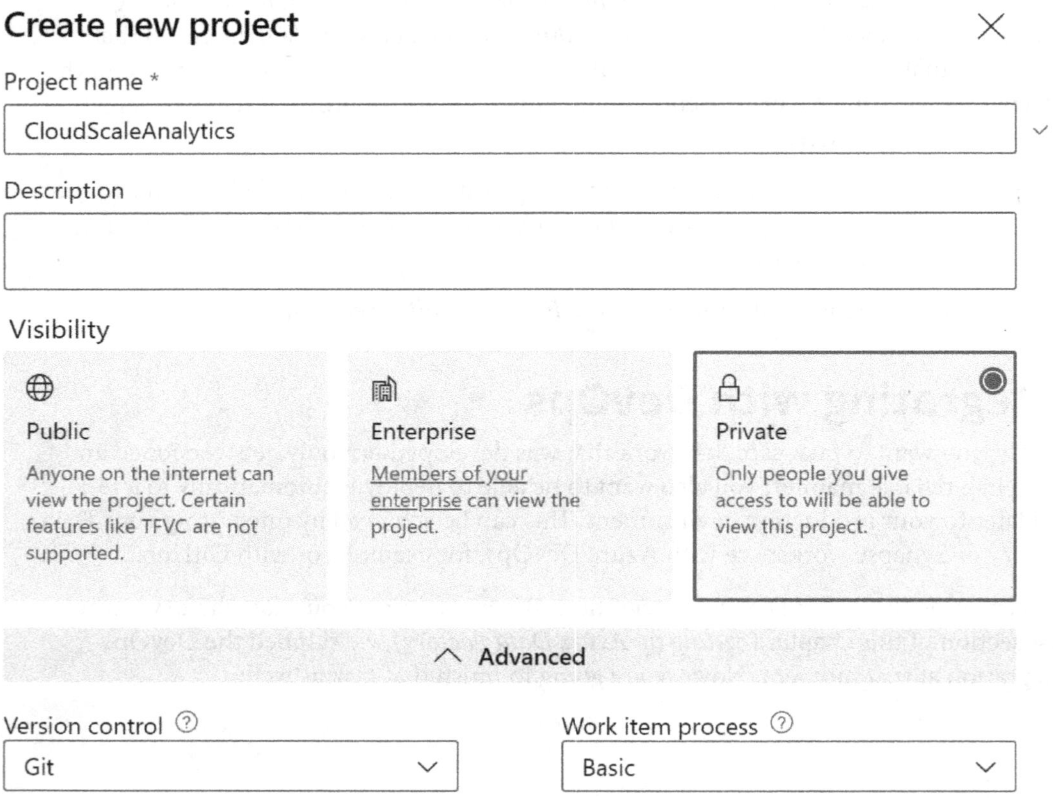

Create new project   ✕

Project name *

CloudScaleAnalytics

Description

Visibility

⊕	🏢	🔒 ◉
Public	**Enterprise**	**Private**
Anyone on the internet can view the project. Certain features like TFVC are not supported.	Members of your enterprise can view the project.	Only people you give access to will be able to view this project.

⌃ **Advanced**

Version control ⑦

Git   ⌄

Work item process ⑦

Basic   ⌄

Figure 5.35 – DevOps settings for your repository

2. Then, navigate to your data factory and either go to the **Management** section and the **Source Control/Git configuration** entry or start the sequence by clicking on **Data Factory** in the upper-left corner of any view and selecting **Set up code repository**.

3. On the **Configure a repository** pane, enter the information required:

   a) For **Repository type**, select **Azure DevOps Git**.

   b) In the **Azure Active Directory** field, please select the directory you are working in.

   c) For **Azure DevOps Account**, please choose where you have just have created a new project.

   d) You can create a new **Repository name** field that will then be created in your project as the root.

   e) Leave **Collaboration branch** as **master**.

   f) You can leave **Publish branch** as the default that is suggested, **adf_publish**.

   g) In **Root folder**, you can leave /.

   h) In the **Import existing resource** option, please tick the **Import existing resources to repository** checkbox.

   i) Finally, for the **Import resource to this branch** option, select **Use Collaboration**.

4. You can start the creation by clicking **Apply**. On the following screen, please create a new working branch called development.

5. Click **Save** and finish the sequence. Your Azure Data Factory is now integrated with Azure DevOps.

When you have finished the integration, your menu bar will have changed slightly. In the upper-left corner, you will now see **Azure DevOps GIT**, and next to it the branch you are actually connected to:

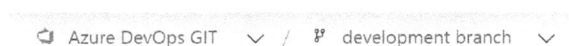

Figure 5.36 – DevOps connection in Azure Data Factory

Working with DevOps

As you are now starting your development work with DevOps integrated mode, you will always first create a branch where you carry out your work. This is called a **feature branch**. Every developer working with you in your environment should do so to avoid overlapping. By creating a feature branch, you are checking out the actual version of the **collaboration branch**. In your case, this would be **master**.

Now that you have an actual version of **master** in your **development** feature branch, you can start your work. Once you are finished with your development, you will create a **pull request** with your work to be delivered to the **master** collaboration branch.

In this phase, there is either another developer or the head of your development team, who will be notified and will check the work to approve or request changes.

Once the approval is there, the changes will be merged into the collaboration branch and can be published to the **publish branch**. From there, the work can be deployed/released either manually or preferably automatically to the next instance in your development life cycle. Typically, this would be the test environment. Once testing is complete and the quality gates are reached, the release can be deployed to the production environment.

The deployment sequence, from the first approval of the reviewer up to the final productive release, can be automated to a high degree with Azure DevOps. The service offers workflow, approval, and pipeline functionalities that will support you in your DevOps setup.

Please refer to the *Further reading* section for more information about Azure DevOps in general and a deep dive document on how to set up approval workflows and pipelines for automated deployment of artifacts.

Summary

In this chapter, you have examined Synapse pipelines/Azure Data Factory. You have learned how to create a data movement pipeline using a wizard, as well as from scratch in the authoring environment. You have seen the orchestration capabilities with the many different activities provided.

You have further implemented your first mapping flow to create transformations your data is going through before it lands in your Data Lake Storage. You have examined wrangling flows and learned the difference between the two data flow components.

We have also examined the IRs and their differences and talked about managed virtual networks and managed private endpoints.

Finally, we have integrated our Data Factory with Azure DevOps and have established source control over our artifacts.

In the next chapter, we are going to dive into another option to transform and process data using one of the main compute components in our modern data warehouse: the **Spark engine**.

Further reading

Please refer to the following links for more information:

- Pipelines and activities: `https://docs.microsoft.com/en-us/azure/data-factory/concepts-pipelines-activities`

- Mapping flows: `https://docs.microsoft.com/en-us/azure/data-factory/concepts-data-flow-overview`

- Language reference mapping flow expressions: `https://docs.microsoft.com/en-us/azure/data-factory/data-flow-expression-functions`

- Wrangling flows: `https://docs.microsoft.com/en-us/azure/data-factory/wrangling-data-flow-overview`

- Language reference power query – M: `https://docs.microsoft.com/en-us/azure/data-factory/wrangling-data-flow-functions`

- Azure IR: `https://docs.microsoft.com/en-us/azure/data-factory/create-azure-integration-runtime`

- Self-hosted IR: `https://docs.microsoft.com/en-us/azure/data-factory/create-self-hosted-integration-runtime`

- SSIS IR: `https://docs.microsoft.com/en-us/azure/data-factory/create-azure-ssis-integration-runtime`

- Data Factory managed VNET and managed private endpoints: `https://docs.microsoft.com/en-us/azure/data-factory/managed-virtual-network-private-endpoint`

- Azure Data Factory DevOps: `https://azure.microsoft.com/en-us/resources/whitepaper-adf-on-azuredevops/`

- Synapse pipelines and Data Factory differences: `https://docs.microsoft.com/en-us/azure/synapse-analytics/data-integration/concepts-data-factory-differences`

6
Using Synapse Spark Pools

In your modern data warehouse project, you may use Azure Data Factory ETL pipelines (see *Chapter 5, Integrating Data into Your Modern Data Warehouse*) to integrate and transform incoming data according to your needs. However, chances are that you are a more code-oriented developer, that you are already very proficient with Spark, or that your transformational needs reach beyond the functionality or the available compute power of Data Factory.

Maybe you need to train and implement machine learning models as part of your project, and you want a Spark engine that can scale to your needs and offers suitable libraries and tight integration with all the other tools that you plan to use on Azure.

This chapter will discuss Synapse Spark pools and how to implement them on Azure. You will learn about their architecture and how jobs are handled when they are dispatched to a cluster. You will examine how to implement notebooks and Spark jobs and integrate additional libraries with your clusters. Finally, we will examine security features and see how to monitor our environment.

This chapter covers the following topics:

- Setting up a Synapse Spark pool
- Examining the Synapse Spark architecture

- Programming with Synapse Spark pools
- Using additional libraries with your Spark pool
- Handling security
- Monitoring your Synapse Spark pools

Technical requirements

To follow this chapter, you will need the following:

- An Azure subscription for which you have at least contributor rights.
- The right to provision a Synapse workspace.
- The right to provision a Synapse Spark pool.
- The right to use Synapse Studio.
- An Azure DevOps Git or GitHub account. This is optional and to be used if you want to integrate your work with a DevOps repository.
- Your Azure Data Factory from *Chapter 5*, *Integrating Data into Your Modern Data Warehouse*.
- Visual Studio Code (optional, if you wish to follow the batch example later in the chapter): `https://code.visualstudio.com/Download`.

Setting up a Synapse Spark pool

Now, let's examine the basic steps to spin up a Synapse Spark pool in this section.

This task is very easy to handle in a Synapse workspace:

1. Please navigate to the **Management** pane and there, in the **Analytics pools** section, select **Apache Spark pools**.
2. In the **Details** pane, click **+ New**. The configuration blade for a new Apache Spark pool is displayed:

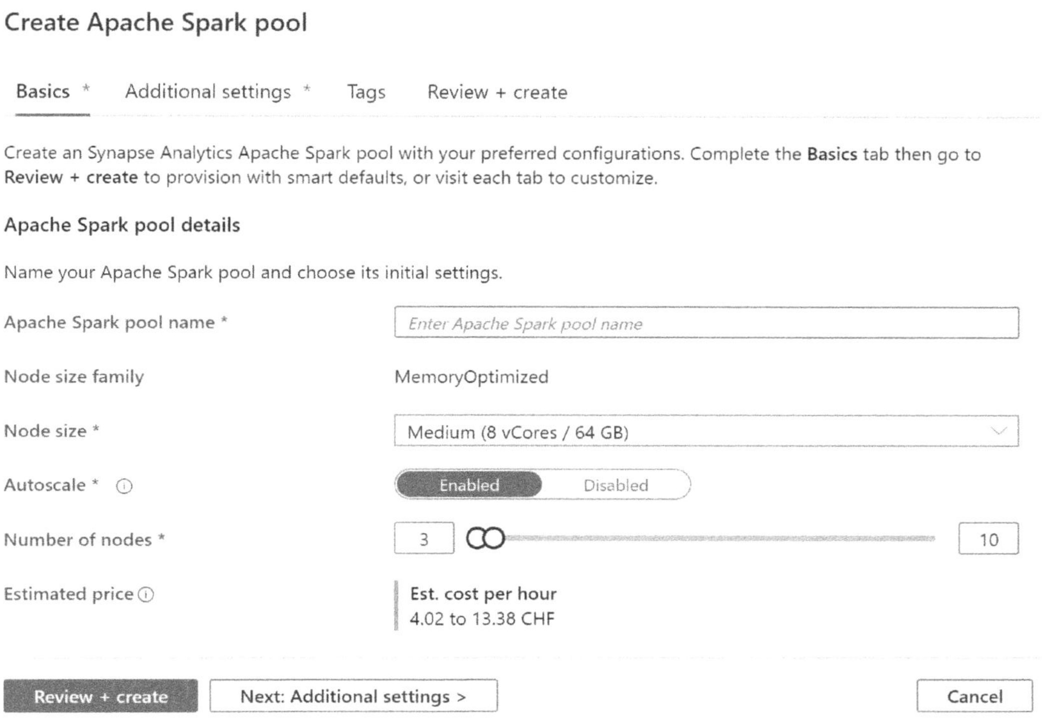

Create Apache Spark pool

Basics *   Additional settings *   Tags   Review + create

Create an Synapse Analytics Apache Spark pool with your preferred configurations. Complete the **Basics** tab then go to **Review + create** to provision with smart defaults, or visit each tab to customize.

Apache Spark pool details

Name your Apache Spark pool and choose its initial settings.

Apache Spark pool name *	Enter Apache Spark pool name
Node size family	MemoryOptimized
Node size *	Medium (8 vCores / 64 GB)
Autoscale * ⓘ	Enabled Disabled
Number of nodes *	3 ∞ 10
Estimated price ⓘ	**Est. cost per hour** 4.02 to 13.38 CHF

Review + create   Next: Additional settings >   Cancel

Figure 6.1 – Create Apache Spark pool – The Basics blade

3. Here you will name your new Spark pool and configure the **node size** value, enable **Autoscale**, and set the lower and upper boundaries for the autoscaling feature, if enabled. The last row in this view shows the potential cost of the lowest and the highest autoscaling setting. Click **Next: Additional settings**.

4. In the upper area of the **Additional settings** blade, you can now configure **Auto-pause** and **Number of minutes idle**, which sets the amount of idle time that will elapse before the cluster pauses. In the **Component version** section, you can control the Spark version and you will see a list of the versions of the other participating components in your cluster. See *Figure 6.2*.

> **Note**
> Microsoft has just put Spark 3.0.1 into preview. The actual version for your productive environments will be 2.4, as you can see in *Figure 6.2*.

5. In the lower area of the blade, you also have the options to add additional packages to your cluster and influence the Apache Spark configuration:

Create Apache Spark pool

Basics *   **Additional settings** *   Tags   Review + create

Customize additional parameters including pause settings and component versions.

Automatic pausing

Configure the pause settings for the Apache Spark pool.

Automatic pausing * ⓘ	● Enabled ○ Disabled
Number of minutes idle *	15

Component version

Select the Spark version for your Apache Spark pool.

Apache Spark *	2.4 ⌄
Python	3.6
Scala	2.11.12
Java	1.8.0_272
.NET Core	3.1
.NET for Apache Spark	1.0
Delta Lake	0.6

Apache Spark configuration

Upload a Spark configuration file to specify additional properties on the Apache Spark pool. This will be referenced to configure Spark applications upon job submission.

File Upload	Select a .txt file 📁

Upload

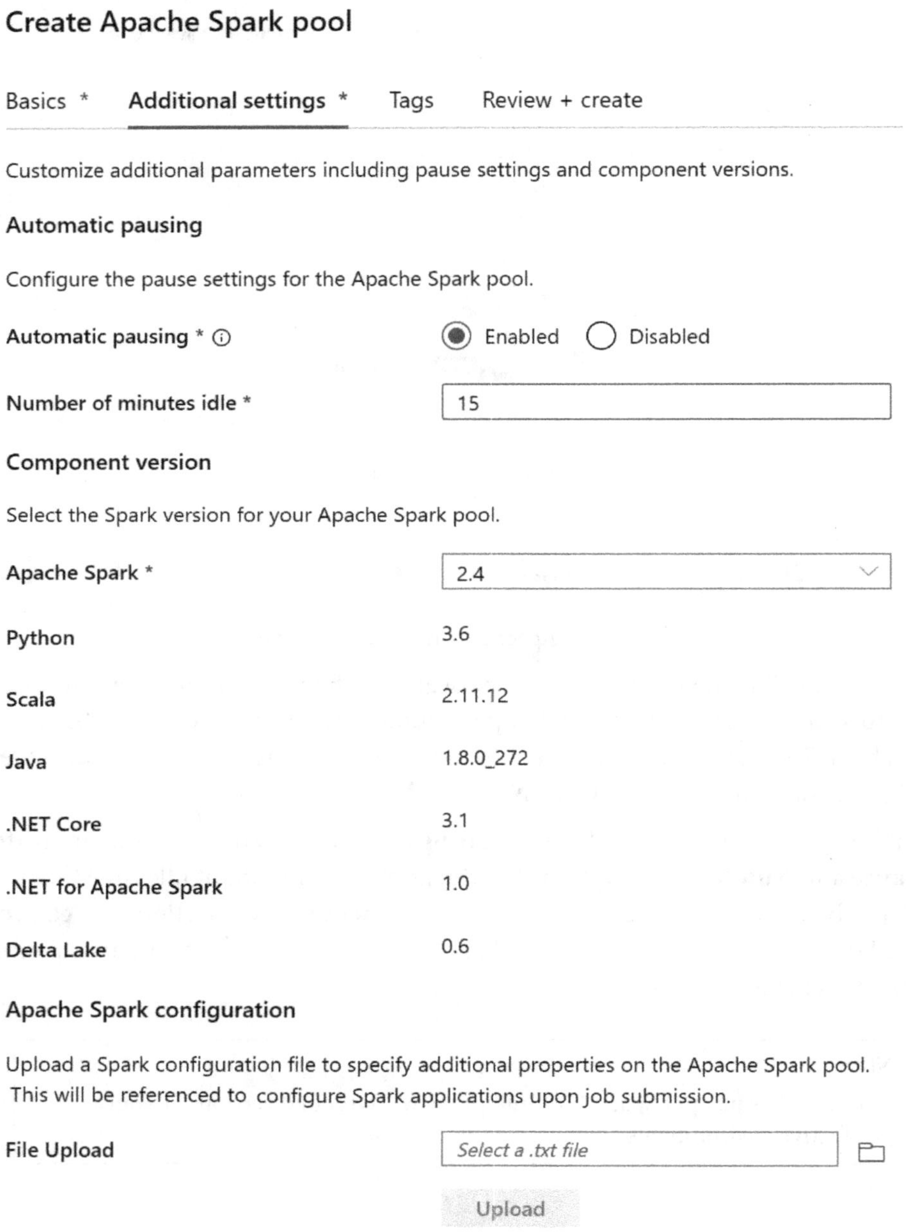

Figure 6.2 – Create Apache Spark pool – Additional settings

6. The **Next: Tags** button will bring you to the penultimate blade in the sequence. This blade will allow you to add key/value pairs to your resource and will give you the ability to track the usage of your resource when you are reporting and analyzing the consumption of your system.

As in the previous chapters, when you now hit **Next: Review + create**, you will be taken to the final overview of your configuration settings before you start the creation of the service. And again, you will find a **Download template for automation** link. Hit **Create** to finally start the provisioning of the Spark cluster.

Bringing your Spark cluster live for the first time

To see your cluster in action, you can now proceed to the **Development** section of your Synapse workspace. Hit + and select **notebook** from the options. When the new notebook is displayed, attach it to your newly created cluster from above. You will find the cluster in the drop-down box right above the notebook. See *Figure 6.3*:

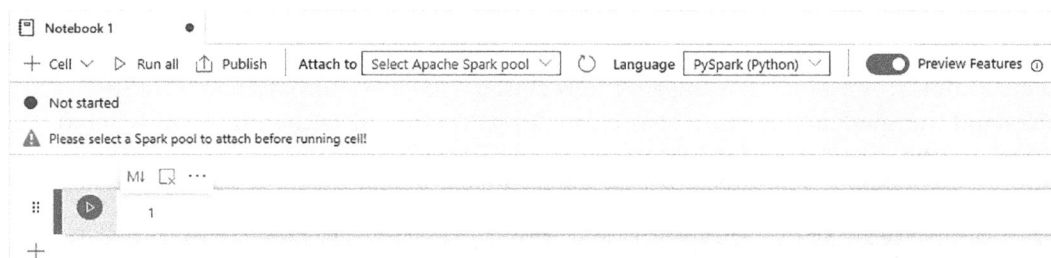

Figure 6.3 – New notebook ready to attach to a cluster

When the new cluster is selected and displayed in the **Attach to** combo box, go to the upper empty command cell of the notebook and type the following:

```
print(sc.version)
```

On the far-left side of the cell, you will find a **Run** button. Click it and run the cell. Your cluster will now start. Please allow around 2 minutes for the cluster to start. You should receive a result as in *Figure 6.4*:

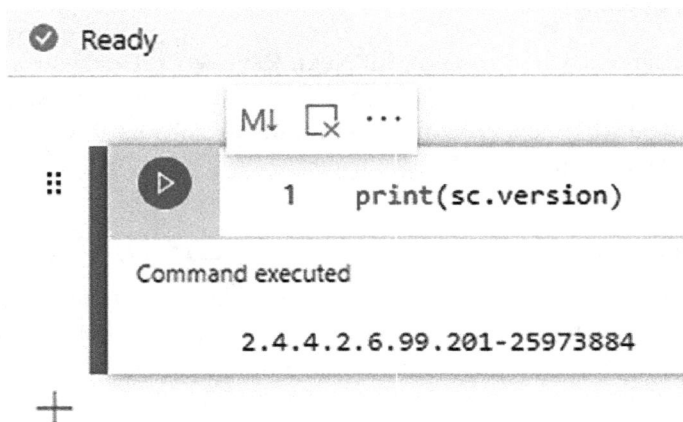

Figure 6.4 – First code run on your first Spark cluster

You can now start developing using your new cluster. However, perhaps you want to understand the Spark architecture before you jump into this adventure. If so, move on to the next section, *Examining the Synapse Spark architecture.*

Examining the Synapse Spark architecture

With Synapse Spark pools, Microsoft adds another scalable parallel processing engine to the Synapse ecosystem. The Microsoft implementation of Spark adds in-memory processing capabilities that support languages such as Python, Scala, Java, and even .NET for Spark and SQL.

The engine comes with built-in compatibility with Azure Data Lake Gen2 and Azure Storage. This enables the Spark Core engine, via the **YARN** layer (which is a JobTracker, resource management, and job scheduling/monitoring tool), to access the data that you have brought to Azure. This way, Spark Core exposes the storage components to libraries such as **Spark SQL** for interactive querying, **MLib** for machine learning, and **GraphX** for graph computation at scale.

Spark implements in-memory computation algorithms that can run your Spark jobs or notebooks in parallel on defined clusters. As mentioned previously, clusters will hold the data to be computed in memory in a distributed manner to guarantee the best possible performance.

In addition to Spark Core, Microsoft has also added the following:

- **Anaconda**: A very popular open source data science and machine learning library collection
- **Apache Livy**: Mainly exposes a REST interface to developers to enable them to submit Spark jobs from anywhere to a Synapse Spark cluster
- **Nteract notebook**: Adds an interactive notebook development experience to Synapse Spark pools as an out-of-the-box functionality

But let's move on to the components of a Spark pool in the next section.

Understanding the Synapse Spark pool and its components

A Spark cluster implements a **driver node**, which controls the whole cluster and creates the so-called **Spark context**, in which an application (code from a notebook or Spark job) is processed.

The application's `main()` function is run on the driver node, and from here the processing of the required calculations is prepared to be run in parallel on the participating **worker nodes**.

The driver node implements the **cluster manager** (**YARN**), which "knows" the participating worker nodes and is responsible for the distribution of the work to those workers.

The final task of the driver is to collect all the intermediary results coming from the worker nodes, put them together into a usable result set, and return them to the application's `main()` function.

The worker nodes are responsible for processing the tasks that are passed from the driver. They access data from the underlying storage component and write back the results of their work to that storage. The datasets that are processed are cached as far as possible into the memory of the workers to achieve the best possible performance. Therefore, the more workers you can use, the better your performance of course:

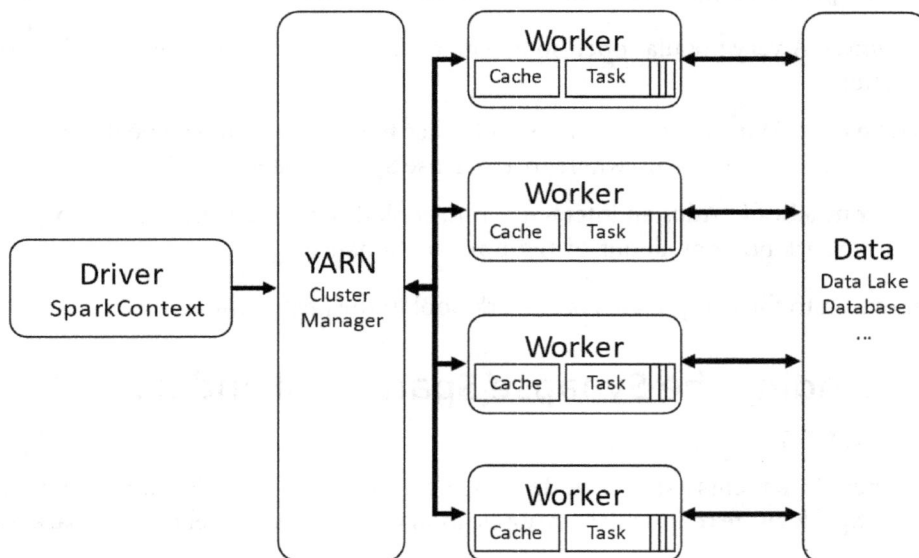

Figure 6.5 – Spark general architecture

In the next section, we will dive into how to run a Spark job in this architecture.

Running a Spark job

When you throw a processing request against a Spark cluster by running a notebook or a Spark job, it will first acquire a SparkContext. This will take care of your code, split it, and run it as a group of processes on the cluster.

YARN, as the cluster manager, will then allocate free resources to the workers. Executors will be reserved, which will run the independent processing tasks coming from the cluster manager. As soon as the allocation of task slots and memory is done, the tasks will receive the code from the SparkContext to run it.

The `main()` function is executed in the SparkContext and controls the parallel execution of the tasks on the workers. These interact with the storage and get the necessary data for the processing before writing back results when applicable. During this processing, the workers cache the data in so-called **Resilient Distributed Datasets (RDDs)** in the memory of the workers.

Once processing is complete, the SparkContext will collect the intermediary results from all the participating workers.

Examining Synapse Spark instances

Now that you have seen how Spark works in general, let's have a look at the architecture of a Synapse Spark instance. Microsoft has made some changes to facilitate configuration and speed up the initialization and startup of Spark clusters; you're also able to save money when you only need a small instance.

This section displays the main difference between Synapse Spark pools and standard Spark clusters: A Synapse Spark pool consists of at least three **virtual machine** (**VM**) nodes. But in comparison to a standard Spark setup, Microsoft doesn't isolate the driver node from the worker nodes. Even the first three nodes in a Synapse cluster can run Spark executors. This already gives us high availability, even in the smallest setups.

Furthermore, the three-run **ZooKeeper** is a service that's responsible for *knowing* the Spark configuration and keeping the cluster in sync. The first two nodes additionally run the **YARN resource manager** and therefore act as the redundant drivers for the cluster. Please see *Figure 6.6*:

Spark Instance

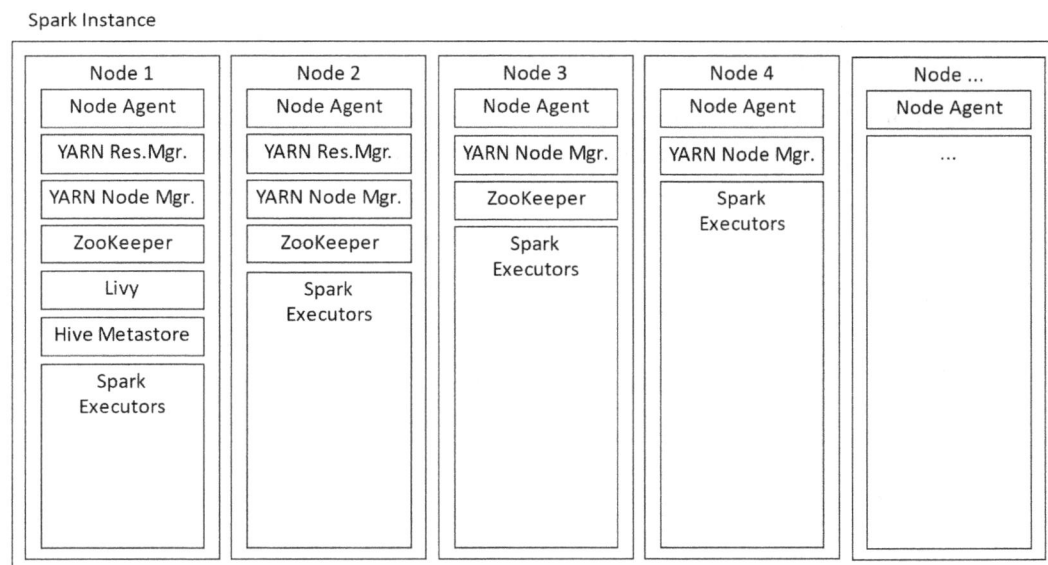

Figure 6.6 – Synapse Spark cluster configuration

The main benefit of the architecture outlined is the fast provisioning time. A Synapse Spark cluster will normally be available within 1 or 2 minutes of a start request being issued. Synapse Spark clusters are equipped with an autoterminate option for when they're idle, and you might want to have a cluster back up and running as quickly as possible to start developing, for your Spark batch job, or your Synapse pipeline (aka Data Factory) job.

As mentioned previously, you will benefit from the alternative Synapse Spark setup even from a cost perspective, especially in smaller scenarios. As you only need three nodes in your Spark instance to run a job, additional nodes aren't required to act as drivers.

Understanding Spark pools and Spark instances

When you configure a **Spark pool** within Synapse, you create metadata entries in the Synapse workspace. Following the creation of the Spark pool, there is not a usable Spark cluster available for your notebooks or jobs. You won't even be charged money. All you have done is described some properties that may be used to spin up compute resources when they are needed.

A **Spark instance** is a physical representation of the configuration that you save when you create a Spark pool. A Spark instance is created when you use a notebook, for example, and run code with a Spark pool attached, just like you did previously (in the *Bringing your Spark cluster live for the first time* section) after you created your first Spark pool.

There can be more than one Spark instance created from a Spark pool. The number of instances that can be run on a pool depends on the resources available for the workspace and the resources required by the jobs that are run in the environment.

Let's now examine the behavior of the environment when resources are consumed:

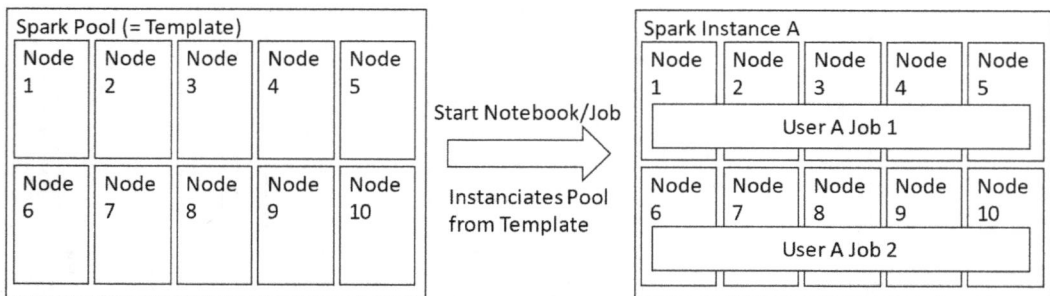

Figure 6.7 – Spark pool with one Spark instance running two jobs

If you configure a Spark pool with a fixed cluster size with 10 nodes and you run 1 job that requires 5 nodes, when you then fire another job that requires another 5 nodes, everything will run smoothly. Synapse Spark will reuse the Spark instance and run both of your jobs.

If a job demands more nodes than the 5 remaining ones, if the request was issued from a notebook, you will get a rejection message, while if the request comes from a job, it will be queued until sufficient resources become available.

In an autoscaling scenario (refer to *Further reading, Autoscaling behavior*), things would look similar, with one difference. The Spark instance for your first job would only cause 10 nodes to be provisioned upfront. The additional 10 nodes are only started and added to the available instance when there is a need for them, for example, when you start your second job.

To consider another scenario, let's say we added a second user to the system who wants to run their job on the Spark pool. In this case, you would again run your notebook (job 1) on a cluster of a fixed size with 10 nodes, consuming 5 nodes for Spark instance A. As soon as your second user's request for 5 nodes hits the Spark pool, another Spark instance, B, will be created and run the second job (job 2). Additionally, you could now even start another job (job 3) using 5 more nodes in your Spark instance, as you have still 5 nodes available there:

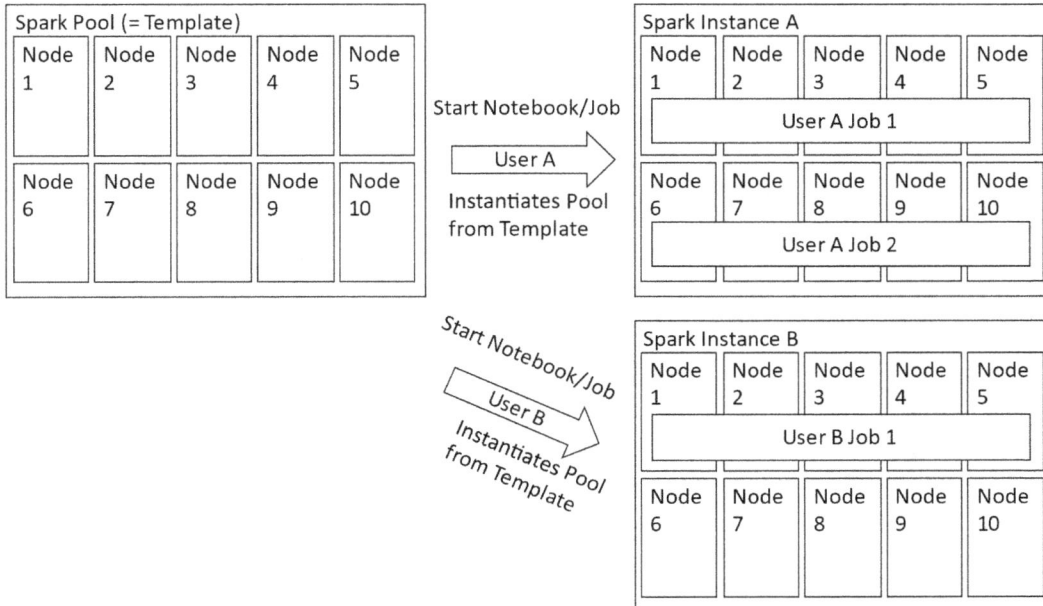

Figure 6.8 – Spark pool with two Spark instances running three jobs

---

> **Notes**
>
> You want your Spark pools to vary in size and the number of nodes depending on the tasks they are used for. For a development environment, for example, you would want to create smaller clusters with smaller VMs, and for a production environment, of course, you would want to create bigger ones.

Understanding resource usage

Depending on the configuration of your Spark pool and the VM sizes configured, a Spark pool will not give away all resources to any of the jobs or notebooks that are run. As Synapse Spark pools support Synapse pipelines (the Synapse instances of Data Factory), there will be limits on how many Synapse pipelines each user can run. The vCores for each cluster will therefore never be allocated completely to one job or one pipeline.

This happens globally on the workspace level as well as on the Spark pool level. If you receive an error message like the one shown in *Figure 6.9*, you have not done anything wrong:

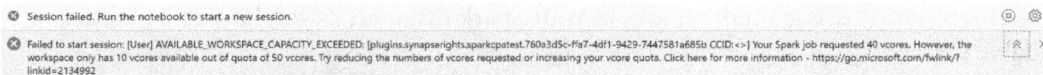

Figure 6.9 – Error message for the workspace quota having been exceeded

If you receive an error message that tells you that your workspace does not have enough vCores left for your job, you need to either increase the quota for the vCores for your workspace (see *Further reading, Requesting a quota increase*) or reduce the number of requested vCores in your session. To do so, click on the gear icon in the upper-right corner of your notebook window and adjust the settings for your session:

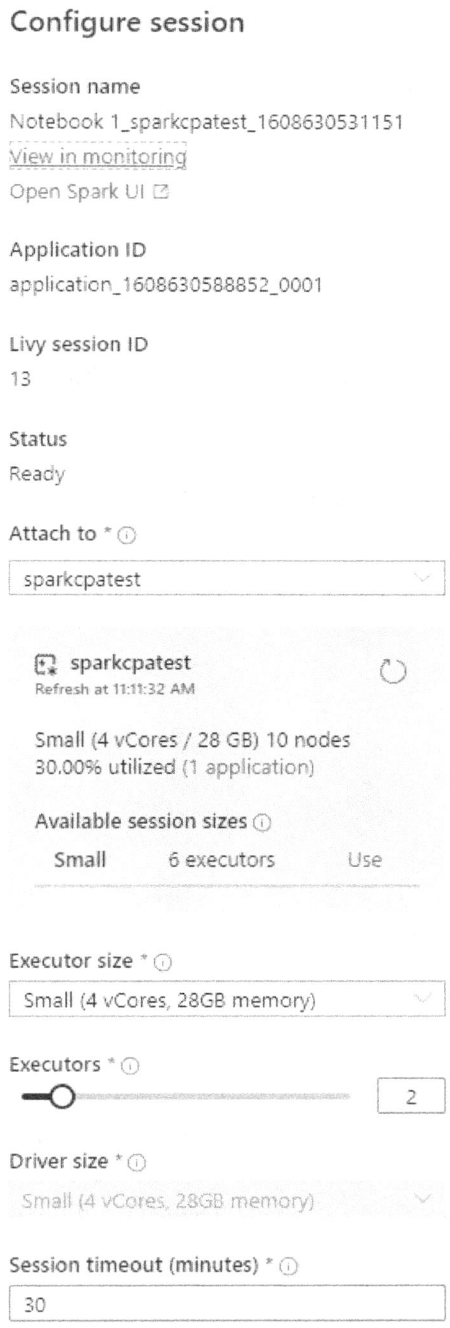

Configure session

Session name
Notebook 1_sparkcpatest_1608630531151
View in monitoring
Open Spark UI ☒

Application ID
application_1608630588852_0001

Livy session ID
13

Status
Ready

Attach to * ⓘ

```
sparkcpatest                                              ∨
```

┌──┐
│ ⌗ sparkcpatest   ↻   │
│ Refresh at 11:11:32 AM   │
│   │
│ Small (4 vCores / 28 GB) 10 nodes   │
│ 30.00% utilized (1 application)   │
│   │
│ Available session sizes ⓘ   │
│   Small   6 executors   Use   │
└──┘

Executor size * ⓘ

```
Small (4 vCores, 28GB memory)                             ∨
```

Executors * ⓘ

━●━━━━━━━━━━━━━━━━━━━━━━━━━━   2

Driver size * ⓘ

```
Small (4 vCores, 28GB memory)                             ∨
```

Session timeout (minutes) * ⓘ

```
30
```

Figure 6.10 – Adjusting session settings

If you receive a similar message related to your Spark pool, this means that there are other jobs already consuming vCores in your Spark pool. Again, you can reduce the requested resources in your session settings.

Programming with Synapse Spark pools

Now that you understand how to provision a Spark pool and how resources are used, let's proceed and examine the different interfaces that you can use to program against a Spark instance.

Understanding Synapse Spark notebooks

Notebooks are the rising star when it comes to interactive data analysis. They offer a step-by-step programming experience where you receive immediate feedback for single code steps. You can enter one line or a block of code into a cell and you can run it directly using an available Spark instance and have the results displayed below the cell.

To create a new notebook, you want to navigate to the **Develop** hub in Synapse Studio. Here, you can hit the + icon in the navigation pane (next to the word **Develop**) and select **Notebook** (*Figure 6.11*):

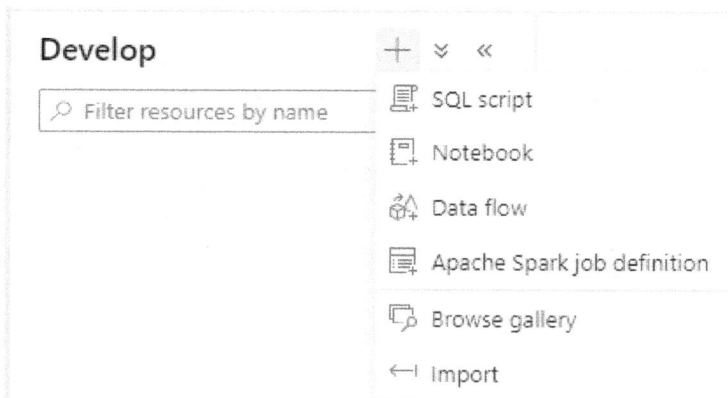

Figure 6.11 – Creating a new notebook

Alternatively, you can right-click on the **Notebooks** section and select **New Notebook**. An empty notebook will be displayed where you can start implementing your code.

But let's now move on to a different method of starting a new notebook where even the first step of data acquisition is already made for you and you can start developing using your selected dataset.

Please proceed to your data hub in Synapse Studio and select the Data Lake Storage Gen2 instance that we created in *Chapter 3*, *Understanding the Data Lake Storage Layer*. If you navigate through the path to the source folder you created, you should find your `airdelays.csv` file from our exercise in *Chapter 5*, *Integrating Data into Your Modern Data Warehouse*. When you right-click on the file, you will have the option to start your notebook with a code snippet and either load the file into a **DataFrame** or create a new **Spark table**:

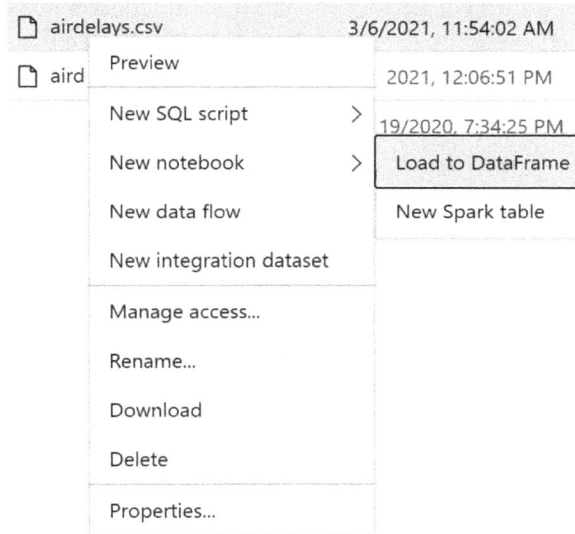

Figure 6.12 – Creating a notebook with code snippets

Please select **Load to DataFrame**. You will receive a new notebook with the following code snippet, and you can start programming directly using your created DataFrame:

```
%%pyspark
df = spark.read.load('abfss://://[YOURDATALAKEFILESYSTEM]@
[YOURDATALAKEACCOUNTNAME].dfs.core.windows.net/
[PATHTOYOURFILE]/source/airdelays.csv', format='csv'
## If header exists uncomment line below
##, header=True
)
display(df.limit(10))
```

The parts of the path that read `[YOURDATALAKEFILESYSTEM]`, `[YOURDATALAKEACCOUNTNAME]`, and `[PATHTOYOURFILE]` are, of course, placeholders that you should change to reflect your setup.

This is a nice bit of support to kick off an analytical session on data in your data lake, isn't it?

If your folder contains files of a similar structure, you can replace the filename with * and leave the .csv part; this will read the whole folder into your DataFrame.

Integrating notebooks with Synapse pipelines

Another example of the additional support you get with notebooks in Synapse Studio is the smooth integration of a notebook that you have developed with Synapse pipelines (the Synapse version of Data Factory).

If you examine the Synapse Studio notebook development surface and you look in the upper-right corner, you will find three buttons there, as follows:

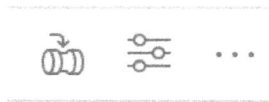

Figure 6.13 – Notebook options

Those buttons will offer the following:

- The three dots offer you the following options:

    a) **Clear output** will eliminate all the outputs of cells that you have run during development.

    b) **Export** will export your notebook to the .ipynb format.

    c) **Copy link** will give you a link to the actual notebook.

    d) **Clone** will create a copy of the actual notebook.

    e) **Delete** – well, this will delete the notebook.

- The **Properties** button in the middle will take you to the **Properties** pane for the notebook. For a newly created notebook, you can enter a name here before the first save. You can add a description and toggle a checkbox that will save the cell's data output/results when you save the notebook or not.

- Upon clicking the **Add to Pipeline** button, you will be presented with a dialog where you can decide to add your notebook to a Synapse pipeline. You can decide to create a new pipeline, or you can add it to an already existing one and put your notebook into a wider context and orchestrate it, for example, together with a Data Copy job to bring the source data for your notebook into your data lake.

Handling code in your notebook

During the development of your Spark application, you will need to navigate in your notebook. You may want to add cells, jump between them, select the programming language, and so on. We'll cover all that in this section.

Selecting the notebook language and using multiple languages

In the menu ribbon of the notebook, you will find the language combo box. This is the spot where you set the primary language for cells that you add to the notebook. The actual language options are as follows:

- Python (PySpark)
- Scala (Spark)
- SparkSQL
- .NET for Spark (C#)

With so-called **magic commands**, you'll be able to use multiple languages in your notebook. When you use either `%%pyspark`, `%%spark`, `%%sql`, or `%%csharp` as the first command in a cell, you toggle the language for that cell.

When you're using multiple languages in your notebook, there is one challenge. A DataFrame that was initially created using PySpark will not be visible to a cell that is using Scala. To accomplish this, you will need to create a temporary table. The Python code for this would look like this:

```
df.createOrReplaceTempView("dftemptable")
```

In a subsequent cell, you might then use the `%%sql` magic command and query the new table by simply running a SQL statement:

```
%%sql
select * from dftemptable
```

But how do you add more cells?

Adding new cells

To add a new cell, you can click the + button in the upper-left corner in the menu above the notebook details. You have two options here: add code or add a text cell. Another option would be to use the + button on the left side under the active cell. Again, you have the option to decide between code or text (Markdown).

There is another way to add a new cell to your notebook. When you have finished a code fragment in an actual cell, you can hit *Shift + Enter*. This will run the code in the active cell and add a new cell below. *Ctrl + Enter* will just run the cell without adding a new one.

Running code

To run code in your notebook, you can either decide to run all cells with the **Run all** button in the menu above the notebook details or run a single cell by clicking on the **Run** icon.

As mentioned previously in the *Adding new cells* section, you can run the code in the active cell by hitting *Shift + Enter* or *Ctrl + Enter*. The first option will also create a new cell, while the second will just run the code.

Another method of running a notebook is to add it to a Synapse pipeline. You can do this as shown in the *Integrating notebooks with Synapse pipelines* section, that is, from within the notebook, or you can add a **Synapse notebook activity** to the Synapse pipeline **Development** pane.

Handling cells

The active cell shows additional buttons in the upper-left corner. You can use these to toggle the active cell between Markdown mode and code cell mode. The button in the middle will clear any output in the cell. With the three dots, you can display additional options such as moving the cell up or down (you can alternatively drag and drop it), active scrolling for the output, hiding or showing the code input, hiding or showing the output, or toggling the active cell as a parameter cell.

In the upper-right corner of the cell, you will find a trashcan symbol. By clicking this, you will delete the cell.

Visualizing data in your notebook

It is important for us to cover data visualization. The notebook experience in Synapse Spark offers the option to instantly convert the tabular output of a cell into a chart. You can use different chart types here, and you can select the attributes to display, the aggregation type for the chart, and even the groupings for the keys displayed.

To display a chart, you will need to use the `display()` function in your code. If you want to switch from table mode to chart mode, go to the upper-left corner of the table that is displayed when a result set is returned by your Spark instance:

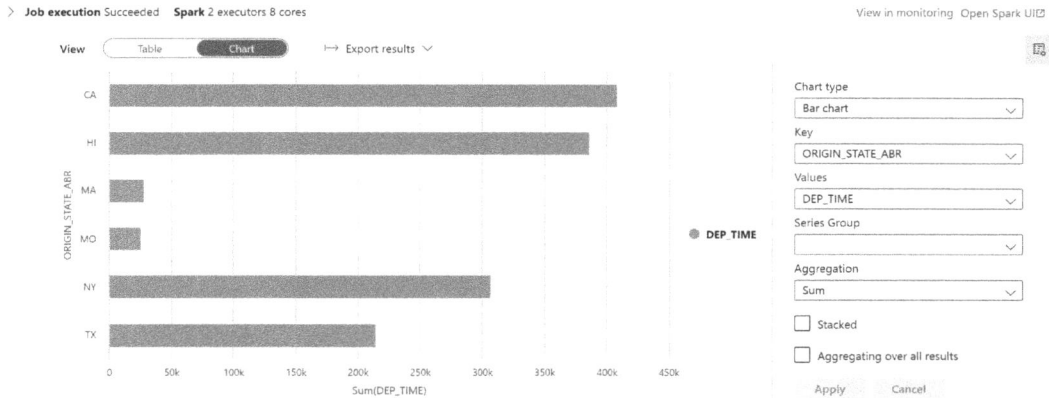

Figure 6.14 – Displaying results as charts in your notebook

Up to now, you have been trying a lot. But you want to keep your work at a certain point, right? Please refer to the next sections about saving your work and how to run Spark applications.

Saving your work

When you have finished working on your notebook, you will want to ensure that it is saved to your repository. You can do this by clicking on **Publish** in the menu ribbon. If you aren't using a DevOps repository, the notebook will just be saved to the Synapse workspace. With a DevOps repository, you will need to first check in your notebook and then issue a pull request into the collaboration branch of your Git repository.

Please refer to *Further reading, Notebook navigation* for more information on notebook handling and development.

Running Spark applications

The interactive analysis is only one facet of working with Spark environments. In a production environment, you might want to move to a batch-oriented strategy where notebooks are part of your pipelines, as shown in the *Integrating notebooks with Synapse pipelines* section.

Another option is to write your applications in your own IDE, such as Visual Studio Code, IntelliJ, or PyCharm, and run them as Apache Spark job definitions, either from a Synapse pipeline, as a batch from within your IDE, or via the Livy REST API.

Let's explore an example. In the following code, you will use the `airdelays.csv` file and create a `.parquet` file that will contain distinct combinations of origin and destination airports:

1. You can use the following code and create a Python file from it. You can copy it to any text editor or to Visual Studio Code. From there, you want to store it, perhaps as `getoriginanddestination.py`, to disk:

```python
from pyspark.sql import SparkSession
spark = SparkSession.builder.
appName("getoriginanddestianation.py").getOrCreate()
df = spark.read.load('abfss://[YOURDATALAKEFILESYSTEM]@
[YOURDATALAKEACCOUNTNAME].dfs.core.windows.net/
[PATHTOYOURFILE]/source/airdelays.csv', format='csv'
, header=True
, delimiter=';'
)
dfout = df.select("ORIGIN", "DEST").distinct()
dfout.write.parquet('abfss://[YOURDATALAKEFILESYSTEM]@
[YOURDATALAKEACCOUNTNAME].dfs.core.windows.net/
[PATHTOYOURFILE]/source/dfout.parquet')
```

The placeholders in the path, `[YOURDATALAKEFILESYSTEM]`, `[YOURDATALAKEACCOUNTNAME]`, and `[PATHTOYOURFILE]`, will, of course, need to be changed to reflect your setup. Please replace them accordingly.

> **Note**
>
> When you're not quite sure where to get the paths of your files from, you might want to use Storage Explorer in Synapse Studio. When you navigate to the file, you just right-click on the file and select **Properties**. You will find an **https:** path and an **abfss:** path. Please use the **abfss:** option.

2. Upload your Python file to your data lake into the source folder you created in *Chapter 5, Integrating Data into Your Modern Data Warehouse*.

3. In Synapse Studio, go to the development hub, hit the + icon, and select **Apache Spark job definition**:

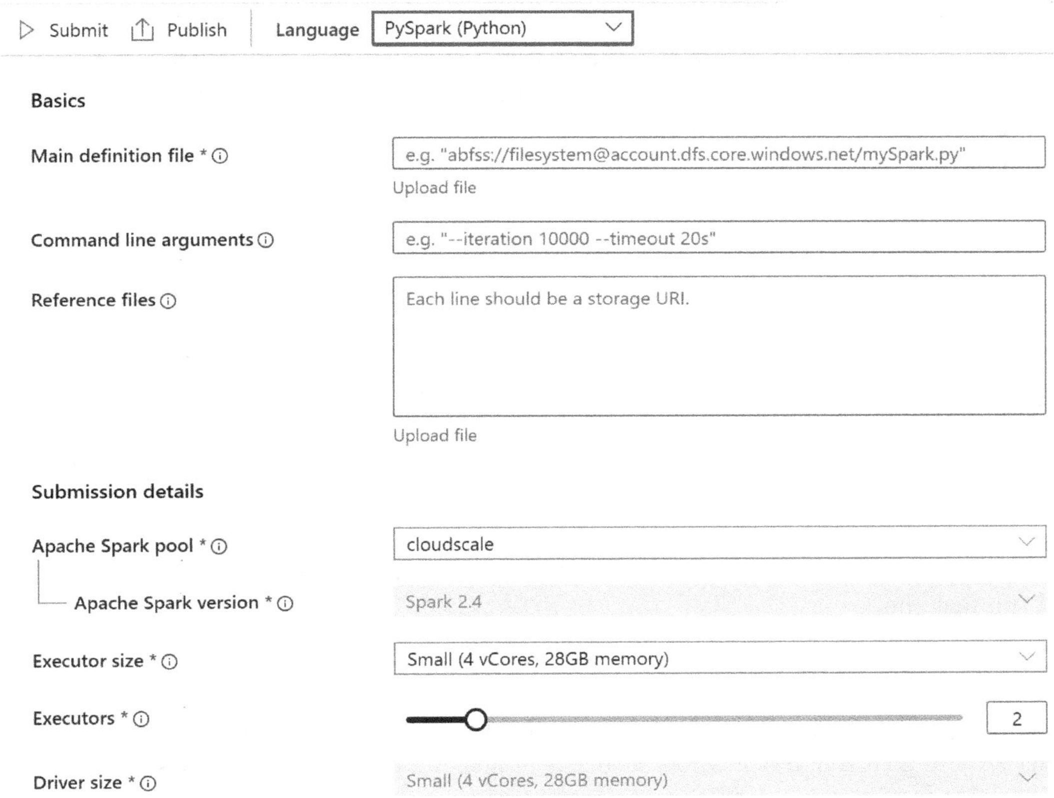

Figure 6.15 – Apache Spark job definition

4. On the right, you will see the **Properties** pane displayed. Name your first Spark job something like `getoriginanddestination`.

5. In the menu ribbon of the configuration blade, select **PySpark (Python)** for the Language combo box.

6. Paste the full **abfss:** path to your Python file into the **Main definition file** field. Refer to the preceding note box for instructions on how to easily get this path copied from the properties of the file.

7. In the **Apache Spark pool** field, select the Spark pool that you created at the start of this chapter.

8. In the **Executor size** field, select **Small (4 vCores, 28GB memory)**.

9. You can leave the remaining fields as their defaults for now.

10. Finally, you need to publish the job definition by clicking on **Publish** or **Publish all**.

11. Your job is now ready and can be submitted. Click **Submit** in the menu above the job configuration. Below the **Configuration** pane, you will get a new **Display** pane with messages regarding your job. Soon after the start of the job, you will get a monitoring URL that will take you to the monitor hub, where you can check the outcome of your job.

12. If everything goes fine, you can check your data lake folder. You should find a new folder named `dfout.parquet` with the following content:

Name	^	Last Modified
🗋 _SUCCESS		12/23/2020, 10:01:58 AM
🗋 part-00000-f2797875-9868-479f-bb5a-bb93077ddd36-c000.snappy.pa...		12/23/2020, 10:01:57 AM

Figure 6.16 – Output of your first Spark job

If you want to check the result, you can right-click the `.parquet` file and either create a new interactive notebook and run the displayed command or create a new SQL script and run that one.

Exploring Visual Studio Code integration

When you are developing Spark applications with Visual Studio Code, for example, you won't want to go through that whole sequence of uploading your code file and submitting your job via the Spark job definition again and again.

With the newest version of the **Spark & Hive Tools extension** for Visual Studio Code, you can stay in the IDE and submit your code to your Synapse Spark pool. To install this extension, perform the following steps:

1. You need to navigate to the **Extensions** section in Visual Studio Code and type `Spark & Hive` into the search field.

2. From the result list, select **Spark & Hive Tools** and hit **Install** in the **Details** window. Reload Visual Studio Code when you are prompted to.

3. You will need to create a work folder to proceed. So, maybe create something such as `C:\AdvancedAnalytics\SynapseSpark` and open it in Visual Studio Code (**File → Open Folder**).

4. To shorten things, you can use the file that you created previously (in the *Running Spark applications* section), move it to your new folder, and open it in Visual Studio Code. It will be displayed in your folder on the left-hand side in the navigation pane.

5. Next, you will need to log in to Azure to be able to use your Spark pool. Please hit *Shift + Ctrl + P* to display the command palette of Visual Studio Code and type and then select **Azure: Sign In** to connect your Visual Studio Code to your Azure subscription. Follow the sign-in procedure.

6. Now, right-click somewhere in the editor and select **Synapse: Set default Spark pool**. The command palette will show you the available Spark pools in your subscription. Select the one that you created at the beginning of the chapter. This action will add an entry to your `.VSCode\settings.json` file.

7. Delete the `dfout.parquet` folder in your data lake before submitting your code to avoid any failure. This has to do with existing files as we are reusing the code from earlier.

8. Finally, you can now right-click in your editor and select **Synapse: PySpark Batch**. This will submit your code to your Spark pool.

9. In the **Output** window below your code, you can now follow the progress.

10. When the job status returns **Success**, visit your data lake folder and check for the `dfout.parquet` file.

Visual Studio Code is, of course, not the only IDE that you can use with your Spark pools. Another option would be the IntelliJ plugin. For a detailed walk-through of how to use it, go to *Further reading, Developing Spark applications with IntelliJ*. You will find additional material about Visual Studio Code there too: *Further reading, Developing Spark applications with Visual Studio Code*.

Benefiting of the Synapse metadata exchange

One of the biggest functional assets of the Synapse workspace is the **metadata exchange**. This will give you the option to re-use tables that have been created by another compute engine in Synapse.

If you have created a Spark table, for example, this will show up in the tables when you query `sys.tables` from a serverless SQL query.

Let's look at this in more detail:

1. Navigate to your data lake and find the `airdelays.csv` file from *Chapter 5,*
   Integrating Data into Your Modern Data Warehouse.

2. Right-click the file and select **New notebook** → **New Spark table**.

3. Adjust the code as follows (`[YOURDATALAKEFILESYSTEM]`,
   `[YOUDATALAKEACCOUNTNAME]`, and `[PATHTOYOURFILE]` will, of course,
   need to be changed):

```pyspark
%%pyspark
df = spark.read.load('abfss://[YOURDATALAKEFILESYSTEM]@
[YOURDATALAKEACCOUNTNAME].dfs.core.windows.net/
[PATHTOYOURFILE]/source/airdelays.csv', format='csv'
, header=True
, delimiter=';'
)
df.write.mode("overwrite").saveAsTable("default.
airdelaystable")
```

4. Run the code.

5. When the notebook has finished, you will see a new table displayed in the **Data** hub
   in your Synapse Studio in the **Spark section** under **Tables**:

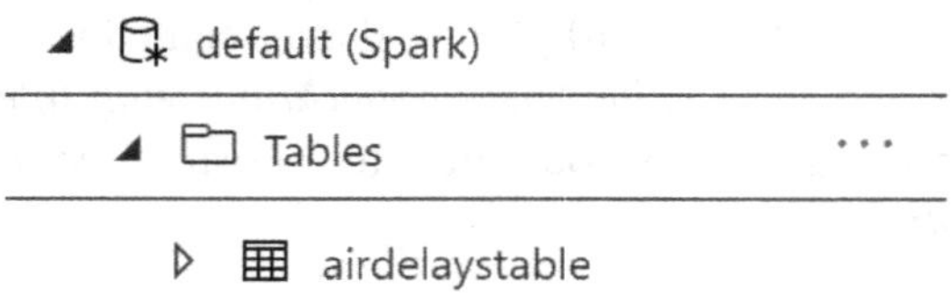

Figure 6.17 – New table in the Spark section

6. You can now navigate to the development hub in Synapse Studio, create a new **SQL query**, and run the following statement:

```
select *
from airdelaystable
```

You have seen how to use the Spark environment in its default setup. But there is more to it when you want to add additional functionality using third-party libraries, for example, or incorporate your own. Please find how to do this in the next section.

Using additional libraries with your Spark pool

There are so many cases where you need to rely on additional functionality from third-party libraries. Synapse Spark supports the addition of libraries to your Spark pool and will make them available when the pool is instantiated. There are different options available for you to use this functionality.

Using public libraries

In the case of PyPi packages, you would create a file named `requirements.txt` and add it to the configuration of your Spark pool. Within this file, you can list all the packages that you want to include upon starting a Spark instance. The format for how you name the packages follows the pip freeze format and will include the package version next to the package name:

```
packagename==1.2.1
```

The `requirements.txt` file can be uploaded to the **Packages** section of the Spark pool properties during creation. You can do this later, too, if you need to.

You'll find the location to upload your file in *Figure 6.16*, **Apache Spark Configuration**:

Create Apache Spark pool

Basics *   **Additional settings** *   Tags   Review + create

Customize additional parameters including pause settings and component versions.

Automatic pausing

Configure the pause settings for the Apache Spark pool.

Automatic pausing * ⓘ   ⦿ Enabled   ◯ Disabled

Number of minutes idle * | 15 |

Component version

Select the Spark version for your Apache Spark pool.

Apache Spark * | 2.4   ⌄ |

Python   3.6

Scala   2.11.12

Java   1.8.0_272

.NET Core   3.1

.NET for Apache Spark   1.0

Delta Lake   0.6

Apache Spark configuration

Upload a Spark configuration file to specify additional properties on the Apache Spark pool. This will be referenced to configure Spark applications upon job submission.

File Upload | Select a .txt file | 🗁

[Upload]

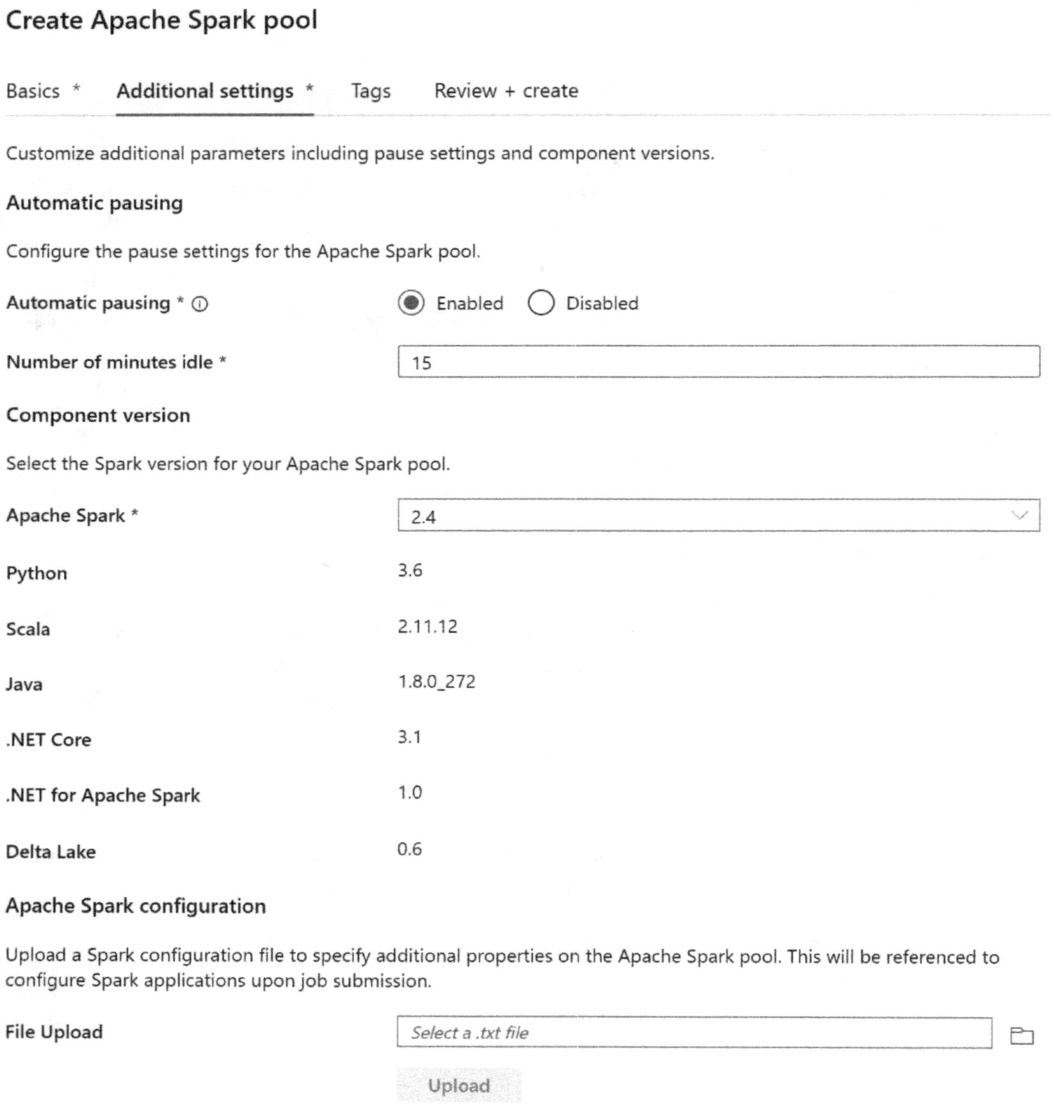

Figure 6.18 – Create Apache Spark pool – Additional settings

You can browse to the file in your filesystem and upload it in this dialog.

When you want to add a `requirements.txt` file to an existing cluster, you need to navigate to the **Management** pane in Synapse Studio and click the three dots button for the Spark pool. Select **Packages** from the context menu. A dialog will be displayed where you can upload the file to the Spark pool configuration:

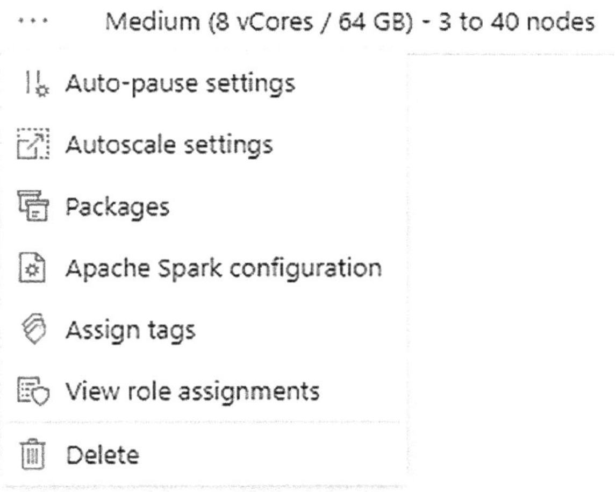

Figure 6.19 – Context menu for your existing Spark pool

Once your Spark instance is created from a pool with additional libraries, you can import them into your code just like any other libraries. Already-instantiated pools can be forced to restart to add new packages to the configuration. In the footer of the upload dialog, you will find a **Force new settings** checkbox. When you check this option, the Spark instance will restart immediately and will end all sessions that are running at that time. You can leave this setting unchecked if you want to wait for all executing sessions to end.

Adding your own packages

Public libraries are a great source for you to extend the capabilities of your system. In an enterprise setup, however, you want to be able to add custom packages with your own functions to your applications. Being able to modularize your code is essential for a resilient and agile architecture.

In the case of PySpark, you will need to create a Python wheel. You will need to install `wheel` in your Python environment:

```
pip install wheel
```

From here you can start creating your wheel file:

1. Create a root folder where you'll place all the .py files that you want to package. Let's try it with a hello world example. Please create a helloworld.py file in your root folder that just displays Hello World!, something like this:

```
print("Hello World")
```

2. In the root folder, you need to create a setup.py file that requires some minimum content as displayed:

```
from setuptools import setup, find_packages

setup(
    name = 'MyWheel',
    version='1.0.1',
    packages=find_packages()
)
```

3. To create your wheel, run the following command from the root folder with your setup.py file:

```
python setup.py bdist_wheel
```

4. If you check your root folder now, you will see two new folders: build and dist. Check the dist folder and you'll find a .whl file. This is the package that you want to upload to your Spark pool.

5. Navigate to the default data lake of your Synapse environment. You will find a path/folder combination as follows:

```
abfss://[YOURFILESYSTEMNAME]@[YOURDATALAKEACCOUNTNAME].
dfs.core.windows.net/synapse/workspaces/
[YOURWORKSPACENAME]/sparkpools/[YOURSPARKPOOLNAME]/
libraries
```

6. Place your .whl file in the libraries folder.

7. When next starting up your Spark instance for this Spark pool template, your additional libraries will be available, and you will be able to import them just like any other libraries. In the hello world case, you would do this, of course, like this:

```
import helloworld
helloworld
```

So far so good. Let's now examine security and how Synapse handles the credentials of a user when accessing storage, for example, in the next section.

Handling security

When you access the data lake storage that was configured during the setting up of the Synapse workspace, you don't need to worry about using the TokenLibrary. The Spark instance will use an Azure Active Directory credential pass-through to access the data in the data lake. This makes it easy for you to integrate your environment and set up detailed control as described in *Chapter 3*, *Understanding the Data Lake Storage Layer*. You have been using this throughout this chapter to access your data lake:

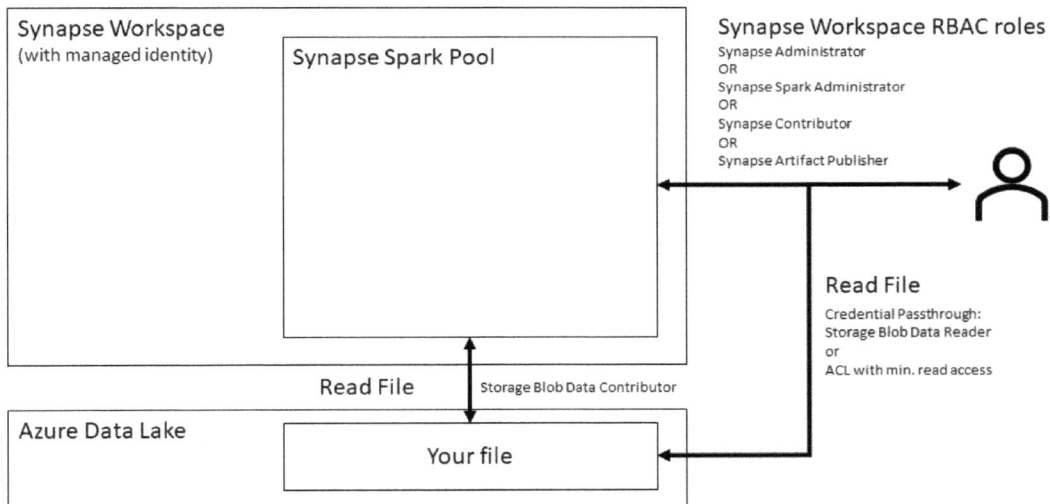

Figure 6.20 – Security setup with credential pass-through

There are other options when it comes to accessing Azure Data Lake Storage Gen2. You might have additional Azure Data Lake Storage Gen2 accounts that you have added as **linked services** to your Synapse workspace. In this case, you have several authentication options when it comes to using the storage:

- If a linked service uses a storage account key, you will need to create a Spark configuration set using `LinkedServiceBasedSASProvider`:

```
%%pyspark
```

```
spark.conf.set("spark.storage.synapse.
linkedServiceName", "[YOURLINKEDSERVICENAME]")
```

```
spark.conf.set("fs.azure.account.auth.type", "SAS")
```

```
spark.conf.set("fs.azure.sas.token.provider.
type", "com.microsoft.azure.synapse.tokenlibrary.
LinkedServiceBasedSASProvider")
```

```
df = spark.read.csv('abfss://[YOURFILESYSTEMNAME]@
[YOURSTORAGEACCOUNTNAME].dfs.core.windows.net/
[YOURPATH]')
```

```
df.show()
```

- If your linked service uses a **managed identity** or a **service principal**, you will need to alter your Spark configuration set as follows:

```
%%pyspark
```

```
spark.conf.set("spark.storage.synapse.
linkedServiceName", "[YOURLINKEDSERVICENAME]")
```

```
spark.conf.set("fs.azure.sas.token.provider.
type", "com.microsoft.azure.synapse.tokenlibrary.
LinkedServiceBasedTokenProvider")
```

```
df = spark.read.csv('abfss://[YOURFILESYSTEMNAME]@
[YOURSTORAGEACCOUNTNAME].dfs.core.windows.net/
[YOURPATH]')
```

```
df.show()
```

- Another option would be to access the additional data lake directly without a linked service in your Synapse workspace:

```
%%pyspark
```

```
spark.conf.set("fs.azure.account.auth.type", "SAS")
```

```
spark.conf.set("fs.azure.sas.token.provider.
type", "com.microsoft.azure.synapse.tokenlibrary.
ConfBasedSASProvider")
```

```
spark.conf.set("spark.storage.synapse.
sas", "[YOURSASKEY]")
```

```
df = spark.read.csv('abfss://[YOURFILESYSTEMNAME]@
[YOURSTORAGEACCOUNTNAME].dfs.core.windows.net/
[YOURPATH]')
df.show()
```

- A fourth solution would be the use of Azure Key Vault. This method will even give you the option to control the credentials from outside the environment and rotate them, for example, without the need to change the code. This is quite a safe method as the developer doesn't need to worry about the credentials; in fact, they don't need to know them at all:

```
%%pyspark
spark.conf.set("fs.azure.account.auth.type", "SAS")
spark.conf.set("fs.azure.sas.token.provider.
type", "com.microsoft.azure.synapse.tokenlibrary.
AkvBasedSASProvider")
spark.conf.set("spark.storage.synapse.
akv", "[YOURAZUREKEYVAULT]")
spark.conf.set("spark.storage.akv.
secret", "YOURSECRETKEY")

df = spark.read.csv('abfss://[YOURFILESYSTEMNAME]@
[YOURSTORAGEACCOUNTNAME].dfs.core.windows.net/
[YOURPATH]')
df.show()
```

For further information about using the TokenLibrary for more linked services, please go to *Further reading, TokenLibrary for other linked services.*

Monitoring your Synapse Spark pools

When you're developing your Spark application, you will sometimes need to get your hands deep into the engine to dig into the details of your jobs and the environment you run them in.

To ascertain details regarding your environment, navigate to the Synapse management hub. Your first stop is the **Apache Spark pools** section. You will see a list of all Spark pools, and by clicking on them, you can get an overview page with information about occupied vCores, allocated memory, and active Spark applications:

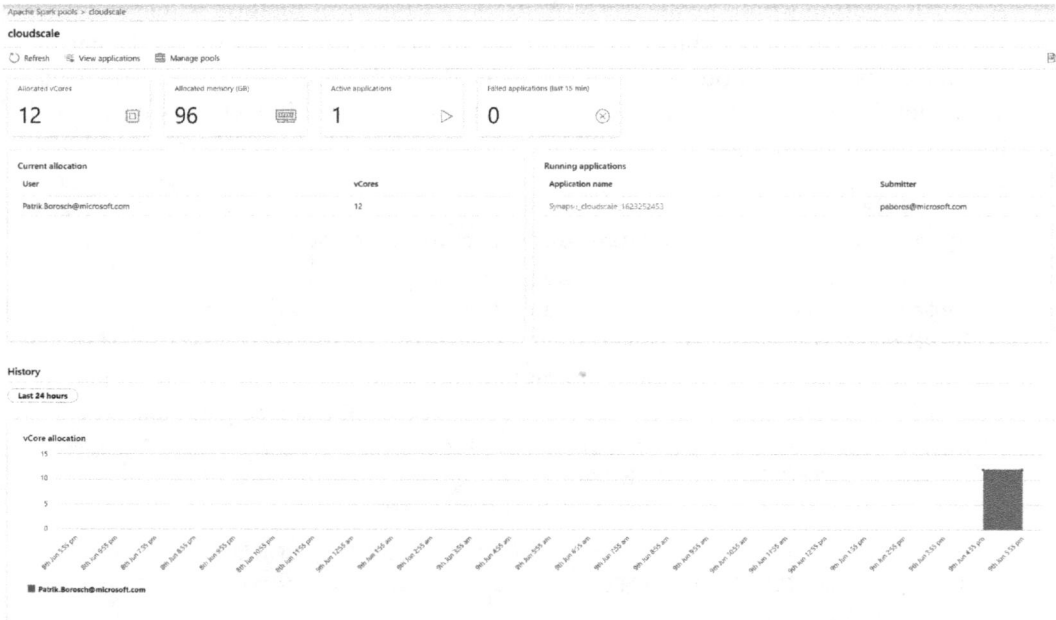

Figure 6.21 – Synapse Spark pools overview

The next level of detail to investigate will then be the application itself. You will find the **Applications** overview in the **Management** pane in Synapse Studio. You will be taken to a list of all applications that are present in your Synapse environment. By clicking on the application name on the line you're interested in, you'll get to the application details. You will have access to an application graph and the diagnostics and logs of the driver and workers that have been used to run the application.

In the drop-down fields in the upper area of the graph, you can toggle between different views, such as **Progress**, **Read**, **Written**, and **Duration**:

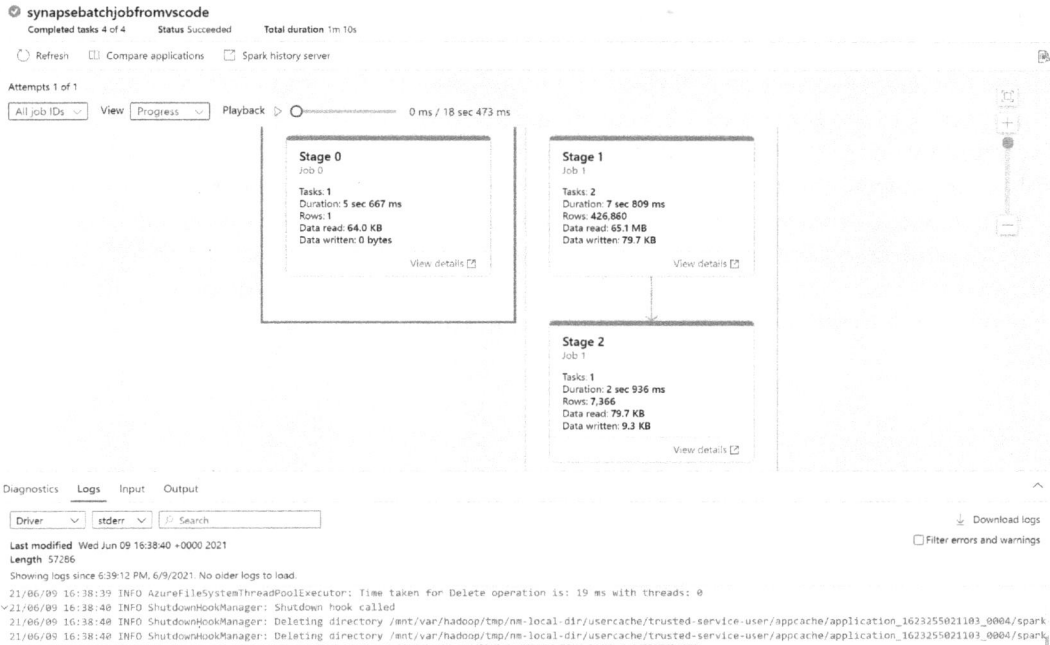

Figure 6.22 – Spark application overview

From here, you have the option to jump to the **history server** of your Spark instance with the context of the application that you have just viewed. The history server will help you to dive into the details of your application:

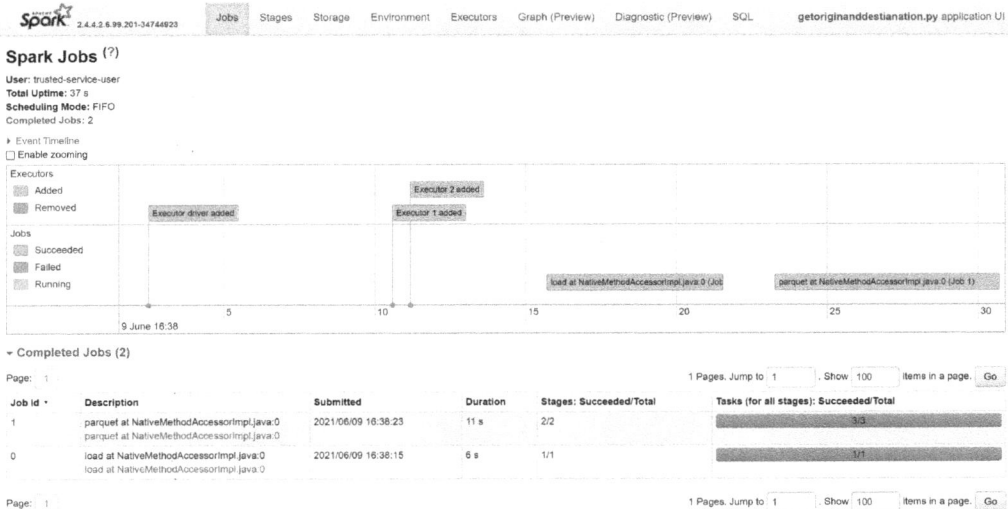

Figure 6.23 – Spark history server

The **Jobs** and **Stages** tabs will help you to understand the different activities that occurred. In the **Storage** section, you'll find insights about the usage of storage components in your job. The **Environment** section will allow you to understand the setup your job was run with. By going to the **Executors** tab, you will get insights into what happened with your drivers and workers. The **Graph** will display a graphical representation of the different stages your job went through. You will be able to play back the complete job history and see the behavior of your job. When you dig into the **Diagnostic** view, you will get details about data and time skew and you'll find an analysis of the usage of your executors. The **SQL** tab on the far right of the display will provide detailed insights into Spark SQL behavior, even into the query plans of your job differentiated by workers:

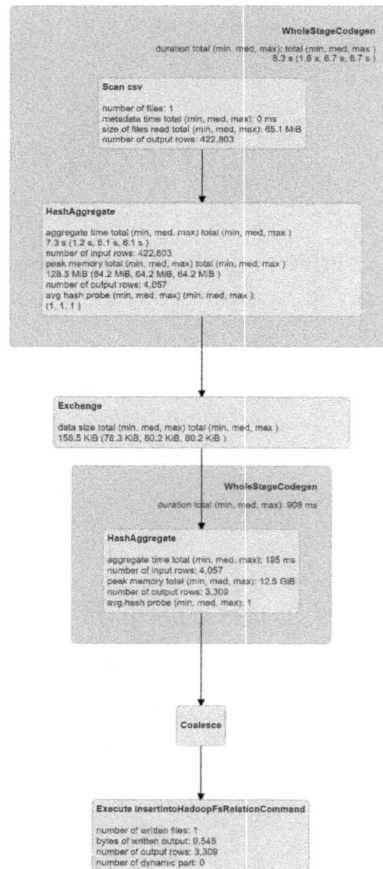

Figure 6.24 – Query details on the SQL tab of the history server

There is so much to learn regarding Synapse Spark pools. Don't forget to check the *Further reading* section.

Summary

In this chapter, you have seen how to provision a Synapse Spark pool. You have learned about Spark's architecture in general and Synapse Spark's architecture.

You have learned about the difference between Synapse Spark pools and Synapse Spark instances. You implemented your first Synapse notebook for interactive analytics and learned how to implement a Spark application that can be run as a batch job.

You have seen how to use a Spark pool from an IDE such as Visual Studio Code and you have investigated how to use additional libraries from public sources and your own libraries.

Finally, you saw how you can interact with storage securely, before learning how monitoring works with your Synapse Spark environment.

In *Chapter 7, Using Databricks Spark Clusters*, you will learn about an alternative Spark environment that Microsoft offers on Azure.

Further reading

- *Requesting a quota increase*: https://docs.microsoft.com/en-us/azure/
  azure-portal/supportability/per-vm-quota-requests#request-
  a-standard-quota-increase-from-help--support

  a) For the **Service type**, select: Azure Synapse Analytics

  b) For the **Quota details**, select: Apache Spark (vCore) per workspace

- *Autoscaling behavior*: https://docs.microsoft.com/en-us/azure/
  synapse-analytics/spark/apache-spark-autoscale

- *Notebook navigation*: https://docs.microsoft.com/en-us/azure/
  synapse-analytics/spark/apache-spark-development-using-
  notebooks

- *Developing Spark applications with IntelliJ*: https://docs.microsoft.com/
  en-us/azure/synapse-analytics/spark/intellij-tool-synapse

- *Developing Spark applications with Visual Studio Code*: `https://docs.microsoft.com/en-us/azure/synapse-analytics/spark/vscode-tool-synapse`

- *TokenLibrary for other linked services*: `https://docs.microsoft.com/en-us/azure/synapse-analytics/spark/apache-spark-secure-credentials-with-tokenlibrary?pivots=programming-language-python#tokenlibrary-for-other-linked-services`

7
Using Databricks Spark Clusters

In the last chapter, *Chapter 6, Using Synapse Spark Pools*, you learned about Spark and the Synapse integrated Spark engine. But what about cases where you only need a Spark cluster to interact with your Data Lake Store? You would, for example, choose Databricks over Synapse Spark pools at this point in time, when you need to work on Spark 3.0 or when you need to implement **Structured Streaming**. R, as a required programming language, will require Databricks as well as the Databricks-specific features of Delta Lake, such as vacuuming and others. Synapse will offer most of these options, too, in the future. But at the moment, they are available only in Databricks.

With Azure Databricks, Microsoft offers a standalone Spark environment that will give you all the aforementioned options and can still integrate with other data services on Azure if needed. And with Databricks, you have the people at your back that invented Spark. The cluster architecture differs slightly from that of Synapse Spark pools and will be a little more complex in terms of handling.

In this chapter, we will examine Azure Databricks. You will see how to provision a Databricks workspace and learn how to use it to create autoscaling Spark clusters. In the *Understanding the Databricks components* section, you will also see how to create Databricks notebooks or run Spark jobs on the clusters to implement machine learning models or simply run ETL jobs on your data lake. You will learn how to create dashboards and how to add additional libraries to the environment. Additionally, you will see how to set up Databricks with VNets and how to implement access control within a Databricks workspace. Of course, we will also examine how to monitor workloads in Databricks.

We will go through these topics in the following sections:

- Provisioning Databricks
- Examining the Databricks workspace
- Understanding the Databricks components
- Setting up security
- Monitoring Databricks

Technical requirements

For this chapter, you will need the following:

- An Azure subscription where you have at least contributor rights or you are the owner
- The right to create a service principal in Azure Active Directory
- The right to provision a Databricks workspace

Provisioning Databricks

Provisioning a Databricks workspace is as easy as the services in the previous chapters:

1. First, navigate to the Azure portal and click **Create a resource**.

2. In the search box, type Databricks and select **Azure Databricks** from the quick results displayed beneath the search. The Databricks info is displayed.

3. Click **Create** and start the provisioning.

4. In the **Basics** blade, fill in or select the values for the input fields. You will need to select the subscription to build your **Azure Data Factory** (**ADF**) and either select an existing resource group or create a new one. See *Chapter 3*, *Understanding the Data Lake Storage Layer*, for a description of resource groups. You want to name your workspace here and assign it to the most suitable region for you. As regards the Pricing Tier, please select the appropriate one. For a first test, you might select **Trial (Premium - 14 Days Free DBUs)** as this won't cost anything. You can then proceed with **Next: Networking >**:

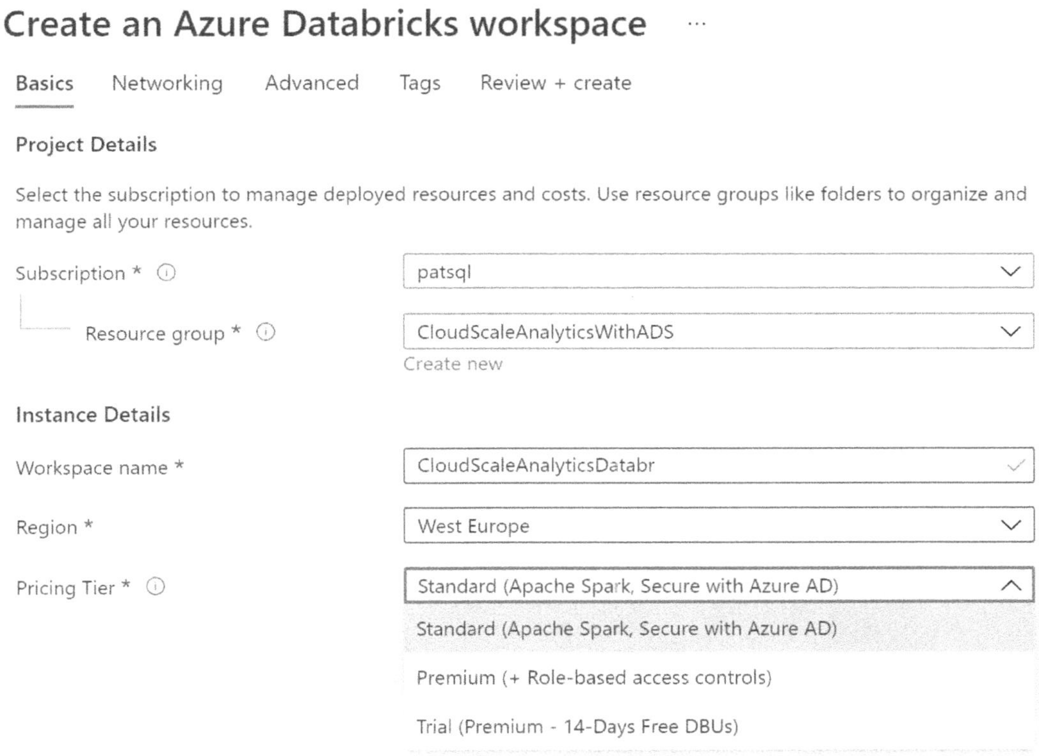

Create an Azure Databricks workspace   ···

Basics   Networking   Advanced   Tags   Review + create

Project Details

Select the subscription to manage deployed resources and costs. Use resource groups like folders to organize and manage all your resources.

Subscription * ⓘ	patsql ⌄
Resource group * ⓘ	CloudScaleAnalyticsWithADS ⌄
	Create new

Instance Details

Workspace name *	CloudScaleAnalyticsDatabr ✓
Region *	West Europe ⌄
Pricing Tier * ⓘ	Standard (Apache Spark, Secure with Azure AD) ⌃
	Standard (Apache Spark, Secure with Azure AD)
	Premium (+ Role-based access controls)
	Trial (Premium - 14-Days Free DBUs)

Figure 7.1 – Create an Azure Databricks workspace – The Basics blade

You will find an overview of the different pricing tiers in the *Further reading*, *Pricing Databricks*, section.

5. In the following blade, you will be able to configure Databricks to be deployed to an available virtual network of your choice. In the **Virtual Network** combo box, you will select the VNet to be used for this. In the fields below it, you can then name the additional subnets that will be created on behalf of Databricks. Please pay attention that the **Classless Inter-Domain Routing (CIDR)** ranges need to be equal to, or less than, 26. This will mean that you can host a maximum of 64 IP addresses in that particular subnet. This time, for simplicity reasons, please select **No** and do not join a VNet now. Please proceed with **Next : Tags >**:

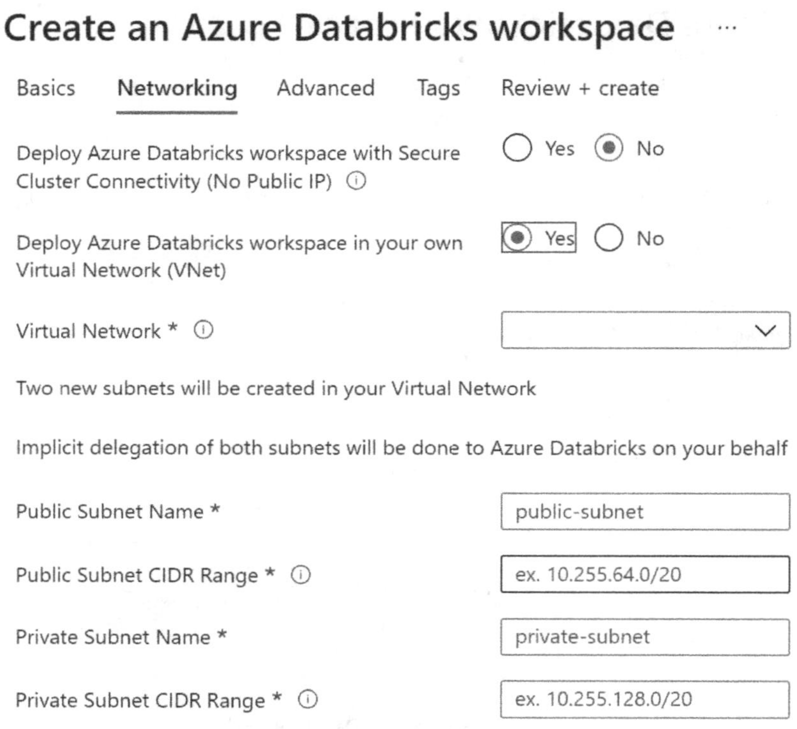

Create an Azure Databricks workspace ···

Basics   **Networking**   Advanced   Tags   Review + create

Deploy Azure Databricks workspace with Secure Cluster Connectivity (No Public IP) ⓘ   ◯ Yes ⦿ No

Deploy Azure Databricks workspace in your own Virtual Network (VNet)   ⦿ Yes ◯ No

Virtual Network * ⓘ   [⌄]

Two new subnets will be created in your Virtual Network

Implicit delegation of both subnets will be done to Azure Databricks on your behalf

Public Subnet Name *   [public-subnet]

Public Subnet CIDR Range * ⓘ   [ex. 10.255.64.0/20]

Private Subnet Name *   [private-subnet]

Private Subnet CIDR Range * ⓘ   [ex. 10.255.128.0/20]

Figure 7.2 – Create an Azure Databricks workspace – The Networking blade

6. Finally, you will again pass the **Tags** blade, where you can add your own tags for better reporting. You have already seen tags during the provisioning of your first Data Lake account in *Chapter 3, Understanding the Data Lake Storage Layer*. Please move to the final blade of the provisioning sequence with **Review + create**.

When the validation of your entries is successful, you can kick off provisioning with **Create**, and as you have seen before, in this case, again, you will be able to download an ARM template when you click on **Download a template for automation**.

When your deployment is complete, you can hit **Go to resource** and start examining the newly created Databricks workspace.

You will do this and more in the next section, *Examining the Databricks workspace*.

Examining the Databricks workspace

Like ADF and Synapse Analytics, too, Databricks follows the concept of a browser-based workspace interface. When your basic deployment has been successful and you navigate to the resource in your Azure portal, you will find the **Launch Workspace** button very prominent on the **Overview** blade of your Databricks service:

Figure 7.3 – The Launch Workspace button on the Overview Blade of the Databricks service

Click on the button to enter your workspace for the first time. You are taken to the Databricks workspace portal:

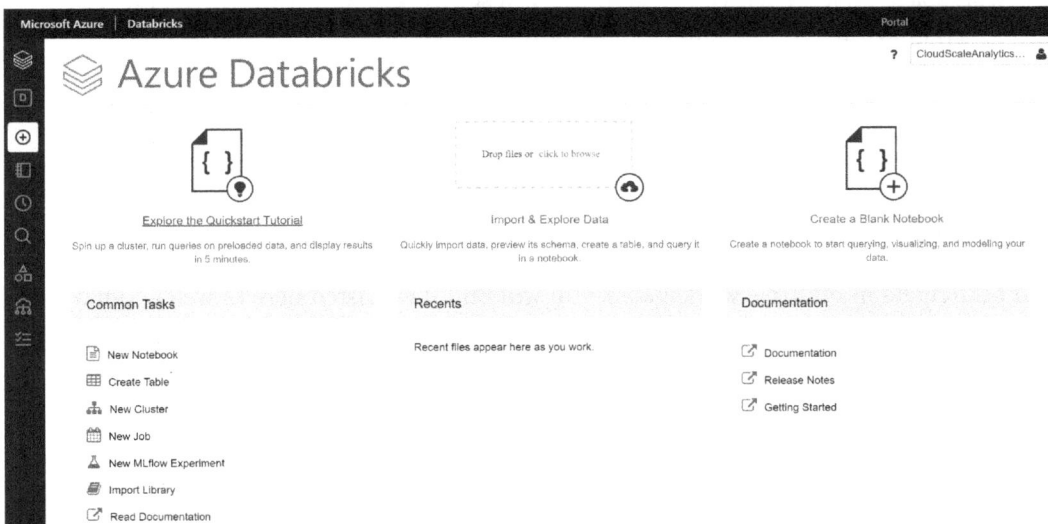

Figure 7.4 – Entering your Databricks workspace

You will find a navigation area on the left side of the screen. Here, you have access to all the different areas of Databricks, such as your workspace, recent artifacts, data environments such as databases and tables, your clusters, Spark jobs, and machine learning models. The final option on the navigation area is **Search**.

Click on **Home** to make yourself familiar with the way folder structures are displayed in the workspace. The navigation area will expand from left to right, into the center of the screen.

In the center of the screen, you'll find different options to get started with Databricks. You'll find a quickstart tutorial, which it's a good idea to go through.

Go through the **Common Tasks** section. Here, you'll see a list of all the typical object types you'll work with in Databricks. You will learn more about the components of Databricks in the next section, *Understanding the Databricks components*.

In the center of the screen, you'll find a list of recently accessed objects, and on the right, you have access to the Databricks documentation.

Above **Recents**, you have the **Import & Explore Data** option. As pointed out there, you can bring data into the workspace and start exploring it using Databricks notebooks. You can even create a table directly from the data in a Databricks database.

In the top-right corner, next to the question mark, you'll find a button with the name of the workspace on it. When you click it, you will see the account with which you are logged in and have access to the following options:

- **User Settings**
- **Admin Console**
- **Manage Account**
- **Log Out**

If you participate in multiple workspaces, you will find them listed here as well and have the option to switch between them.

Understanding the Databricks components

In the last chapter, *Chapter 6, Using Synapse Spark Pools*, we examined the basic Spark architecture, and Databricks also follows those rules. You will find driver and worker nodes that will process your requests. And we shouldn't forget that Databricks was the first to deliver autoscaling Spark as a Service, which will even take the compute environment down as soon as an idle time threshold is reached.

Although Databricks is based on Apache Spark, it has built its own runtime, optimized for usage on Azure. When you spin up a cluster, for example, different sessions will reuse the same cluster and will not instantiate it as with Synapse Spark pools.

Creating Databricks clusters

This section will take you through the provisioning process of a Databricks cluster. You will see the different node sizes and the options that you have, such as autotermination and autoscaling, when you create your compute engine here.

But let's see how this works. Let's create your first Databricks cluster and use it:

1. Navigate to your newly created workspace and either use the Clusters hub in the navigator area on the left or hit **New Cluster** in the **Common Tasks** area in the center of the workspace screen. The following screen is displayed:

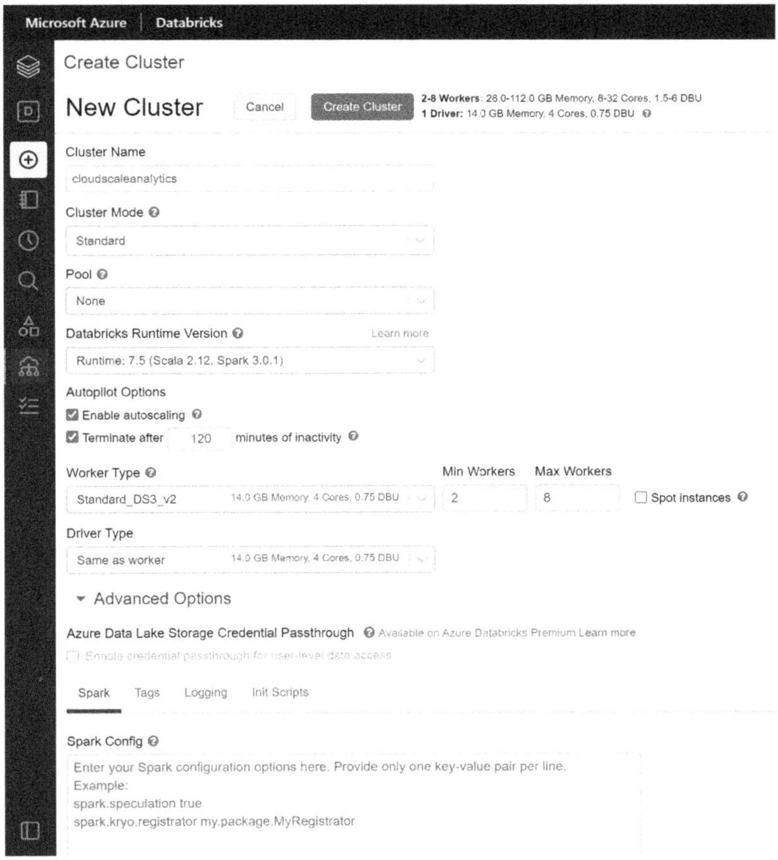

Figure 7.5 – New Cluster dialog

2. First, fill in the **Cluster Name** field for the new cluster.

3. Next, you can select **Cluster Mode**. Your options are as follows:

    High Concurrency: Aimed at data scientists and business analysts and is optimized for multi-user usage. Supports SQL, Python, and R.

    Standard: Targets data engineers for single-user usage and is best used for batch mode jobs in ETL scenarios. Supports Scala, Java, SQL, Python, and R.

    Single Node: Targets developers who do not need bigger clusters and want to save money.

4. Below **Cluster Mode**, you can select a pool to join. When you join a pool, you can profit from a collection of already started instances. These will be available for you when you start your cluster. This can speed up the start time for your cluster to be available far more quickly.

5. **Databricks Runtime Version** gives you access to different Scala and Spark versions if you need different compatibilities. When you check the combo box, you will also find GPU support in the runtime.

6. When you now check **Autopilot Options**, you can enable autoscaling by ticking the checkbox. The second checkbox gives you access to the autotermination option for your Databricks cluster. When you tick the checkbox, you can then select the number of minutes as a threshold for inactivity before the cluster is taken down.

7. The **Worker Type** and **Driver Type** drop-down boxes offer you access to a vast (30+) collection of different VM types (compute or memory-optimized, with and without GPU support, and with up to 64 virtual CPUs and 432 GB of memory per VM) that you can use for your Databricks cluster.

    Worker Type offers additional settings if you enable autoscaling mode. You can set the lower and upper boundaries for the cluster to scale.

8. In **Advanced Options**, you can configure additional settings, such as for your Azure Data Lake Storage Credential Passthrough. If you want to use this, you will need to select **Azure Databricks Premium** when you provision your account.

9. Additionally, you can add configuration options on the **Spark** tab, in the **Spark Config** section. Below that, you can configure **Environment Variables** for your cluster.

    In the lower section where you found the Spark settings, you have additional tabs where you can add tags to your cluster, control the logging path, and provide additional init scripts for the cluster startup.

Create a standard cluster, naming it as you wish. Tick the **Enable Autoscaling** option, with **2** as the minimum and **8** as the maximum number of workers. Select **120 minutes** as the the autotermination option, leave the rest as their defaults, and then click **Create Cluster** on the upper ribbon.

When the cluster is up and running, the screen will change a little and show you not only the configuration, but a bunch of tabs on the upper ribbon giving you access to all the details regarding your newly built environment:

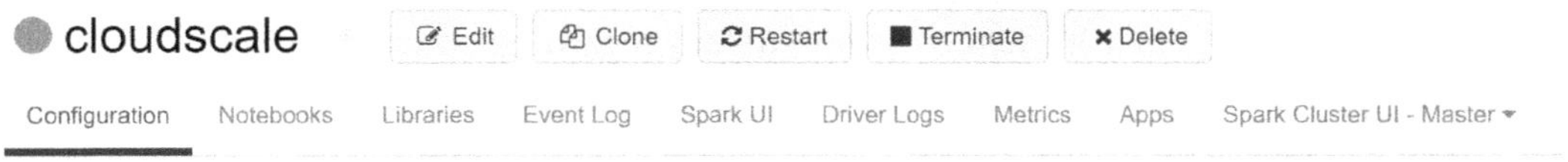

Figure 7.6 – Cluster details and controls

Let's check the buttons:

- With the **Edit** button, you can get back into the configuration of the cluster and change the actual configuration. You can change nearly every option there apart from **Cluster Mode**. Following a change, you need the cluster to restart with **Confirm** and **Resize**.

- **Clone** will just create an identical copy of the actual cluster. This comes in handy when you quickly want to create a new cluster with similar configurations to an existing one. But please note: cloning does not copy the permissions on a cluster, nor will the installed libraries or the attached notebooks be available in the clone of an existing cluster.

- **Restart** just boots an existing cluster. If you change an attached library, for example, and want the change to take effect, you will need to restart the particular cluster.

- **Terminate** will stop the cluster. With Databricks clusters, just like Synapse Spark pools, you are in a fully managed world. When you have finished your work, you will terminate the cluster and from that moment on, you won't be charged for the environment.

- **Delete** will wipe the cluster from your workspace.

- The tab ribbon below the buttons provides access to different details regarding the actual cluster.

- On the **Notebooks** tab, you can get an overview of all the notebooks that have the actual cluster attached. You will find **Status**, **Last Command Run**, and **Location** details.

- The **Libraries** tab does not only show all the attached libraries; it gives you access to the library installer for the actual cluster. This is one of the entry points when you need to add libraries to the environment.

- The **Event Log** tab, as you will have already guessed, gives you access to the cluster's event log. You can filter the log via the combo box in the upper-left corner to dive into the details quickly.

- **Spark UI** will take you to the Apache Spark interface with access to all the details that your Spark engine can provide. There, you'll find jobs, stages, storage consumption, an environment overview, details regarding your executors, SQL statements used, JDBC/ODBC server-side information, and a **Structured Streaming** tab with in-depth information about your streaming jobs.

- **Driver Logs** provides an overview of everything that happens on the driver nodes of your cluster.

- On the **Metrics** tab, you will find snapshot reports that display all kinds of metrics regarding your cluster:

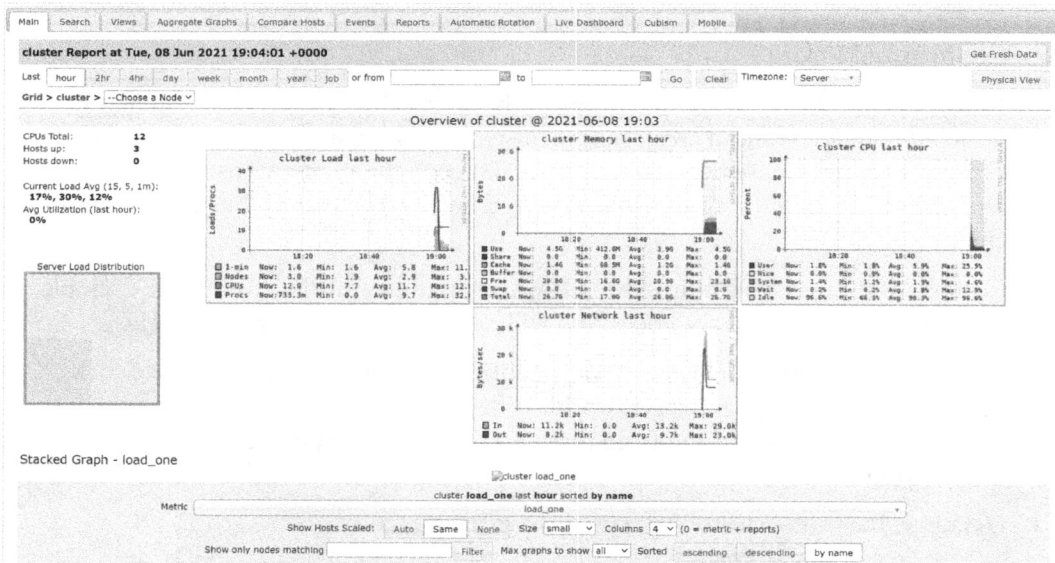

Figure 7.7 – Cluster metrics snapshot report

- The **Apps** tab allows you to configure a Bash terminal in your cluster and to set up RStudio for use with your actual cluster.

- Finally, the **Spark Cluster UI** combo box on the far right of the ribbon gives you access to the Apache Spark Cluster UI with the option to select either the **Driver** or the available **Worker** nodes to be displayed:

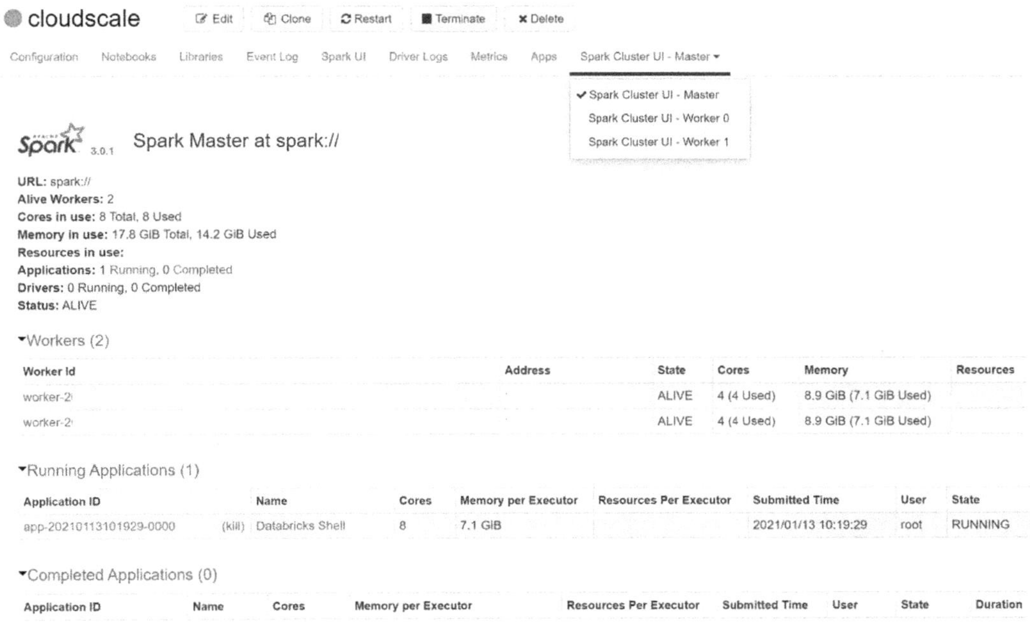

Figure 7.8 – Spark Cluster UI Master view

There is a lot to know here but don't let the options confuse you. You will be able to start in any case and build your knowledge while using the service.

Managing clusters

When you check the navigation area in the Databricks workspace on the left, you will find a **Clusters** node. When you click on it, you'll be presented with a list of all the available clusters in your workspace:

- **All-Purpose Clusters**: These are used with notebooks, Spark jobs, and dashboards, and can be used and shared by more than one data scientist, for example.

- **Job Clusters**: These are optimized for use with Spark jobs. They will be stopped as soon as your job is done.

- **Pools**: These are more or less clusters with "warm" nodes that are available quickly if needed.

From here, you have access to all the available clusters in your workspace. You can create new ones, review the status of existing clusters, filter the view, and search for a particular one.

The list view also shows you the actual number of nodes that a particular cluster uses and already gives you some information on the configurations:

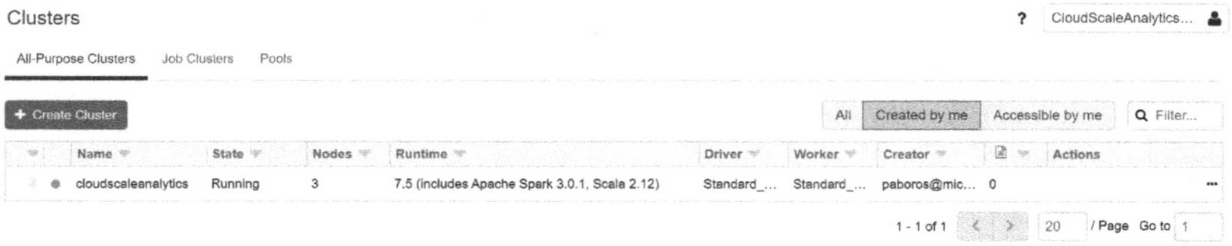

Figure 7.9 – Clusters node view

Do you remember the **Create Cluster** dialog? There was an option where you could select a pool of VMs to be used for your cluster. When you click on the **Pools** tab in this view, you can provide pools of VMs of a particular size to be available for faster cluster creation.

Using Databricks notebooks

Now that you have a running cluster available, let's create your first Databricks notebook and use the available cluster.

Navigate to your **Databricks** workspace and either select **New Notebook** in the **Common Tasks** section on the home screen or the more prominent **Create a Blank Notebook** icon on the right side to start your first notebook:

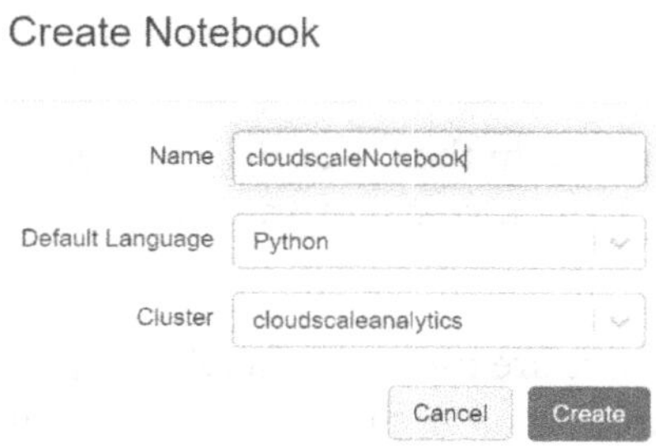

Figure 7.10 – Create Notebook dialog

A small dialog will prompt you to fill in the **Name**, **Default Language**, and **Cluster** boxes. In the language field, you will find the language that Databricks expects in a notebook, for example, as the default language. You can change this language between cells in a notebook by adding a so-called **Magic Command** to the first line of the cell. We have talked about this concept already in *Chapter 6, Using Synapse Spark Pools*.

If you have created your notebook with Python as the default and you reach a point where you'd rather continue using SQL, you need to create a temporary table with the data you want to use in SQL. This works exactly the same as in the Synapse Spark pool example in *Chapter 6, Using Synapse Spark Pools*. Moving to the next cell in your notebook, you can then type %sql into the first line of the cell. Databricks will then expect SQL code in all of the following lines in this particular cell.

Possible magic commands to mix languages in Databricks include the following:

- %sql
- %python
- %r
- %scala

Databricks allows additional magic commands such as %sh, which allows you to run shellcode in a notebook, and %fs, which will give you access to the dbutils filesystem commands of Databricks. Another available magic command is %md, which allows you to include images, text, mathematical formulas, or equations for documentation purposes.

Name your new notebook and select **Python** (it shows up already) as the default language for the notebook. In the **Cluster** drop-down box, you will already see the cluster that you created in the *Creating Databricks clusters* section.

An empty notebook, perhaps with some hints popping up, will be displayed with some controls that are easy to understand and use:

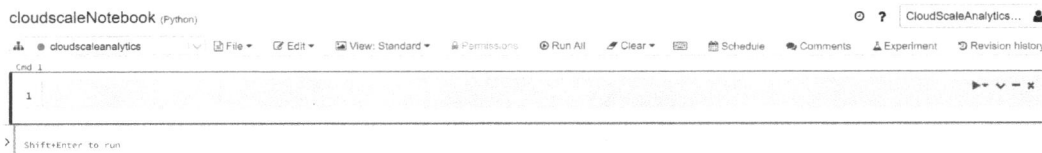

Figure 7.11 – First, empty notebook

When you now check the drop-down box right beneath the notebook name in the top-left corner, you will see the cluster this notebook is attached to. When you click the down arrow next to the cluster name, you will get a brief overview of the cluster with some options to detach, restart the cluster, or jump to the Spark UI or the driver logs of the cluster:

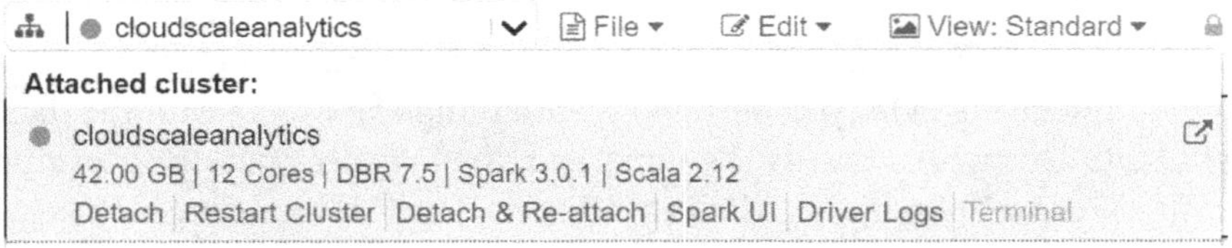

Figure 7.12 – Cluster info in the notebook

To run your first piece of Python code, you could, for example, enter the following into the first empty cell of the notebook and hit *Ctrl + Enter*:

```
print('Hello World')
```

This will run the code and will display the following right under the cell:

```
Hello World
```

We won't deep dive into the world of Spark programming. That would fill many other books on its own.

But let's examine your options when you are interactively analyzing your data using Databricks notebooks.

You have already used your first shortcut in your notebook. *Ctrl + Enter* will run the code in the actual cell and the results will be displayed right under the cell. *Shift + Enter*, alternatively, will do the same, but will additionally create a new cell below the one you're working in and jump to that one so that you can smoothly write your next lines of code and interact with your data.

Alternatively, you can hit the small **Run** button in the upper-right corner of the active cell:

Figure 7.13 – Cell context menu

A small context selection will appear giving you the options to either **Run Cell**, **Run All Above**, or **Run All Below**:

Figure 7.14 – Run Cell context menu

Additionally, the down arrow in this context menu offers you a lot of other options, such as copying the active cell or cutting it, but also adding other cells above or below, moving the active cell in the notebook, and more:

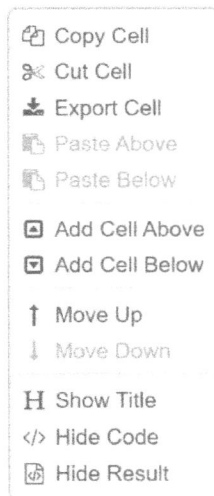

Figure 7.15 – Cell Edit context menu

The two additional options in that context menu are to collapse the cell (or expand it when it's collapsed) with the - button (which turns into a + when the cell is collapsed) and using the **x** button to delete the active cell.

When you check the ribbon below the notebook name, you will find additional menu items for interaction with your notebook.

The **File** menu provides access to all menu commands that you need for the notebook file itself:

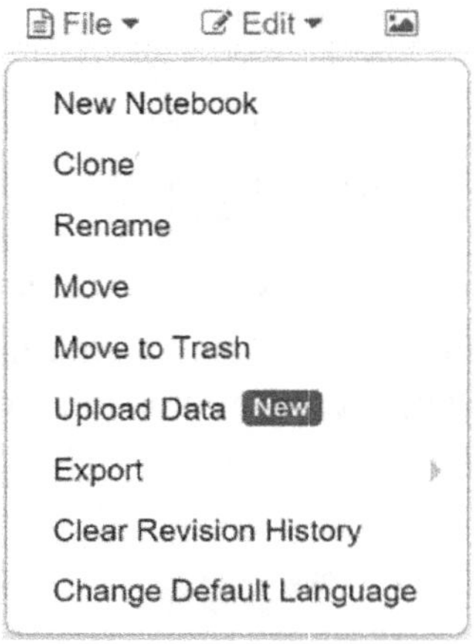

Figure 7.16 – Notebook File options

Another extended menu for interacting with the cells in the notebook is the **Edit** menu. Here, you will find all the necessary commands for editing your notebook and the cells it contains:

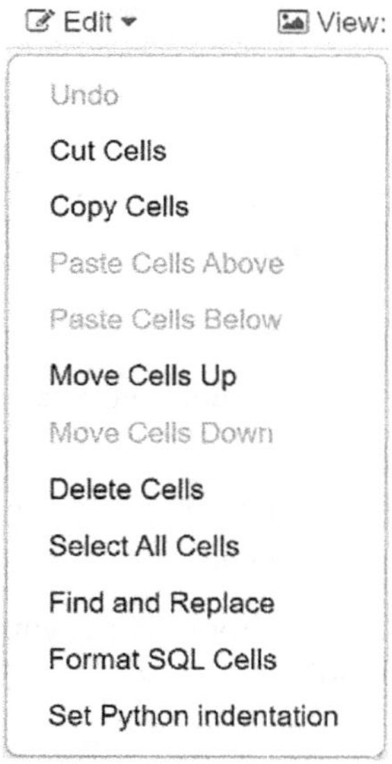

Figure 7.17 – Notebook Edit options

When you check the contents of the **View** menu, you will find options to adjust how the notebook is displayed and the command to create a new dashboard:

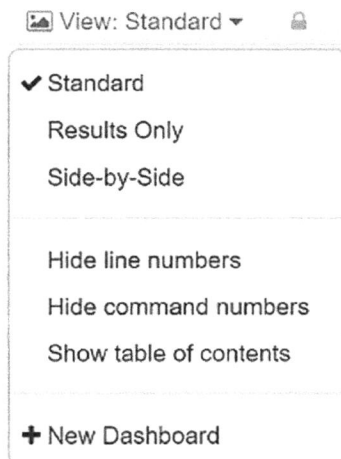

Figure 7.18 – Notebook View options

The **Permissions** menu will be available in this view when you provision Databricks in the premium plan. In this case, you decided to do so, so you will be able to control permissions on the notebook. You will be able to influence who can read, run, edit, and manage parts of or the whole notebook and who has no permissions. The options you can control in the notebook are as follows:

- **View cells**
- **Comment**
- **Run via %run or Notebook Workflows**
- **Attach and detach Notebooks**
- **Run commands**
- **Edit cells**
- **Change Permissions**

The **Run All** button next to the **Permissions** menu does just that. When you click it, all cells in the notebook will be executed in a sequence from top to bottom.

The **Clear** menu offers you options in terms of clearing off either all the results in your notebook, the state, or a mixture of the two:

Figure 7.19 – Notebook Clear options

However, more options are available in this view. On the far-right side of this ribbon, you will find a small keyboard symbol that leads you to all the shortcuts available for notebook development as well as some more menu items:

- **Schedule**: This option allows you to schedule the active notebook as a Spark job. When you hit **+New** on the following screen, you can create a schedule for your notebook and have it running regularly as a job.

- **Comments**: Comment on your code in the notebook to make it easier for others to understand or to comment on the results or similar.

- **Experiment**: Implement MLflow experiments.

- **Revision History**: Get a full history of all the versions of your notebook. From there, for example, you can additionally configure a Git repository to store your work.

Above these items, you will find another gimmick when you click on the small clock symbol. It will show a list of recent activities as a log, which comes in handy when you want to review what you have done.

Connecting your Databricks notebook to your data lake

Now let's go on and connect to your data lake from *Chapter 3*, *Understanding the Data Lake Storage Layer*, and read the file you created earlier. In comparison to the tightly coupled Synapse Spark pools, Databricks needs you to do some setup steps before you can go and read in your data from the data lake:

1. Create a service principal for your Databricks environment. Databricks will use this one to access your data lake. Within your data lake, this service principal will need the RBAC Storage Blob Data Contributor. You can find a link to a tutorial to create a service principal in the *Further reading*, *Creating a Service Principal*, section. Copy the **Tenant ID** (this may appear under **Directory** (tenant) ID), **App ID** (this may appear as **Application** (client) ID), and the **Client Secret** values for later use to a text file. The tenant and app IDs will be available on the **Overview** blade of your service principal. The client secret must be created under **Certificates & Secrets**. You will need to create a new secret. When you see it displayed right after creation, copy the secret. You won't be able to see it again. If you forget to do so, you will need to create a new one.

2. Go to your notebook and enter the following Python code into the first cell. Take care that you replace the values in brackets ([]) with the values that you collected in the first step or from your Data Lake account before you run the cell.

   Note the second command, dbutils.fs.mount(), which mounts the particular Data Lake folder to your cluster:

```
configs = {"fs.azure.account.auth.type": "OAuth",
"fs.azure.account.oauth.provider.type": "org.apache.
hadoop.fs.azurebfs.oauth2.ClientCredsTokenProvider","fs.
azure.account.oauth2.client.id": "[AppID]",

"fs.azure.account.oauth2.client.secret":
"[ClientSecret]",

        "fs.azure.account.oauth2.client.
endpoint": "https://login.microsoftonline.com/
[TenantID]/oauth2/token",        "fs.azure.
createRemoteFileSystemDuringInitialization": "true"}

dbutils.fs.mount(source = "abfss://[YOURFILESYSTEM]@
[YOURDATALAKE].dfs.core.windows.net/[YOURFOLDERPATH]",

mount_point = "/mnt/data",

extra_configs = configs)
```

3. Now you can go on and create a data frame as the next step, from your `airdelays.csv` file from the last exercise:

```
dfairdelays = spark.read.format('csv') \
    .options(header='true', inferschema='true') \
    .load("/mnt/data/[YOURFOLDER]/airdelays.csv")
display(dfairdelays)
```

4. The content of the file should now display beneath the code as a result set:

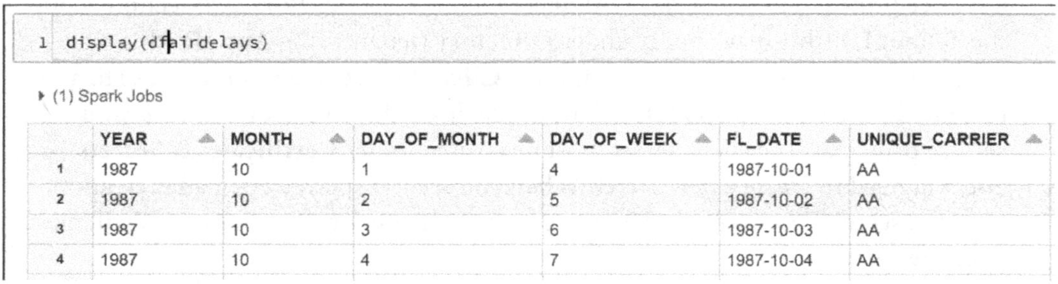

```
1  display(dfairdelays)
```

▶ (1) Spark Jobs

	YEAR	MONTH	DAY_OF_MONTH	DAY_OF_WEEK	FL_DATE	UNIQUE_CARRIER
1	1987	10	1	4	1987-10-01	AA
2	1987	10	2	5	1987-10-02	AA
3	1987	10	3	6	1987-10-03	AA
4	1987	10	4	7	1987-10-04	AA

Figure 7.20 – Results in the notebook

When you now check the result set that is displayed, you will find additional icons in the lower-left corner of the cell: a table, a chart, and an export icon. With the table and the chart icon, you can toggle the output between a tabular and chart view. When you experiment a little with **Plot Options** for this one, you will quickly be able to display a useful chart:

```
1   display(dfairdelays.select('YEAR', 'MONTH', 'DAY_OF_MONTH', 'UNIQUE_CARRIER'))
```

▸ (3) Spark Jobs

Aggregated (by count) in the backend.

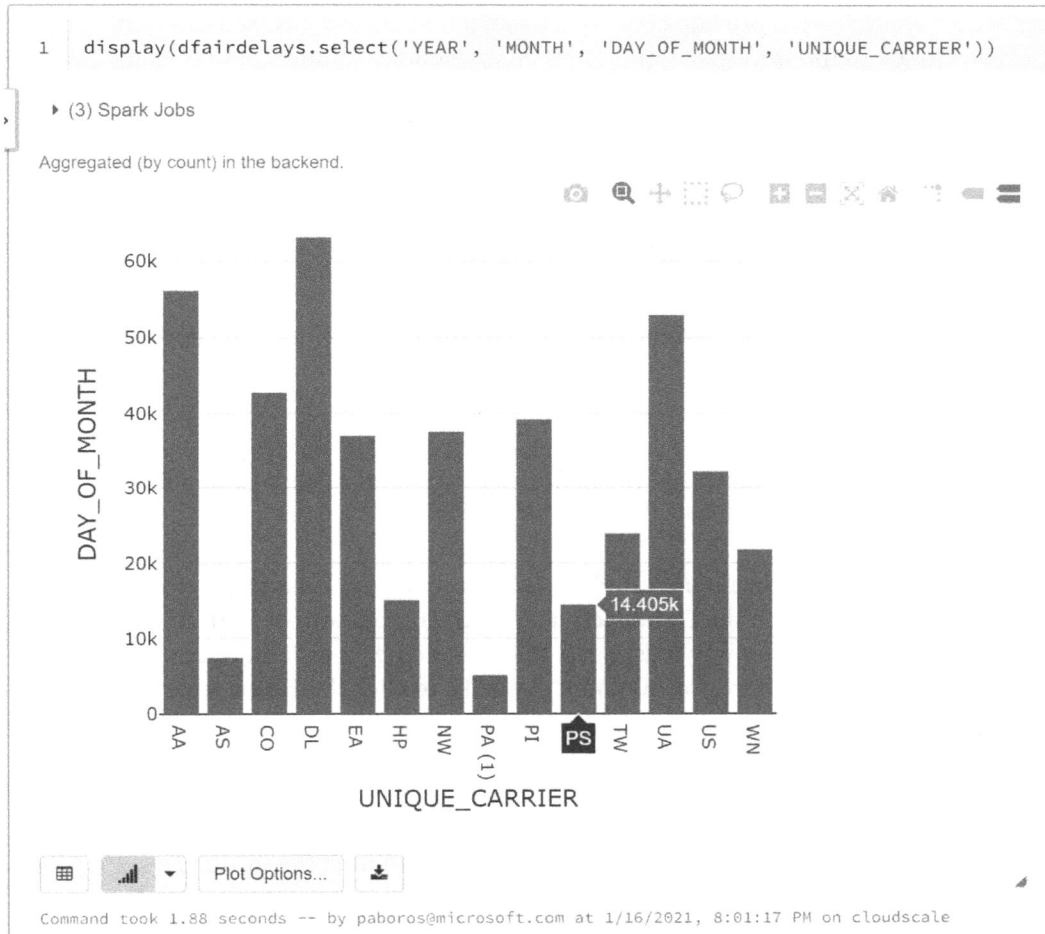

Command took 1.88 seconds -- by paboros@microsoft.com at 1/16/2021, 8:01:17 PM on cloudscale

Figure 7.21 – Chart output

Now, let's move on and examine the option of running Spark batch jobs against a Databricks cluster in the following section, *Using Databricks Spark jobs*.

Using Databricks Spark jobs

Interactively analyzing data using Databricks notebooks, as in the preceding section, is one of the most prominent activities that Databricks is used for. When you have reached a point in your analysis where you need a recurring routine to wrangle data and perform ETL jobs or use a machine learning model with your data, another strength of this platform kicks in: jobs.

There are different ways to provide code for a batch job in Databricks. You can write a notebook, store it in the workspace, and call it on a schedule. Another option would be to provide a JAR file that contains Java code and run that in a scheduled batch. A third option would be to use `spark-submit` to throw a JAR or a Python file at a cluster from outside the environment and run it.

Let's examine the three.

Scheduling a notebook as a batch job

There are two ways to do this from the UI. The first would be from the notebook directly.

> **Tip**
>
> Why are we not examining Data Factory/Synapse Pipelines? You will find a Databricks section with several options for orchestrating Databricks notebooks or jobs there, which addresses this in a wider context.

Navigate to the notebook that you created in the last section. You can find it, for example, on the main screen (click on the **Databricks** icon in the navigation area), in the center area under **Recents**. Another way to navigate there would be to click on **Recents** in the navigation area, and a third way to find it would be to click the **Home** icon in the **Navigation** area and select it from the folder there.

Hopefully, you have found your notebook now and it is open in front of you. Let's adjust it a little to avoid errors, on the one hand, and to have a more real-world example, finally.

Comment out the first cell where we created the configs and fired the `dbutils.fs.mount()` command. As the data lake is already mounted in the workspace, this would cause an error if we were to try and run it again. In Python, you comment a line with `'#'`.

Then you might add the following line as a new cell:

```
dbutils.fs.rm("/mnt/data/[YOURTARGETFOLDER]/airdelays.parquet",
True)
dfairdelays.write.parquet("/mnt/data/[YOURTARGETFOLDER]/
airdelays.parquet")
```

This will use `dbutils.fs.rm()` to remove the file if it already exists and then output the content of your `airdelays.csv` file to Parquet format into your target folder on your data lake.

When you are ready with the extension of your notebook, click **Schedule** in the upper-left area. From the right side of the screen, Databricks opens the **Schedule job** dialog:

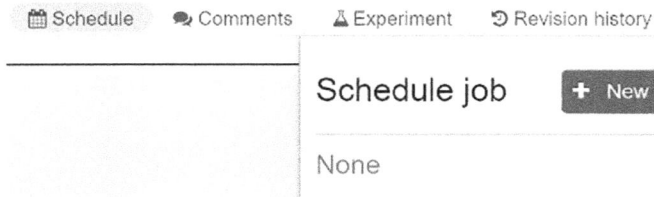

Figure 7.22 – Schedule job dialog

If there are schedules already available, they will show up here. Click **+ New** and continue:

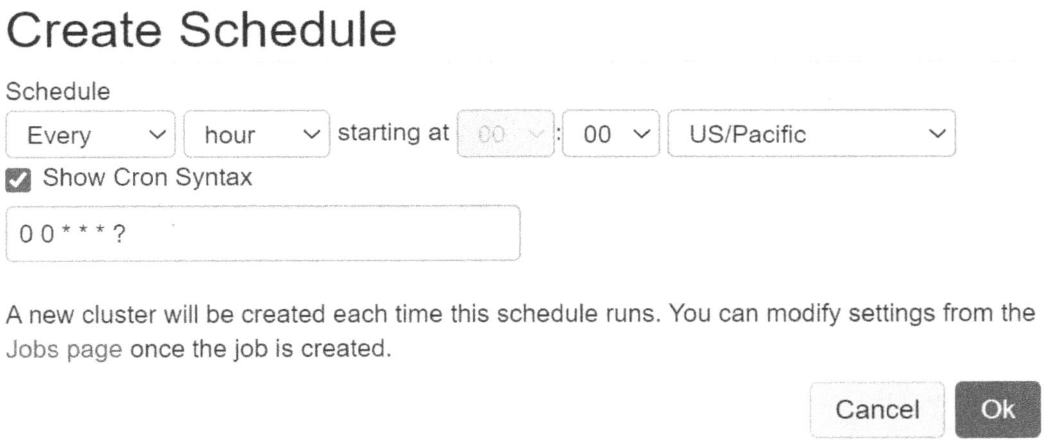

Figure 7.23 – Create Schedule dialog

In the **Create Schedule** dialog, you can now fix the intended schedule for your recurring job and start it right away.

Your new job will now appear in the **Jobs** list:

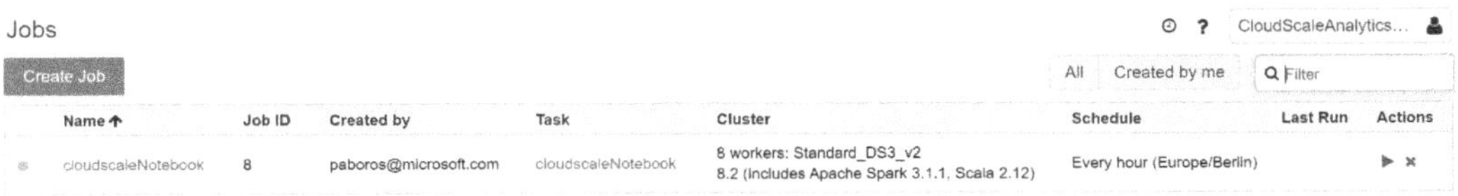

Figure 7.24 – Jobs list

From here, you can now dive into the details of your job when you click on **Name** on the left, for example:

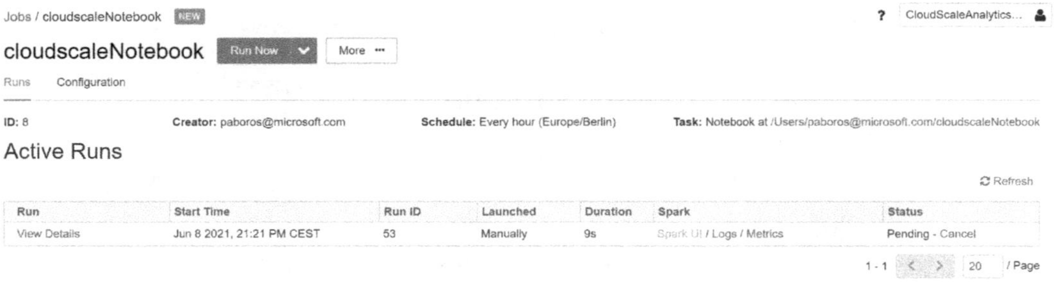

Jobs / cloudscaleNotebook [NEW]   ?   CloudScaleAnalytics... 👤

cloudscaleNotebook [Run Now ▼]   [More ⋯]

Runs   Configuration

ID: 8   **Creator:** paboros@microsoft.com   **Schedule:** Every hour (Europe/Berlin)   **Task:** Notebook at /Users/paboros@microsoft.com/cloudscaleNotebook

Active Runs

                                                                                                  ⟳ Refresh

Run	Start Time	Run ID	Launched	Duration	Spark	Status
View Details	Jun 8 2021, 21:21 PM CEST	53	Manually	9s	Spark UI / Logs / Metrics	Pending - Cancel

1 - 1 < >   20 / Page

Figure 7.25 – Job details

With the **All Jobs** link, you can navigate back to the job list.

What you will have noticed is that the job does not fire immediately as it will be scheduled for midnight at the earliest. But when you check the **Action** column on the far right of the jobs list, you will find a play button. This one will kick off your job right away. Let's do so and see what happens.

When you now click again on the job name, you will see your job being performed. Hopefully, soon you will find it with the status **Succeeded**. It would be interesting to check your data lake now to establish whether there is a new Parquet file in your target folder.

> **Note**
>
> The output created in the notebook, the cell where you call the `display()` function, will be skipped when you run the notebook as a batch job.

Using JAR files in jobs

With this alternative, you would provide JAR files containing the job logic, calculations, ML integration, and so on.

To create a job definition based on a JAR file in the UI, you need to upload your JAR during job creation:

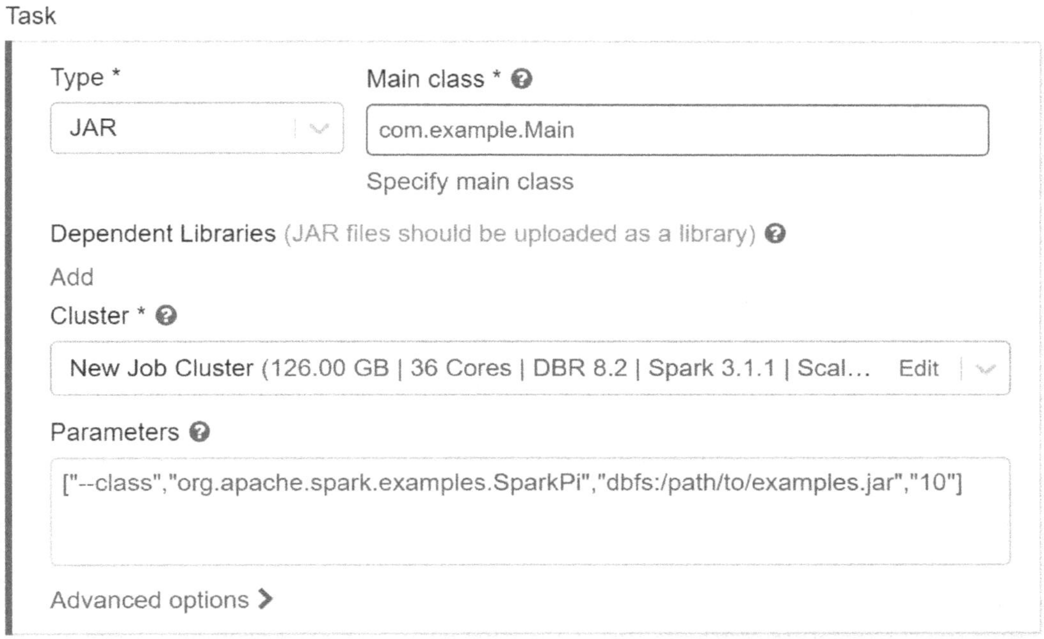

Figure 7.26 – Uploading a JAR file during job creation

Once you have dropped your file into the dialog, you need to set a reference to **Main class** in the job creation dialog. Name additional parameters in the **Arguments** field in POSIX-shell parameter format (in the API, you would use a JSON array).

Please note that you need to use the shared SparkContext API for your routines to attach to the SparkContext for your job. As Databricks provides the SparkContext, your program will fail if you try to invoke the SparkContext yourself with the following:

```
new SparkContext()
```

Therefore, you would do the following to avoid that error:

```
val SparkContext = SparkContext.getOrCreate()
val SparkSession = SparkSession.builder().getOrCreate()
```

Additionally, you shouldn't use one or more of the following code fragments to end your SparkContext:

```
SparkContext.stop()
System.exit(0)
sc.stop()
sys.addShutdownHook(jobCleanup)
```

These can result in unintended results in your code, and the shutdown hooks cannot be relied on because Spark container management is different in Azure Databricks.

Please observe another limit: In Databricks Runtime 6.3 and above, you can set the Spark configuration to avoid job outputs being directed to `stdout`. There is a limit of 20 MB, at which point it will cause an error and your job will fail. You can set `spark.databricks.driver.disableSatalaOutput` to `True` to avoid returning results to the client.

Using spark-submit

When you want to run a Python script or a JAR using `spark-submit` via the UI, you can start configuring a job as above. You can navigate to the **Jobs** hub in the Databricks workspace and start creating a new job. From the task links in the job creation dialog, you will then select **Configure spark-submit**. The dialog will give you an example of how to set the parameters for the call:

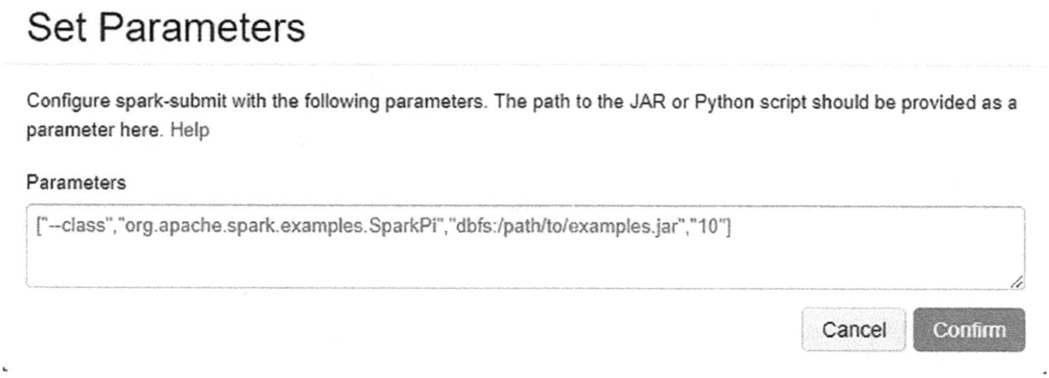

Set Parameters

Configure spark-submit with the following parameters. The path to the JAR or Python script should be provided as a parameter here. Help

Parameters

```
["--class","org.apache.spark.examples.SparkPi","dbfs:/path/to/examples.jar","10"]
```

Cancel   Confirm

Figure 7.27 – Configuring spark-submit

> **Note**
> You can run 1,000 jobs per workspace concurrently and you will see the error `429 Too Many Requests` when you exceed that number. The second limit is the number of jobs you can create per hour, per workspace – 5,000.

Adding dependent libraries to a job

It is important to note that you should always name all dependent libraries when you create new job definitions. When a job gets executed on an existing cluster, or when you have configured libraries to be installed on any cluster created (see the *Adding libraries* section), this now guarantees that they will be loaded before a job starts its execution.

Refer to the *Further reading, Creating Spark Jobs via the REST API*, section for additional material on how to programmatically create and run jobs calling the Databricks REST API.

Setting advanced options

When you extend the **Advanced** section by clicking on the down arrow next to it, you will be offered additional options to control the job run, such as **Alerts**, where you can set email addresses to be notified about the start, the success, or the failure of your job.

The **Maximum concurrent runs** setting for a job will limit the possible concurrent runs of a job definition.

With **Timeout**, you control the threshold, after which the job will be canceled and will fail.

Retries is the setting for how often the job will be restarted if it fails, to retry running it successfully. You can set a wait time between two attempts in that dialog too.

Creating Databricks tables

Databricks offers you two different types of tables that are available in different contexts.

When you create a global table, it will be stored in the Hive metastore of Databricks. This table will be available to all clusters that you spin up.

You can create global tables either via the UI or programmatically from a notebook or a Spark job.

A local table is created in a job or a notebook. You would use the `createOrReplaceTempView()` command, which we have already seen in the *Using Databricks notebooks* section, and before that in *Chapter 6, Using Synapse Spark Pools*, when we created local tables to make data frames available in two different languages in a notebook. Remember magic commands?

Creating a global table in the UI

A global table in Databricks represents a table object that is available to all clusters in your workspace. In comparison, you can create local tables, too, that are only then available to the cluster where they are created. A global table is registered in the Hive metastore of your workspace:

1. When you navigate to the data hub on the main screen, you will first select a database from the menu displayed. In your case, you would find the default database there. To create a new database, you could invoke a new notebook and use a SQL statement: `create database [YOURDATABASENAME]`.

2. At this point, you won't yet have any tables available in your environment. In the top-right corner of the dialog, you'll find the **Create Table** button.

3. When you select **Upload File**, you have the option to select a folder in **Databricks File System** (**DBFS**) as the target directory.

4. Below the target folder selector, you can throw in a file (CSV, JSON, or AVRO) to upload:

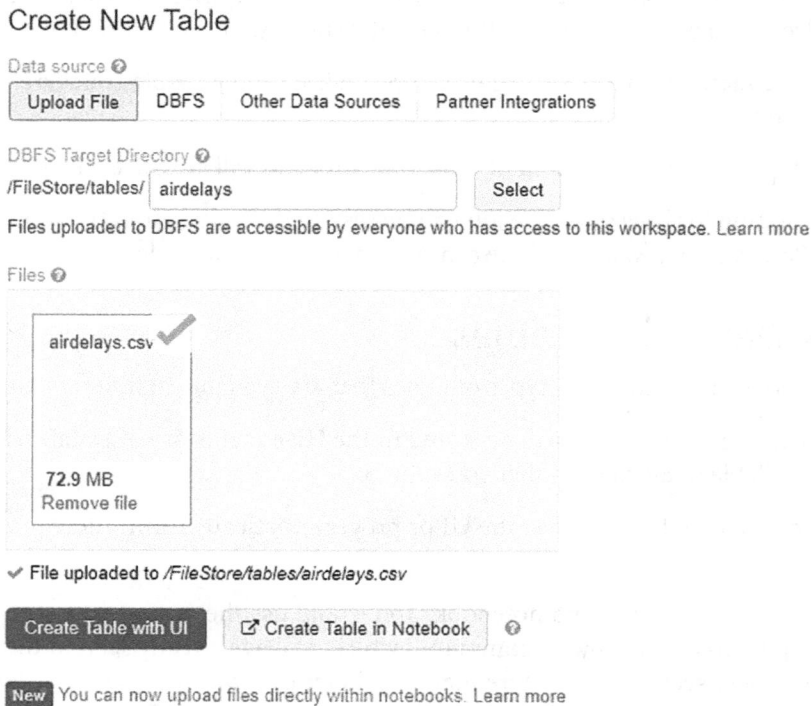

Create New Table

Data source

| Upload File | DBFS | Other Data Sources | Partner Integrations |

DBFS Target Directory

/FileStore/tables/ | airdelays | Select

Files uploaded to DBFS are accessible by everyone who has access to this workspace. Learn more

Files

airdelays.csv ✓

72.9 MB
Remove file

✓ File uploaded to /FileStore/tables/airdelays.csv

Create Table with UI | Create Table in Notebook

New You can now upload files directly within notebooks. Learn more

Figure 7.28 – Create New Table dialog

5. Once the file has been uploaded, you can either create your table using the UI or you can do so in a notebook.

6. Let's continue using the UI:

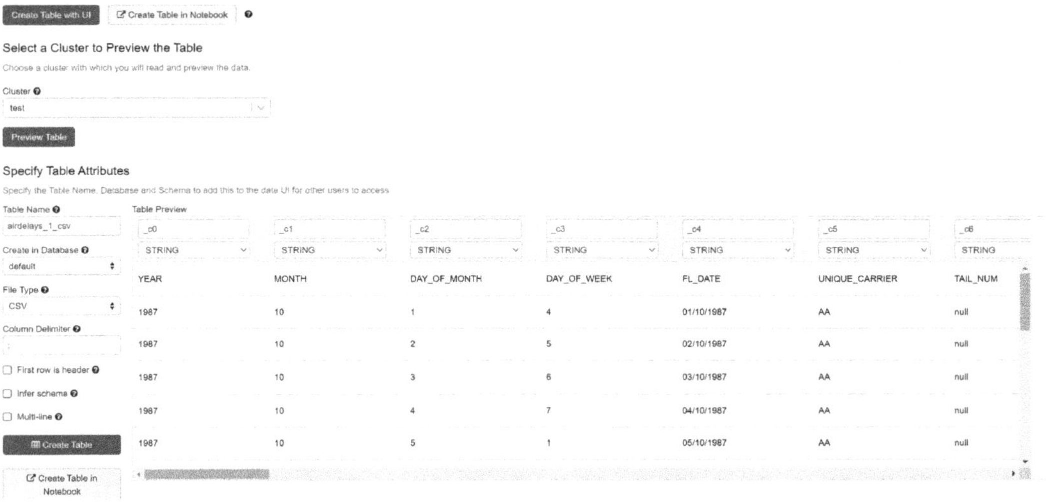

Figure 7.29 – Creating a table in the UI

7. You will need to select a cluster to process the contents of the file to preview your table first. Select the cluster that you created previously, in the *Using Databricks notebooks* section.

8. Now you can add a table name and select a database to store the table in. **File Type** and **Column Delimiter** are suggested based on the content of the file that you refer to.

9. Below the input fields, you will find three checkboxes:

    First row is header: This option will use the first row of your file as headers for the columns.

    Infer schema: This option will try to guess the datatypes for your columns.

    Multi-line: This option will enable multiple lines in cells of your table, for example, when you use JSON files with arrays.

10. Now, click **Create Table** and finish the sequence to make the data available in your database. Databricks will show you the schema and some sample data once your table has finally been created:

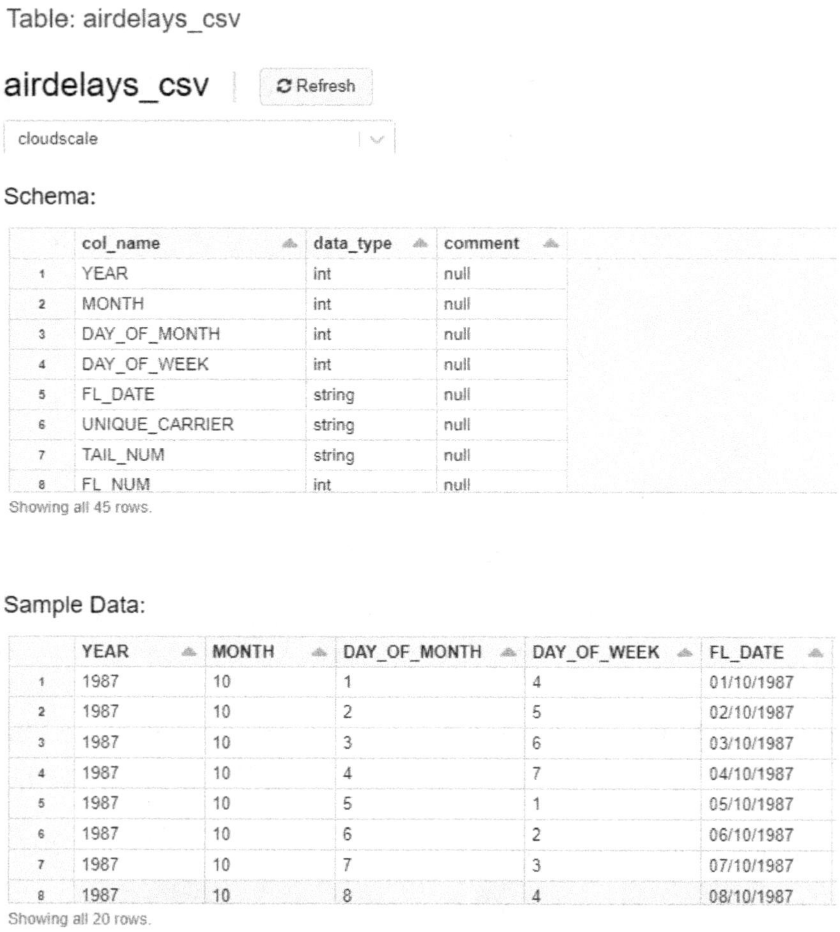

Table: airdelays_csv

airdelays_csv | ⟳ Refresh

cloudscale

Schema:

	col_name	data_type	comment
1	YEAR	int	null
2	MONTH	int	null
3	DAY_OF_MONTH	int	null
4	DAY_OF_WEEK	int	null
5	FL_DATE	string	null
6	UNIQUE_CARRIER	string	null
7	TAIL_NUM	string	null
8	FL_NUM	int	null

Showing all 45 rows.

Sample Data:

	YEAR	MONTH	DAY_OF_MONTH	DAY_OF_WEEK	FL_DATE
1	1987	10	1	4	01/10/1987
2	1987	10	2	5	02/10/1987
3	1987	10	3	6	03/10/1987
4	1987	10	4	7	04/10/1987
5	1987	10	5	1	05/10/1987
6	1987	10	6	2	06/10/1987
7	1987	10	7	3	07/10/1987
8	1987	10	8	4	08/10/1987

Showing all 20 rows.

Figure 7.30 – Newly created table

Now let's access this table using a new Python notebook.

11. Enter one of the following code fragments into the first cell of your notebook (don't forget to replace [YOURTABLE] with the name of your table).

Use this first command if you want to use SQL:

```
%sql
select * from [YOURTABLE]
```

Alternatively, use the following command to code in Python. And remember, you can, of course, interchangeably use the languages available in Spark using the %-magic commands and the language that you want to use:

```python
%python
df = spark.sql('select * from [YOURTABLE]')
display(df)
df = spark.table('[YOURTABLE]')
display(df)
```

Here is the example written in Scala:

```scala
%scala
val df = spark.sql("select * from [YOURTABLE]")
display(df)
val df = spark.table("select * from [YOURTABLE]")
display(df)
```

The DBFS option will offer two options as well to either use the UI or a notebook to create a table based on files stored in DBFS.

12. When you select **Other Sources and Partner Integrations**, you will need to use a notebook to create your table. Databricks will open a notebook with template code, which you can adjust to your needs and use to create your table:

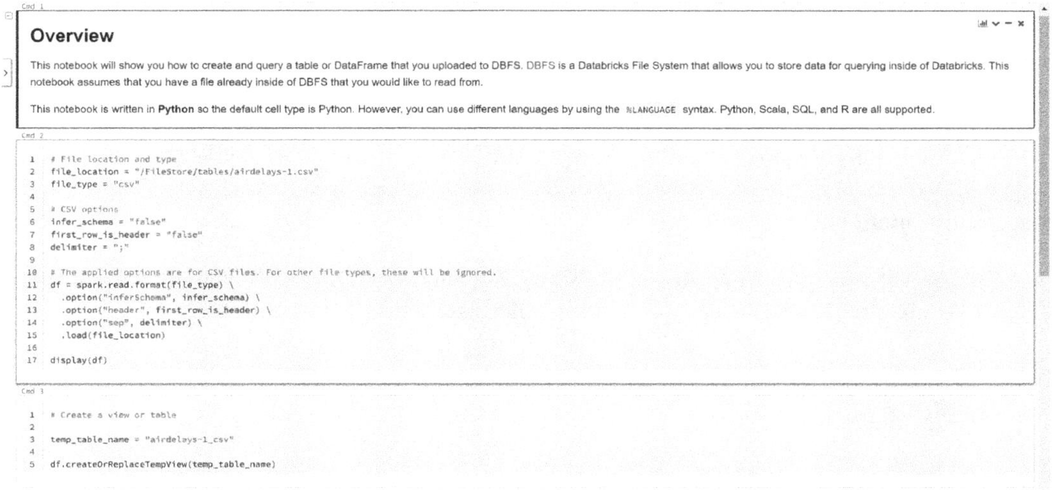

Figure 7.31 – Notebook template to create a Databricks table

Now that you have seen how to prepare your environment and how to implement code, let's examine a topic that will add value big time to your data handling in the *Understanding Databricks Delta Lake* section.

Understanding Databricks Delta Lake

Delta Lake is an extension developed by Databricks to implement a so-called data lakehouse on top of a data lake. Delta Lake implements features to achieve transactions that conform with **Atomic, Consistent, Isolated, Durable (ACID)**, enable time travel on multiple versions of data rows, and enforce schema conformity when adding rows to a Delta Lake file. Additionally, Delta Lake will enable you to unify streaming and batch insertion in one target table. Also, and this is important for many users, Delta Lake adds upserts and deletes for rows in files via versioning.

The Databricks Delta Engine introduces a collection of performance optimization measurements to increase efficiency for big ETL processes as well as ad hoc and interactive queries. These optimizations are mostly built-in and do not require you to write additional code to implement them as they automatically happen when you use the engine.

Delta Engine includes features for optimizing file management: Auto Optimize (writes and compaction), caching, dynamic file pruning, isolation levels, Bloom filter indexes (indexing techniques), join performance, and data transformation.

As Delta Lake is a vast topic in itself, we are not going to cover it in detail here in this book. Please refer to the *Further reading, Introducing Delta Lake*, section for a deep dive.

Having a glance at Databricks SQL Analytics

Recently, Databricks added SQL Analytics as a preview to its offering. Pretty similar to the Synapse serverless SQL pools, Databricks SQL Analytics offers a SQL surface on top of a storage component such as your data lake.

This offering will add SQL endpoints, the ability to create dashboards to visualize data, and alert you to the functionality collection of the Databricks environment.

You can integrate it with the following Azure data services:

- Synapse SQL pools
- Cosmos DB
- Data Lake Store Gen2
- Azure Storage Accounts Blob Store

> **Note**
> Databricks SQL Analytics requires you to run a premium plan Databricks environment.

If you want to read more about Databricks SQL Analytics, please refer to the *Further reading, Databricks SQL Analytics*, section.

Adding libraries

There are tons of machine learning, ETL, plotting, and other libraries, such as OpenCV for video and image processing, and many more available on the internet to be used in your code.

Databricks offers options to add them to your clusters to have them available after the cluster starts. You can even choose to have libraries automatically available for any cluster that is created in your workspace.

To add a library (perhaps the Python wheel that we created in *Chapter 6, Using Synapse Spark Pools*), go to the main screen in your Databricks workspace and search for **Import Library** under **Common Tasks**.

Here, you can choose to either upload the library file directly in the dialog, put the library in your DBFS, or get it from PyPI, Maven repositories, or CRAN.

So, let's use the Python wheel as mentioned above:

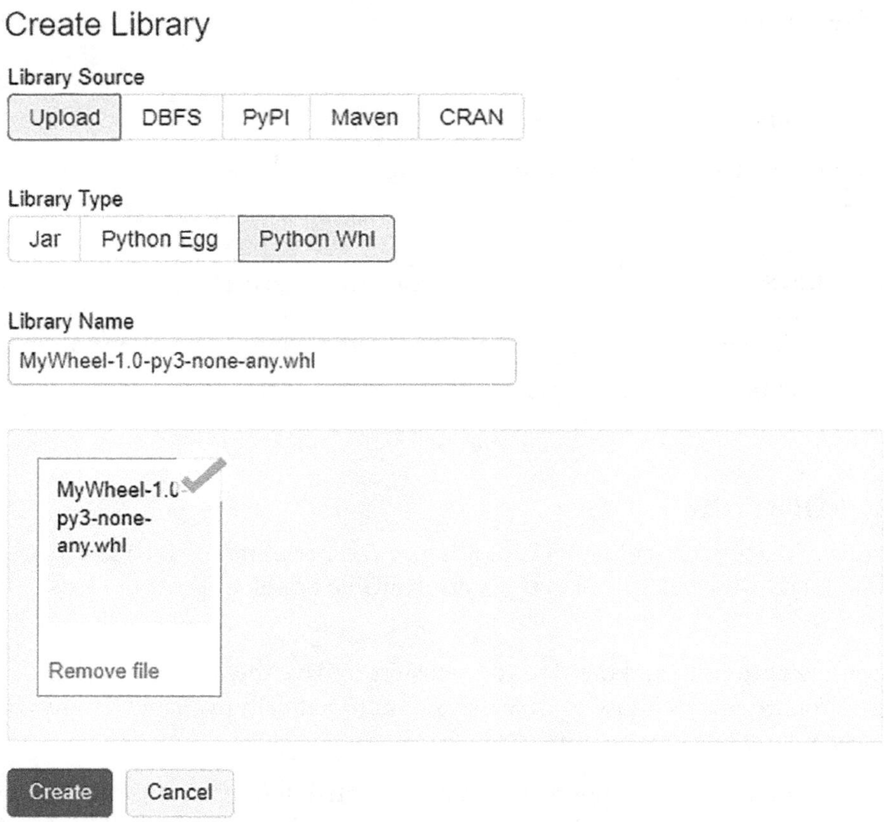

Create Library

Library Source

| Upload | DBFS | PyPI | Maven | CRAN |

Library Type

| Jar | Python Egg | Python Whl |

Library Name

MyWheel-1.0-py3-none-any.whl

MyWheel-1.0-
py3-none-
any.whl

Remove file

Create   Cancel

Figure 7.32 – Uploading the Python wheel

When the file upload is done (the green check symbol appears on the filename), click **Create**. On the next screen, you can then attach the newly uploaded library to any cluster of your choice, choose to install it automatically on all clusters in the workspace, uninstall it from clusters, or delete it entirely.

You need to tick the checkbox on a particular row and then click an action on the screen:

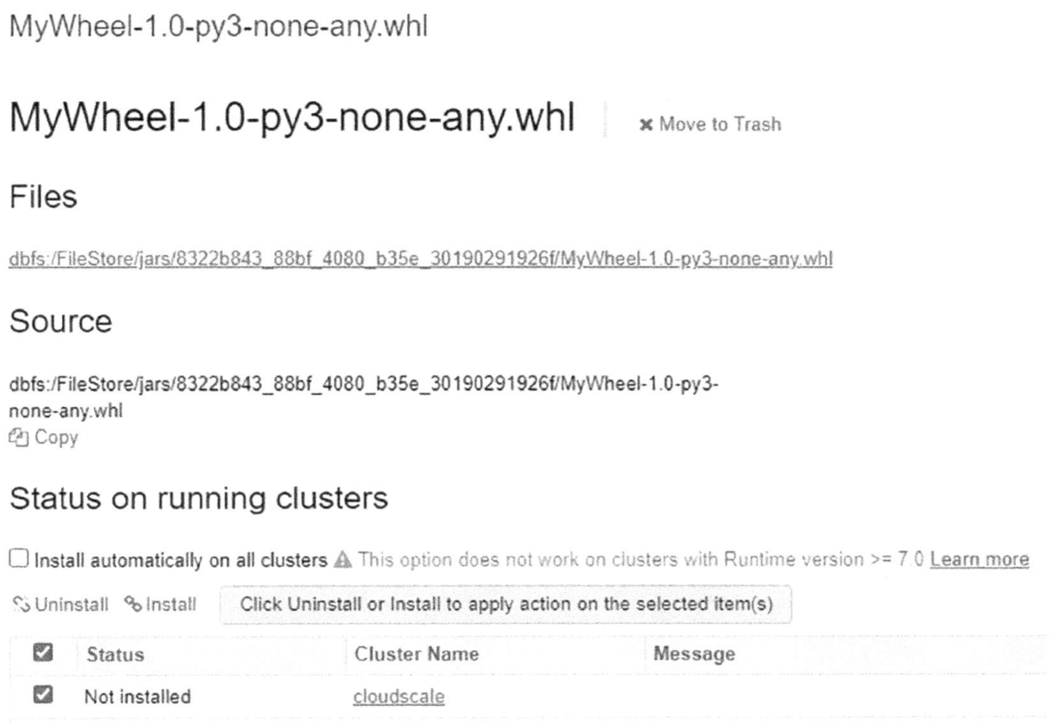

MyWheel-1.0-py3-none-any.whl

MyWheel-1.0-py3-none-any.whl ✕ Move to Trash

Files

dbfs:/FileStore/jars/8322b843_88bf_4080_b35e_30190291926f/MyWheel-1.0-py3-none-any.whl

Source

dbfs:/FileStore/jars/8322b843_88bf_4080_b35e_30190291926f/MyWheel-1.0-py3-none-any.whl
Copy

Status on running clusters

☐ Install automatically on all clusters ⚠ This option does not work on clusters with Runtime version >= 7 0 Learn more

Uninstall Install | Click Uninstall or Install to apply action on the selected item(s) |

| ☑ | Status | Cluster Name | Message |
| ☑ | Not installed | cloudscale | |

Figure 7.33 – Library details

Now you have maybe added your own libraries or third-party ones and you have made progress with your analysis in your notebook. Why not start to visualize your data? Let's see how in the next section, *Adding dashboards*.

Adding dashboards

When you analyze data using notebooks, as mentioned in the *Using Databricks notebooks* section, chances are that you will create visualizations from your data. One feature in Databricks allows you to collect plottings that you have created during your work in notebooks and add them to dashboards. You can then publish these dashboards to allow other users to browse the collected graphs.

Dashboards can be updated when new data arrives. This can be done on demand, but also on a schedule. You would need to schedule the notebook(s) the graph(s) originate from.

To add a visual to a dashboard, you can use the **Show in Dashboard** menu button in the upper-right corner of the cell that shows the chart. This is the button that looks like a small chart next to the **Play** button. If no dashboard is available as yet, the menu will only show + **Add to New Dashboard**:

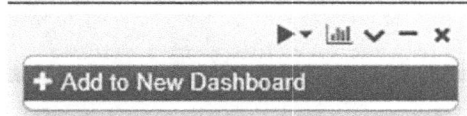

Figure 7.34 – The Add to the Dashboard menu option

You can try it and use the notebook from the *Using Databricks notebooks* section. Perhaps you have managed to create a visual from the displayed data. Click the button and then click **Add to New Dashboard**.

A new dashboard will open in a new tab in your browser. From there, you can name it and configure some presentation options and you can start presenting it in fullscreen mode by clicking the **Present Dashboard** button:

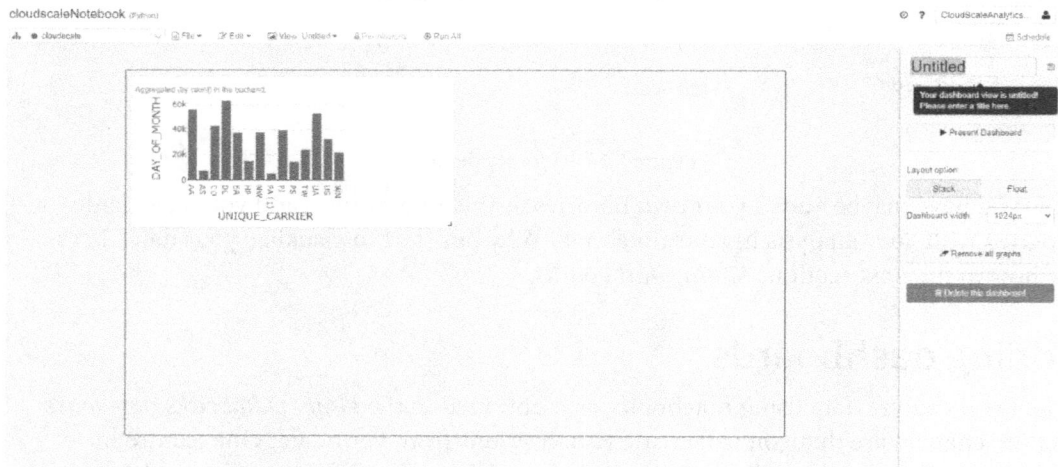

Figure 7.35 – New dashboard

In the next section, we need to talk about a topic that you may or may not like. But it is a very important one. Please check out the *Setting up security* section.

Setting up security

The two most important aspects when you examine the security mechanisms of Databricks are networking, on the one hand, and access controls, on the other.

Examining access controls

Access control lists are, as you have already seen in *Chapter 3*, *Understanding the Data Lake Storage Layer*, a fine-grained method for controlling who can see what and do what in the environment. If you have chosen to provide a premium plan workspace, you can set up access control lists for the following:

- **Workspaces**: Within workspaces, you can set ACLs on a finer level for **Folders**, **Notebooks**, and **MLflow Experiments**. The different object types will show up with several different abilities that you can control. You can find a detailed overview via the link in the *Further reading*, *Workspace Access Control*, section. You can configure **No Permissions**, **Read**, **Run**, **Edit**, and **Manage** for the different abilities in the different artifacts.

- **Clusters**: For clusters, you can set **No Permissions**, **Can Attach To**, **Can Restart**, and **Can Manage** as permissions for abilities such as **Attach notebook to cluster**, **View Spark UI**, and others. You will have the **Can Manage** permission on any cluster that you create. You can find the abilities and more detailed information in the *Further reading*, *Cluster Access Control*, section.

- **Pools**: The permissions for pools are **No Permissions**, **Can Attach To**, and **Can Manage** for the abilities **Attach Cluster to pool**, **Delete Pool**, **Edit Pool**, and **Modify poll permissions**. You can find more information in the *Further reading*, *Pool Access Control*, section.

- **Jobs**: The available job permissions are **No Permissions**, **Can View**, **Can Manage Run**, **Is Owner**, and **Can Manage**. You can find abilities such as **View job details and settings**, **View results, Spark UI, logs of a job run**, and others, along with more information on job access controls, in the *Further reading*, *Job Access Control*, section.

- **Tables**: Table access controls work a little differently to the others and follow a SQL-like approach. You can grant and deny access rights to objects (tables, databases, views, and functions). These can be **SELECT**, **CREATE**, **MODIFY**, **USAGE**, **READ_METADATA**, **CREATE_NAMED_FUNCTION**, **MODIFY_CLASSPATH**, or **ALL_PRIVILEGES**. There are two different ways to enable table access control. You can select a SQL-only table access control or switch to Python and SQL table access control. You can read more information in the *Further reading*, *Table Access Control*, section.

- **Secrets**: Access controls for secrets can be set to **MANAGE, WRITE,** or **READ.** You will need to run Bash commands to create and view the ACLs for your secrets.

In the next section, *Understanding secrets*, you'll find a short introduction to secrets as well as find additional information about access controls for secrets in the *Further reading*, *Secret Access Controls*, section.

Understanding secrets

Secrets and secret scopes are a way to store credentials for data source access securely in your Databricks workspace. You can reference secrets in your notebooks or jobs to avoid using credentials directly in the code. As an alternative to keeping a secret in your workspace, you can decide to store your secrets in **Azure Key Vault** and set the scope accordingly.

Secrets and Databricks secret scopes are created using the Databricks CLI. Scopes can alternatively be set up in the Databricks UI. To use the UI, you need to go to `https://[YOURDATABRICKSINSTANCE]#secrets/createScope`. (Pay attention to the URL as it is case sensitive.)

Check the *Further reading*, *Secrets*, section for a link to documentation relating to secrets.

Understanding networking

Databricks will always create a virtual network for your workspace. This is where all the clusters will be created and it represents the data plane. It will always be created in the customer subscription and resides next to the customer's data store to ensure data will not travel outside of the customer's subscription.

You can choose to create the **Data Plane** of your Databricks workspace in your self-managed virtual network. This is referred to as VNet injection. This can be done during the provisioning of your Databricks workspace. Remember the provisioning process described in the *Provisioning Databricks* section and *Figure 7.2 – Create an Azure Databricks workspace – The Networking blade*. There will be two subnets created: one public subnet that will be used to communicate with the control plane, and one private subnet whose purpose is to communicate with your data sources.

You can peer other networks within your subscriptions with your Databricks data plane's private subnet and control network routes between your Databricks environment and other Azure data services such as databases and storage. This can be extended to on-premises networks as well and will enable secure communications.

The **Control plane**, on the other hand, will be created with a Microsoft-managed subscription. This is where the web application runs that you use as the workspace interface, where you create your notebooks, where your jobs are controlled, and where the cluster manager is run:

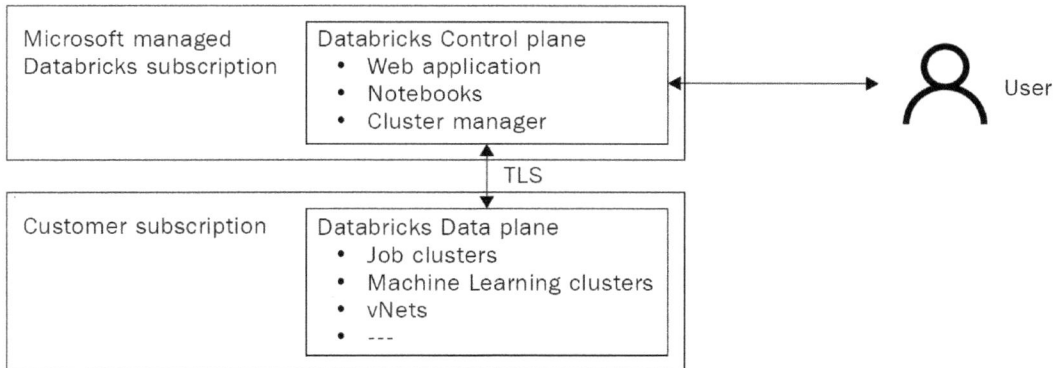

Figure 7.36 – Databricks control and data planes

You can check the *Further reading, Virtual Networks*, section for a deep dive into virtual networks in the Azure Databricks documentation.

Monitoring Databricks

Azure Monitor is the Azure-wide log collection service that enables you to collect, analyze, and correlate logs from Azure, but also from on-premises applications. With Azure Monitor, you will be able to analyze not just one particular service, but bring together information from a wider context, and with this, develop a new level of understanding and insights.

As Azure Databricks is not (yet) natively integrated with Azure Monitor, your applications will need to use an additional library to inject your log events into the Log Analytics workspace of Azure Monitor. Microsoft provides a GitHub repository where you can download and build the required library to be used in your code. You can find the link to the documentation in the *Further reading, Monitoring Databricks*, section.

Summary

This chapter took us into the world of Databricks. You provisioned a Databricks workspace and examined it. In the workspace, you created a new Spark cluster and learned how to manage it.

You created a Databricks notebook, ran it interactively, and saw how to visualize data in your notebook. You also saw how to create a batch job from your notebook and learned about other alternatives for running code as a batch in your environment.

In the section that followed, you learned about Databricks tables and we examined additional capabilities, such as using Delta Lake to manage your data in your environment.

We saw how to add additional functionality using third-party libraries and how to create dashboards from your data.

Finally, we examined security features, such as access controls and secrets, and learned about networking features and how to integrate with Azure Monitor.

There are many more topics related to Azure Databricks that would have exceeded the capacity of this book. If you want to deep dive further into MLflow on Databricks, for example, or how to integrate versioning, you will find these topics in the *Further reading*, *Understanding MLflow*, and *Further reading*, *Databricks Versioning*, sections.

Now let's move on to *Chapter 8*, *Streaming Data into Your MDWH*, and dig into real-time processing with Azure data services.

Further reading

- **Pricing Dataricks**: https://azure.microsoft.com/en-us/pricing/details/databricks/
- **Creating a Service Principal**: https://docs.microsoft.com/en-us/azure/active-directory/develop/howto-create-service-principal-portal
- **Creating Spark Jobs via the REST API**: https://docs.microsoft.com/en-us/azure/databricks/dev-tools/api/latest/jobs
- **Introducing Delta Lake**: https://docs.microsoft.com/en-us/azure/databricks/delta/
- **Databricks SQL Analytics**: https://docs.microsoft.com/en-us/azure/databricks/scenarios/sql/

- **Understanding MLflow**: `https://docs.microsoft.com/en-us/azure/databricks/applications/mlflow/`

- **Workspace Access Control**: `https://docs.microsoft.com/en-us/azure/databricks/security/access-control/workspace-acl`

- **Cluster Access Control**: `https://docs.microsoft.com/en-us/azure/databricks/security/access-control/cluster-acl`

- **Pool Access Control**: `https://docs.microsoft.com/en-us/azure/databricks/security/access-control/pool-acl`

- **Jobs Access Control**: `https://docs.microsoft.com/en-us/azure/databricks/security/access-control/jobs-acl`

- **Table Access Control**: `https://docs.microsoft.com/en-us/azure/databricks/security/access-control/table-acls/`

- **Secret Access Control**: `https://docs.microsoft.com/en-us/azure/databricks/security/access-control/secret-acl`

- **Secrets**: `https://docs.microsoft.com/en-us/azure/databricks/security/secrets/`

- **Virtual Networks**: `https://docs.microsoft.com/en-us/azure/databricks/administration-guide/cloud-configurations/azure/`

- **Monitoring Databricks**: `https://docs.microsoft.com/en-us/azure/architecture/databricks-monitoring/`

- **Databricks Versioning**:

  `https://docs.microsoft.com/en-us/azure/databricks/notebooks/azure-devops-services-version-control`

  `https://docs.microsoft.com/en-us/azure/databricks/notebooks/bitbucket-cloud-version-control`

  `https://docs.microsoft.com/en-us/azure/databricks/notebooks/github-version-control`

8
Streaming Data into Your MDWH

More and more analytical projects need to show real-time or near real-time data, that is, data that is coming from online systems such as shops and trading platforms or IoT telemetry. You want to collect and analyze that data maybe even right as it hits your system. IoT data might give you input about the status and potential failure of machines on your shop floor, or you may just seek to display online data of your production. Shop telemetry could inform you about potential customer churn, or trading events might be checked for fraudulent behavior. There are multiple use cases as well as options to implement them on the Microsoft Azure platform.

This chapter will inform you about **Azure Stream Analytics (ASA)** and the configuration-based approach that this service offers. ASA is a fully managed PaaS component. You will learn how to set up the service and how to connect to sources and targets. You will learn about SQL queries with windowing functions and pattern recognition to process and analyze incoming events. Furthermore, you will learn how to add reference data to your analysis and how to use machine learning for lookup functions. Finally, you will build an online dashboard with Power BI that shows data as it pours into your system.

Additionally, you will examine other options for how to implement stream processing beyond the capabilities of ASA using Structured Streaming with Spark.

These sections are waiting for you in the chapter:

- Provisioning ASA
- Implementing an ASA job
- Understanding ASA SQL
- Using Structured Streaming with Spark
- Security in your streaming solution
- Monitoring your streaming solution

Technical requirements

To follow along with this chapter, you will need the following:

- An Azure subscription where you have at least contributor rights or you are the owner
- The right to provision ASA
- A Synapse Spark pool (optional if you want to follow the additional options)
- A Databricks cluster (optional if you want to follow the additional options)

Provisioning ASA

Now let's provision your first ASA job on Azure:

1. To create your ASA environment in your Azure subscription, please navigate to your Azure portal and hit **+ Create a resource**, and then type `Stream Analytics` in the following blade. From the quick results below the search field, select **Stream Analytics job**. On the following details blade, you can get a first glance at your options with ASA. Please hit **Create** on the blade to start the provisioning sequence.

2. On the following blade, you will need to enter some basic information required to create your ASA job:

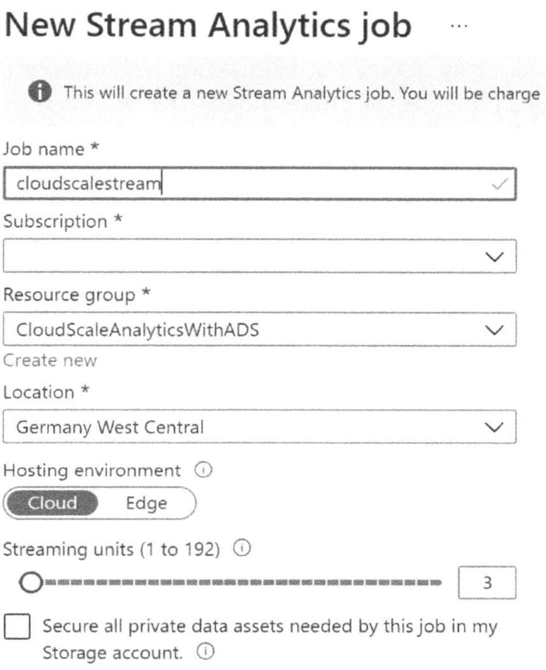

New Stream Analytics job ···

ⓘ This will create a new Stream Analytics job. You will be charge

Job name *

cloudscalestream ✓

Subscription *

⌄

Resource group *

CloudScaleAnalyticsWithADS ⌄

Create new

Location *

Germany West Central ⌄

Hosting environment ⓘ

(**Cloud**   Edge)

Streaming units (1 to 192) ⓘ

○▬▬▬▬▬▬▬▬▬▬▬▬▬▬▬▬▬▬▬▬▬▬▬▬ 3

☐ Secure all private data assets needed by this job in my
Storage account. ⓘ

Figure 8.1 – Basic information provisioning in ASA

Please enter a job name and select the target subscription. You can either select an existing resource group or create a new one and select the data center location for your ASA job.

The **Cloud** hosting environment will create an ASA job in your subscription as a cloud service. If you select **Edge**, the ASA job will be containerized and deployed to an IoT gateway edge device.

The **Streaming Units (SUs)** that you can configure on this blade will determine the available compute for your streaming job.

The checkbox below all the options can trigger additional inputs where you can optionally select a storage account of your choice to store all data assets related to your ASA job. Otherwise, this data (metadata, checkpoints, job configurations, and so on) is stored with your ASA job.

3. Please hit **Create** to start the provisioning of your configuration.

4. Once your deployment is complete, you can hit **Go to resource**.

Now let's start examining the newly created ASA job in the next section, *Implementing an ASA job*.

Implementing an ASA job

ASA offers you a convenient way to create streaming analysis on a configuration basis. This means you do not need to code the environment, the engine, the connection, logging, and so on. The service will take care of all these tasks for you (to see an example, refer to the *Integrating sources* and *Writing to sinks* sections that follow). The only thing you will need to code is the analytical core of your streaming job. To ease things for you, this is done using a SQL dialect that was tailored for this task (see *Understanding ASA SQL*).

After the provisioning of your new resource, you are taken to the following overview blade:

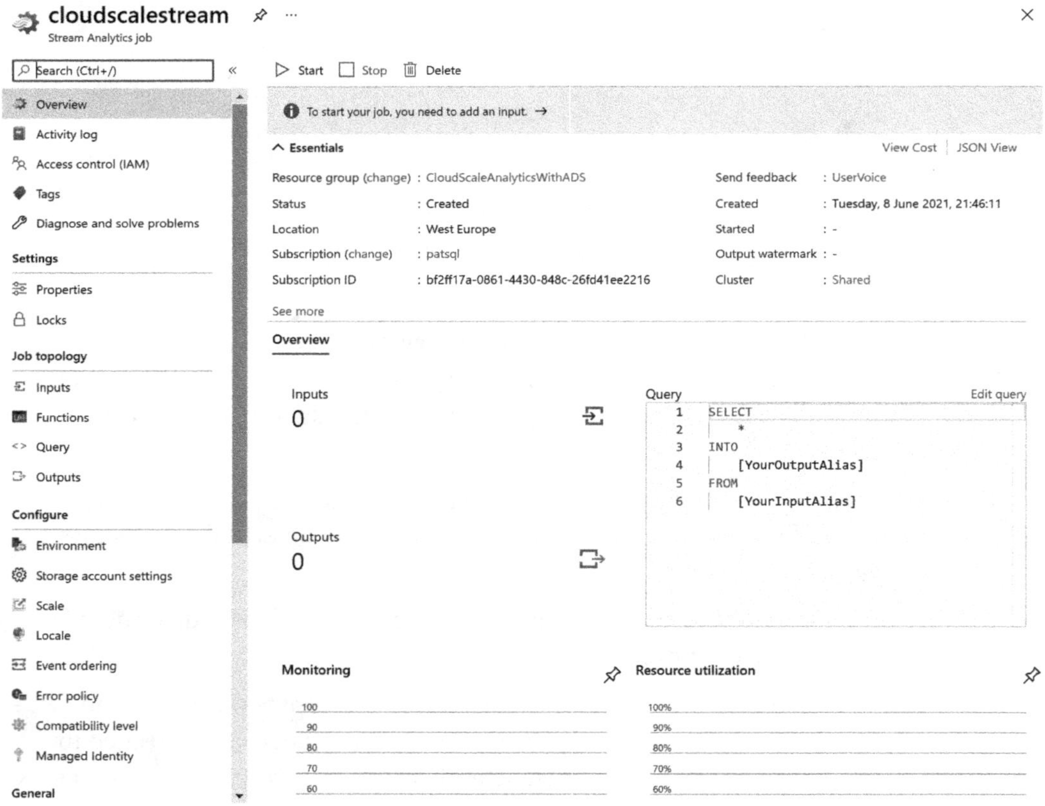

Figure 8.2 – Overview blade of the ASA job

You can already see three of the most important areas of your ASA job:

- **Inputs**: This will show all the configured source connections available in your job.

- **Outputs**: This will show all the configured target connections available in your job.

- **Query**: This will show all the SQL statements that your job will use to process and route the data.

If you examine the navigation blade, you'll find the additional options that we are going to use and configure in the following sections. But let's move on and start creating the first source connection in the next section, *Integrating sources*.

Please see *Figure 8.2* for an overview of how these fit together for your ASA job.

Integrating sources

The following Azure services can be configured as sources in your ASA job:

- **Event Hubs**: A PaaS service for caching, buffering, and processing millions of events in real time as they are sent to the Azure platform. Event Hubs is used as the "address" to which applications can send events to be further processed on the Azure platform. See the *Further reading, Event Hubs* section for more information.

- **IoT Hub**: A PaaS service specialized in IoT implementations. Optimized to scale and process billions of telemetry events, implement secure connections between devices and the Azure platform, and establish bi-directional communication between the cloud platform and devices. See the *Further reading, IoT Hub* section for more information.

- **Blob Storage**: See *Chapter 3, Understanding the Data Lake Storage Layer.*

- **Azure Data Lake Gen2**: See *Chapter 3, Understanding the Data Lake Storage Layer.*

For the easy setup of your first streaming job, let's use the data that you have available already in Data Lake:

1. Please first navigate to your newly created ASA job. Click **Inputs** either on the main blade or in the **Overview** section.

2. On the inputs list, please click **Add stream input** and select **Blob storage/ADLS Gen2**. The configuration blade will pop up from the right:

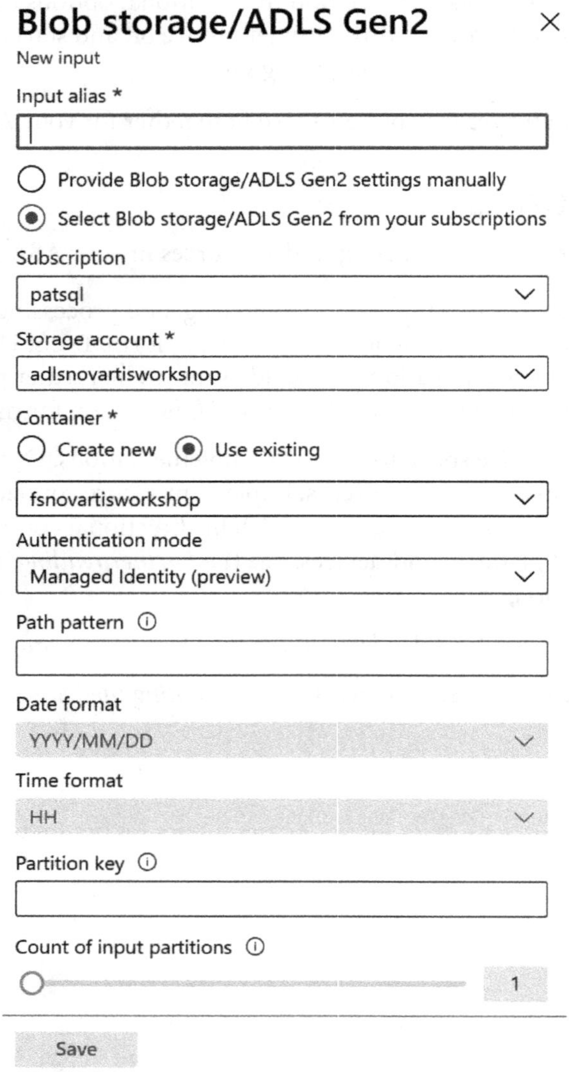

Blob storage/ADLS Gen2 ✕

New input

Input alias *

[   ]

○ Provide Blob storage/ADLS Gen2 settings manually
◉ Select Blob storage/ADLS Gen2 from your subscriptions

Subscription

[patsql   ∨]

Storage account *

[adlsnovartisworkshop   ∨]

Container *
○ Create new   ◉ Use existing

[fsnovartisworkshop   ∨]

Authentication mode

[Managed Identity (preview)   ∨]

Path pattern ⓘ

[   ]

Date format

[YYYY/MM/DD   ∨]

Time format

[HH   ∨]

Partition key ⓘ

[   ]

Count of input partitions ⓘ

○━━━━━━━━━━━━━━━━━━━━━   [1]

[Save]

Figure 8.3 – Storage source configuration

3. You can now start entering an alias for your input connection. Name it something such as `airdelaystreaminginput`.

    If you choose to select the configuration data from your subscription, you can select from the available services from your subscription in the **Subscription**, **Storage account**, and **Container** fields (or create a new one). You can, of course, enter the values manually. Please select your subscription and for the storage account, select the one created in *Chapter 3*, *Understanding the Data Lake Storage Layer*. In the **Container** drop-down field, please select the filesystem that you created earlier.

    For **Authentication mode**, please select **Connection String**. The **Storage Account Key** field underneath this will automatically be populated.

    In the **Path pattern** field, you can now enter the path and name of your `airdelays.csv` file that will be used as streaming input. The two fields below this, **Date format** and **Time format**, can be used to resolve the date and time portions of the path pattern. Please leave them for now.

    Partition key and the **Count of input partitions** slider can be very useful for creating input partitions to parallelize streaming jobs for higher throughput. Please leave these for now.

    If you now scroll down a little, you can influence the input format in the **Event serialization format** field (**JSON**, **Avro**, **CSV**, or **Other** (Protobuf, XML, proprietary)) and the **Delimiter** (**comma (,)**, **semicolon (;)**, **space**, **tab**, or **vertical bar(|)**) drop-down box. Please select **CSV** and **semicolon (;)** and in the following **Encoding** field, leave it as **UTF-8**.

    You don't need to select an event compression type (**None**, **GZip**, or **Deflate**).

4. Please hit **Save** to finish and store your new input. ASA will test and then save the connection.

5. To check the source, you might click on the **Sample data from input** button on the far right in the row of your new input to the right of the delete button (the trashcan symbol). In the following dialog, please set the date of the last modified time of your file. Otherwise, the sample won't be successful. In **Duration**, please set **Days** to 1 to ensure you find some data. Click **Sample** in the footer. The sampling starts and you'll be notified with a small dialog dropping in from the upper-right corner of your window. If you miss the notification, you can always check for it by clicking on the bell symbol in the top-right corner of your Azure portal. You'll find the message that your sample data is ready, and you can click to download it. If you follow the link and you're able to download the sample, your input works. Please try this as we will use this data later to test the analytical query.

Configuring Event Hubs

When you are configuring Event Hubs, you will find slightly different options to configure:

- **Event Hub Namespace**: The namespace is the parent collection where event hubs are grouped.

- **Event Hub Name**: You can use an existing one or create a new event hub from this dialog.

- **Event Hub consumer group**: You can use an existing one or create a new event hub consumer group. Consumer groups store the read state for the read events per group and allow you to have different groups of event consumers reading events at different times and with different rhythms. An event hub will always have a `$Default` consumer group.

- **Authentication Mode**: You can use a managed identity or connection string when you authenticate your ASA job against the event hub.

As IoT hubs are built based on event hubs, they will have a lot of similarities but differences too. Let's check them out in the next section, *Configuring IoT Hub*.

Configuring IoT Hub

The IoT Hub settings differ slightly from Event Hubs:

- **IoT Hub**: The IoT hub from which you need to read events.

- **Consumer group**: Consumer groups store the read state for events per group and allow you to have different groups of event consumers reading events at different times and with different rhythms. IoT Hub will always have a `$Default` consumer group.

- **Shared access policy name**: A shared access policy provides a set of permissions under a name.

- **Shared access policy key**: You will need to set the key according to the shared access policy name.

- **Endpoint**: If you select **Messages**, you will consume device-to-cloud messages. If you select **Operations Monitoring**, you will be able to read device telemetry and metadata.

But let's proceed with our example and provide a sink for our data stream in the next section, *Writing to sinks*.

Writing to sinks

Every ASA job writes to at least one sink. Let's examine your options here.

The following services can be used as sinks in an ASA job. Please find the different properties in this list:

- **Event Hubs**.

- **SQL databases**: Azure SQL Database tables as sinks. Can come in very handy when used with reporting.

- **Blob storage/Azure Data Lake Storage Gen2**: See *Chapter 3, Understanding the Data Lake Storage Layer*.

- **Table storage**: Uses a storage account table storage.

- **Service Bus topic**: Azure Service Bus offers messaging services for asynchronous messaging. A Service Bus topic offers a publish and subscribe pattern for a service bus and can be used as a sink from ASA. Please find more details in the *Further reading* section.

- **Service Bus queue**: In comparison to a Service Bus topic, the queue offers a first-in/first-out messaging queue for one or more, in this case, "competing" consumers. Please find more details in the *Further reading, Service Bus* section.

- **Cosmos DB**: NoSQL component in Azure.

- **Power BI**: The Microsoft reporting and dashboarding solution. Power BI offers the option of streaming datasets that can catch ASA streams. You will need to have a PowerBI.com account and you will need to first authenticate with the PowerBI.com service.

- **Data Lake Storage Gen1**: Legacy Azure Data Lake Storage technology is provided for backward compatibility.

- **Azure Functions**: Compute component for serverless functions.

- **Azure Synapse Analytics**: See *Chapter 4, Understanding Synapse SQL Pools and SQL Options*.

For detailed overviews, please check the *Further reading, Sinks* section.

Please proceed now to set up a sink according to the input that you have created to a target folder in your Data Lake Storage. Maybe name it something such as `airdelaystreamingtarget`.

Understanding ASA SQL

The main processing in your ASA job will be done using SQL to implement the analytical rules you want to apply to your incoming data.

Compared to data warehouse batch-oriented processing, stream processing observes a constantly delivered chain of events. The processing, therefore, will need different approaches as you will, for example, aggregate values over a certain recurring time frame. This is called windowing. The ASA SQL dialect implements a collection of windowing functions that will support you in doing this.

But before we dive into the magic of windowing functions and ASA, let's first finish our basic ASA job and kick it:

1. Please select **Query** from either the navigation blade or the **Overview** blade and select **Edit query**:

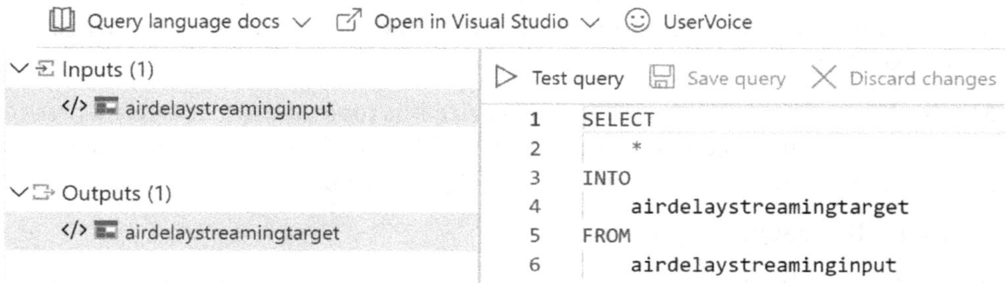

Figure 8.4 – ASA query editor

2. In the editor, please enter your ASA query. Please replace the displayed query with the following:

```
SELECT
    *
INTO
    airdelaystreamingtarget
FROM
    airdelaystreaminginput
```

3. Of course, please check the two aliases and if you have named them differently, please use the names that you have used.

4. Once you're done adjusting the query, you can click **Save query** in the editor window. There is the option to test your query with some sample data that you can get from the source. We're going to have a look at this later.

5. For now, please save the query and return to the **Overview** blade of your ASA job. You can find some controls in the upper area of the overview:

▷ Start   ☐ Stop   🗑 Delete

Figure 8.5 – Run controls for ASA jobs

6. Please press start and kick your ASA job for the first time. From the right, the **Start job** dialog pops up:

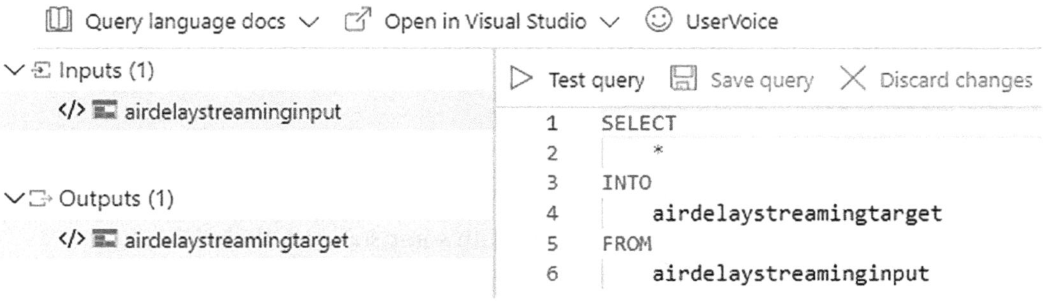

📖 Query language docs ⌄   ↗ Open in Visual Studio ⌄   ☺ UserVoice

⌄ ⎐ Inputs (1)   ▷ Test query   💾 Save query   ✕ Discard changes
    </> ▣ airdelaystreaminginput   1   SELECT
                                        2   *
                                        3   INTO
⌄ ⎘ Outputs (1)   4   airdelaystreamingtarget
    </> ▣ airdelaystreamingtarget   5   FROM
                                        6   airdelaystreaminginput

Figure 8.6 – Start job dialog

7. Please use **Custom** and select the creation date of your source file in the **Start time** fields. Click **Start** in the footer when you are ready.

    If you use **Now**, the job would expect events in your source with a date and time equal to or after the starting time.

    When last stopped will resume the job from where it was stopped and will pick up the event right after the last-saved timestamp.

ASA will internally store the state of your job to enable you to pick up where you left off.

Once you have started your job, it will take some seconds before it kicks off. You can see a small running bar beneath the bell symbol, the notification symbol, in the upper-right ribbon of your Azure portal:

Figure 8.7 – Notification symbol in the Azure portal

If you click the notification button, you will find all kinds of messages related to your Azure environment:

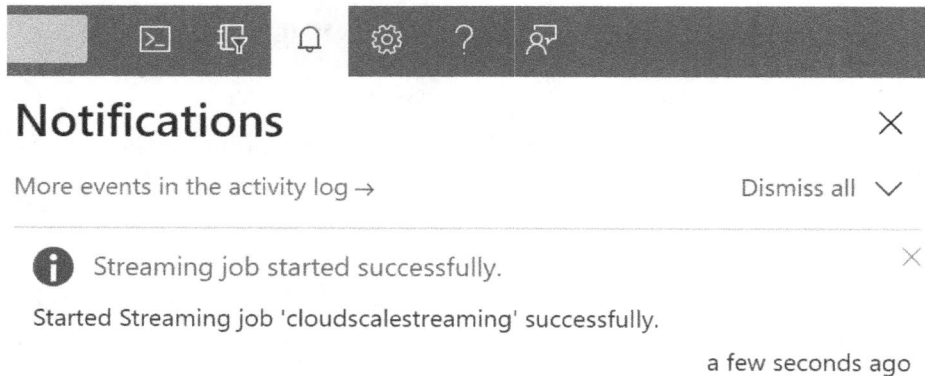

Figure 8.8 – Notifications in your Azure subscription

Once your job is running, you will be presented with some statistics about it in the **Overview** window. **Input Events**, **Output Events**, **Runtime Errors**, and **SU Utilization** are shown in two charts:

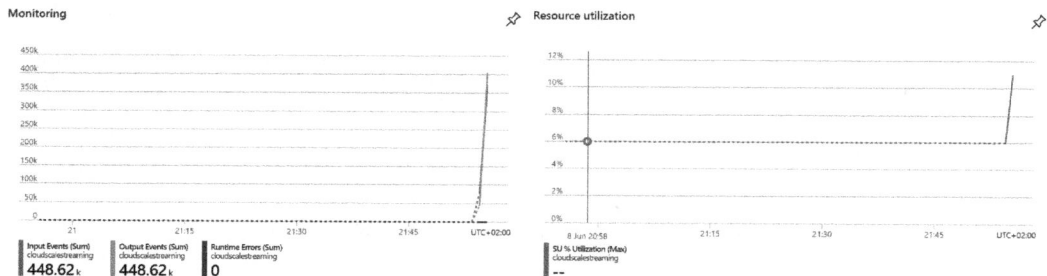

Figure 8.9 – Job statistics

Please proceed now to your target folder in Data Lake and check the output of your job. You should have created a file with all the columns that came from the source file. If you scroll to the far-right end of the file, you will find additional information added: **EventProcessedUtcTime**, **PartitionID**, **BlobName**, and **BlobLastModifiedUtcTime**.

If there is no data, you can check the graph shown in *Figure 8.9* on the left to see whether there have been some runtime errors. The notification list (see *Figure 8.8*) will list errors if they occur. So, if your job was not started successfully, you will find a message there.

In this case, please go through the settings of your input and output and maybe even check the query. Sometimes the input and output get mixed up.

Understanding windowing

As mentioned previously, stream computing adds additional options to your processing logic. In a batch-oriented world, you have a certain input dataset. It can be read, filtered, and aggregated in one go and is then written to the target.

In a streaming world, the input keeps flowing into your system. It has a start but not necessarily an ending date and time. Therefore, we need to observe certain time frames. Maybe the analogy of a traffic census can help with understanding the challenge:

Figure 8.10 – Traffic census windowing example

You are recording the information for the census and will count cars and their types and colors passing by for an agreed amount of time. This time frame constitutes a window.

Now, this is a very basic example of a window. There can be quite a few differences when we think about the time frames and possible overlaps that windows can have.

Understanding tumbling windows

Maybe you want to count your cars every 10 minutes. The 10 minutes will then be your window. If you don't overlap the windows and want to keep the window time frame fixed, you are talking about a tumbling window:

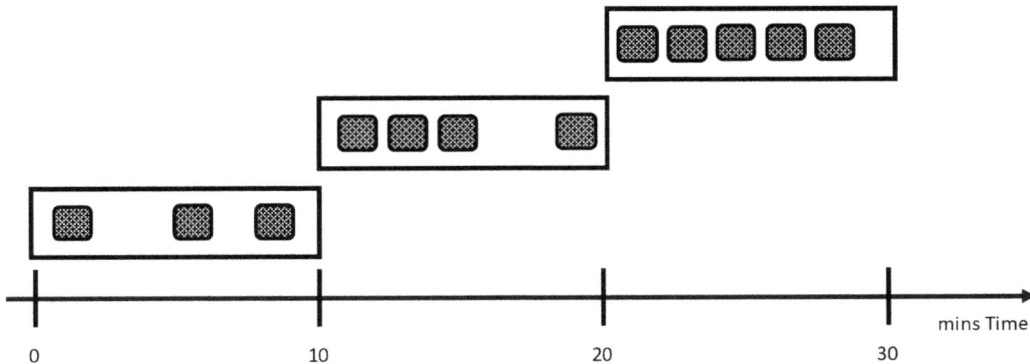

Figure 8.11 – Tumbling window

Later, you'll find some code examples about windowing functions in the *Using window functions in your SQL* section.

Understanding hopping windows

If you need to analyze overlapping time frames – let's say for every 2 minutes you want to count the cars that passed by in the last 4 minutes – you are looking at so-called hopping windows:

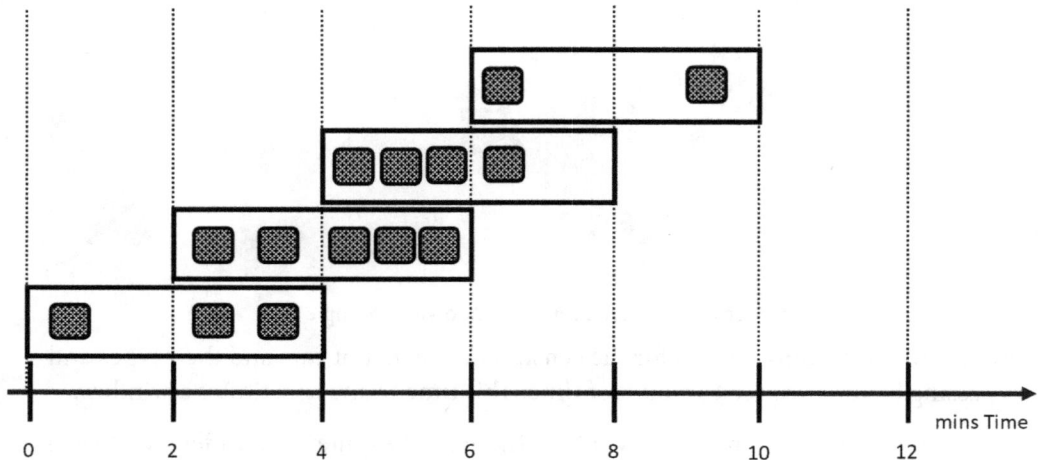

Figure 8.12 – Hopping windows

Understanding session windows

Maybe you need to analyze a set of events that are close in time to each other. Let's say you want to count cars that drive 30 seconds behind each other. In this case, your *session* starts when the first two cars appear that are 30 seconds or less behind each other. The session will only end when there has been no car following the *last* one for 30 seconds or when the duration configured to the function has been reached. The next session will then follow the same rules but doesn't need to last the same amount of time as the first one. Session windows might produce very different time frames for the session:

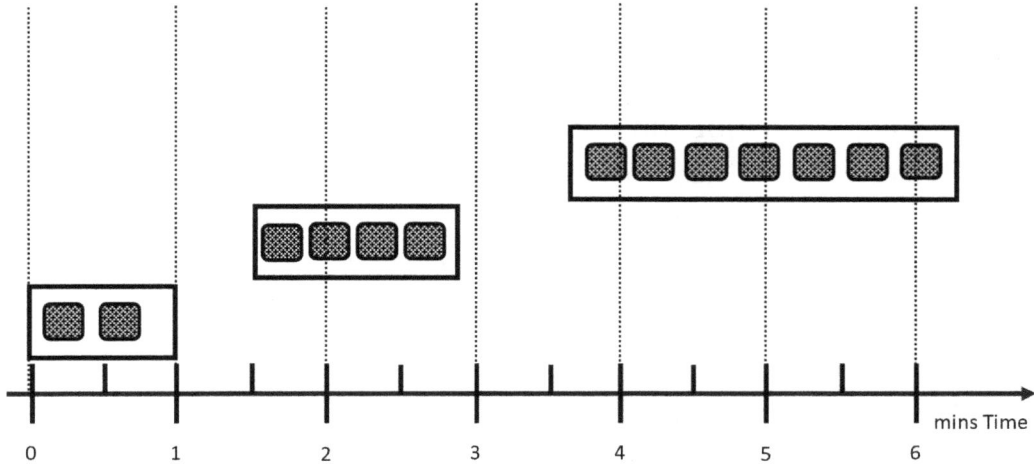

Figure 8.13 – Session windows

Understanding sliding windows

On the one hand, sliding windows consider all the possible windows of a given length. As this might lead to an infinite number of windows, ASA will only return the windows that actually caught a change.

On the other hand, sliding windows add another dimension to the analysis. You can add conditions to the groups that you are analyzing. Let's say you want to know whether there have been more than five cars per type in the last 10 minutes. To extract this information from your stream, you will use a sliding window:

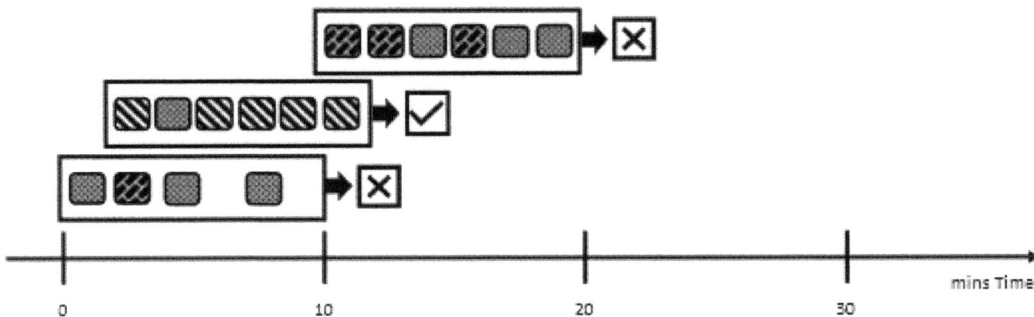

Figure 8.14 – Sliding windows

The example in *Figure 8.14* shows that for the second window, the condition of five cars per type in the last 10 minutes has been satisfied. This condition is false for the two other windows.

Later, you'll find some code examples about window functions in the *Using window functions in your SQL* section.

Understanding snapshot windows

There is yet another type of window that checks exactly one point in time:

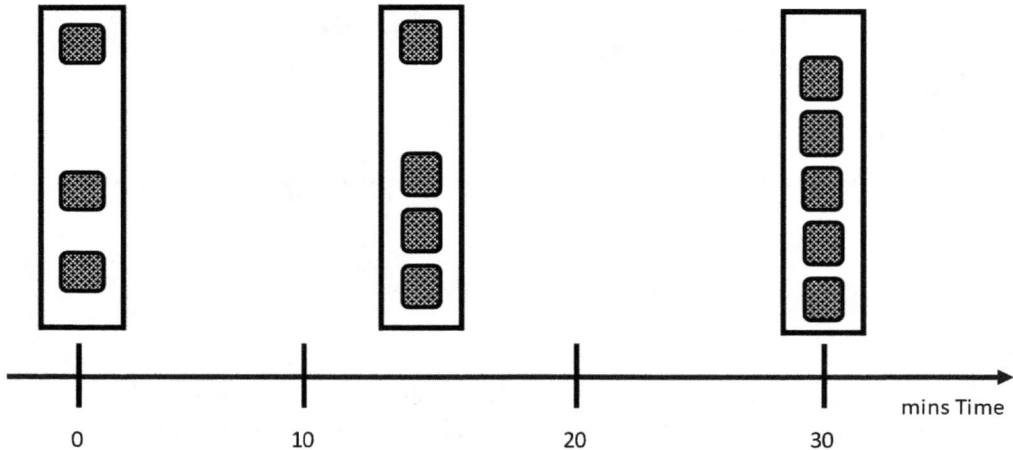

Figure 8.15 – Snapshot window

Using window functions in your SQL

To be able to use windowing functions with your stream, you would need a timestamp coming from your input. In the SQL statement, you will flag this timestamp with the TIMESTAMP BY statement.

Windowing functions are aggregate functions. This means they are always implemented in the GROUP BY clause of your SQL statement:

- For example, the tumbling window from *Figure 8.11* would be implemented as follows:

```
SELECT
    CensusStation,
    COUNT(*) as Amount
FROM
    Cartraffic
TIMESTAMP BY
    ObservedT
GROUP BY
```

```
    CensusStation,
    TUMBLINGWINDOW(minute, 10)
```

- The hopping window from *Figure 8.12* would be implemented like this:

```
SELECT
    CensusStation,
    COUNT(*) as Amount
FROM
    Cartraffic
TIMESTAMP BY
    ObservedT
GROUP BY
    CensusStation,
    HOPPINGWINDOW(minute, 4, 2)
```

- If you want to implement a session window, you would again need three parameters set:

```
SELECT
    CensusStation,
    COUNT(*) as Amount
FROM
    Cartraffic
TIMESTAMP BY
    ObservedT
GROUP BY
    CensusStation,
    SESSIONWINDOW(seconds, 30, 180)
```

In this case, we have added a maximum 3-minute duration (180 seconds) for the observation.

- The sliding window will need another SQL clause that we implement to *filter* the grouped results – the HAVING clause:

```
SELECT
    CensusStation,
    CarColor,
    COUNT(*) as Amount
```

```
FROM
    Cartraffic
TIMESTAMP BY
    ObservedT
GROUP BY
    CensusStation,
    CarColor,
    SLIDINGWINDOW(minute, 10)
HAVING COUNT(*) > 5
```

- The snapshot window, finally, is not using a particular window function. It *just* groups by a timestamp that you want to explore:

```
SELECT
    CensusStation,
    COUNT(*) as Amount
FROM
    Cartraffic
TIMESTAMP BY
    ObservedT
GROUP BY
    CensusStation,
    System.Timestamp()
```

Please check the *Further reading* section for a deep dive into the documentation.

For an example that runs a small app and sends data to an event hub from where you can consume a real stream, please check the GitHub repo.

Delivering to more than one output

The SQL query of your ASA job is not tied to one output only. You can use it to deliver data from your input to several different outputs using different queries with different granularities.

Think of it as a configurable routing mechanism where you create the suitable datasets for any target that you need to deliver to. Maybe you want to land every event in its raw form into Data Lake. This is often referred to as the **cold path**.

At the same time, you need to display some aggregated numbers on a Power BI dashboard, which is the **hot path**.

Additionally, you might need to process the input data using a machine learning model to detect fraud or predict machine failure. The results of such a prediction and only the results need to be sent to another event hub where it will be used to trigger an alert in another system. This is another branch of the hot path.

All these routes can be set up in your ASA job query. Let's use some of the query examples from previously and put them together. You don't need to add a terminator between the queries; just put them behind each other. Let's stick to the assumptions just formulated.

First, we just drop every event into a data lake, then the second query adds a tumbling window over 10 minutes (just like previously) and writes the aggregated numbers to a Power BI streaming dataset that feeds a dashboard visual.

The example tumbling window from *Figure 8.11* would be implemented as follows:

```
SELECT
    CensusStation,
    COUNT(*) as Amount
INTO
    POWERDISTREAMINGDS
FROM
    Cartraffic TIMESTAMP BY ObservedT
GROUP BY
    CensusStation,
    TUMBLINGWINDOW(minute, 10)
```

Have you managed to implement a window function in your ASA job?

Adding reference data to your query

In many cases, the data coming from your input will not satisfy the needs of your required analysis. You will need to add data from other sources, such as a master data database, for example, or files that contain data that you need to enrich the query to create the information needed in your analysis.

In ASA inputs, you can add reference data as inputs that can be additionally used to be joined into your analytical queries. These inputs can be derived from the following:

- Blob storage or Data Lake Gen2
- SQL databases

The references are literally joined into the query using a `JOIN` clause and you have the option to use an inner or left outer join.

If we follow our example from the census station and maybe try to add the fuel type and potential average air pollution from a car type file, we could, for example, derive the average air pollution emitted at the census station:

```
SELECT
    t1.Cartype,
    SUM(t2.mgNOx/60) as SumNOx
FROM
    Cartraffic as t1 TIMESTAMPED BY ObservedT
JOIN
    CarStats as t2
ON
    t1.Cartype = t2.Cartype
GROUP BY
    t1.Cartype,
    TUMBLINGWINDOW(minute, 10)
```

Please check the *Further reading, Reference data* section for additional information on how to use reference data and join it to your input.

Using joins for different inputs

Joins are not just used to add reference data to your stream. You can also use the `JOIN` clause of the ASA SQL dialect to join two inputs together. You will need a `TIMESTAMPED BY` statement for each of the sources to synchronize the stream and sort the right events together. You won't need to explicitly add an `ON` clause to your statement to bring the two inputs together on the timestamp unless you need additional join logic on attributes of the two streams.

Please check the *Further reading, Joins* section for information about joining inputs. You might also want to examine the topic of temporal joins, such as joins with `DATEDIFFS` and temporal analytical functions such as `FIRST`, `LAST`, or `LAG` with `LIMIT DURATION`.

Implementing pattern recognition

ASA SQL implements a function that might come in handy when you want to find and react to patterns hidden in your input data. MATCH_RECOGNIZE enables you to implement even complex **regular expression** patterns and use them with your incoming data stream.

MATCH_RECOGNIZE will support you in recognizing patterns within a configurable time frame and over several rows when you want to find a repeating pattern in your data.

Let's examine a simple example for a pattern matching in our census case for car colors to display the usage. In this case, you would produce outputs when you find two red cars within 1 minute in your input stream:

```
SELECT
    *
INTO
    DataLakeOutput
FROM
    Cartraffic TIMESTAMPED BY ObservedT
MATCH_RECOGNIZE (
    LIMIT DURATION(minute, 1)
    PARTITION BY CensusStation
    MEASURES
        Last(RedColor.CensusStation) AS CensusStation,
        Last(RedColor.CensusTracker) AS CensusTracker,
        Last(RedColor.CarType) AS RedColorCartype
    AFTER MATCH SKIP TO NEXT ROW
    PATTERN (Red{2,} Blue*)
    DEFINE
        Red AS Red.Color = 'Red'
        Blue AS Blue.Color = 'Blue'
) AS MATCHINGSTREAM
```

You will find a link to the MATCH_RECOGNIZE function documentation in the *Further reading, MATCH_RECOGNIZE* section.

There is also a link to a great list of typical queries and usage patterns in ASA for you in the *Further reading, Typical query usage* section.

Adding functions to your ASA job

If you examine the **Job topology** section in the navigator of your ASA job, you will find the **Functions** entry. You can implement additional functionality for your ASA job.

Maybe you want to decode the payload of your incoming event because it is binary-coded following a special pattern. Another option would be to add an Azure Machine Learning service or a model from Azure Machine Learning Studio to score your input or to predict any circumstance based on incoming data.

Functions can be implemented from the following:

- **The Azure Machine Learning service**: You can create a function within Azure Machine Learning Studio and deploy it as a web service.

- **Azure Machine Learning Studio**: You can use Azure Machine Learning (classic) and create a callable web service that will be implemented here.

- **JavaScript user-defined functions** (**UDFs**): You can add your own JavaScript UDF code to your ASA job in this way. An editor window will open where you can enter your function.

- **JavaScript user-defined aggregators** (**UDAs**): You can add a custom JavaScript UDA using this option. An editor will open where you can add and edit your custom aggregation function.

As a fifth option, you can develop **C#** user-defined functions with Visual Studio Code and use them in your SQL query. With Visual Studio Code, you can then deploy the whole setup to your ASA job.

Please find details about implementing and using functions in the *Further reading*, *Functions* section.

Understanding streaming units

ASA does its processing in memory. Therefore, you want to make sure your job is always equipped with the right amount of SUs. These represent the compute resources of ASA and form a combined factor of CPU and memory.

If you check the chart on the right in *Figure 8.9*, you will find a percentage of memory utilization of SUs of your ASA job. This is quite an important one to observe in your ASA job. You want to keep the utilization percentage of your ASA job always below 80%. If your job runs out of memory, it will fail. You won't see the CPU consumption here, so you might want your SUs to be a little bit ahead of your jobs. Maybe you want to set an alert to be notified when your job exceeds a threshold on its way up over 80%.

As a rule of thumb, you would start with six SUs and observe the performance of your job. Don't partition from the beginning! You need to experiment with your data, the complexity and number of steps of your queries, and the number of partitions of your job to balance the right SU settings. Please find a link to more information in the Microsoft documentation about the limits and how to calculate a baseline setting for SUs for your job in the *Further reading, Optimizing Azure Stream Analytics* section.

Partitioning

Adding partitions to your ASA job will help improve the performance, especially when your input data is already partitioned. When you examine the partition key in the source settings (see the *Integrating sources* section), you can select the attribute from your source and determine the number of partitions for your job. If you manage to synchronize your input and output partitions, this can have a big impact on the throughput of your job.

Additionally, you have the option to use the `PARTITION BY` statement in your query logic to even synchronize the job logic with the input and output.

Synchronizing your partitions means that you need to have the same amount of input partitions as you have output partitions. When your job writes to Azure Storage, the storage account will inherit the partition settings. Please check the *Further reading, Optimizing Azure Stream Analytics* section for more information on partitioning.

Resuming your job

As mentioned already, ASA will always store checkpoint information internally to be able to resume your jobs in the case of failure. Functions that support a so-called stateful query logic in temporal elements are as follows:

- The window functions in the `GROUP BY` clause
- Temporal joins
- Temporal analytical functions

If your job needs recovery, it will be able to restart from the last available checkpoint.

> **Note**
>
> There are rare cases when ASA will not be able to store checkpoints for a job recovery. When Microsoft needs to update the service where your ASA job is running, checkpoint data is not stored. You will need to additionally take care of your input data and its retention time.
>
> As a rule of thumb, you might minimally have a retention time in your source data that fits the window size of your ASA job. Please find more information about job recovery and replay times in the *Further reading* section.

Using Structured Streaming with Spark

If you are more the kind of developer that loves to code and you are a fan of Spark, maybe you want to have a look at Structured Streaming with Spark. This might be an interesting alternative for you.

Spark clusters are a widely used engine to implement streaming analytics using one of the available programming languages, such as Python or Scala. With the massive scalability of Spark clusters in Azure services such as Synapse or Databricks, you will be able to implement an environment that can grow with your needs and deliver the necessary performance.

Next to performance, there is the extensibility of Spark clusters that is a factor to consider. You will be able to combine streaming algorithms with the capabilities of Spark and programming languages such as Python (PySpark), Scala, or R.

Take Kafka as input for your streaming analysis, for example. Kafka is an event streaming platform that is quite widely used. ASA does not yet offer a connector to Kafka; therefore, you will need to find another way. Spark offers a Kafka library that you can implement into your streaming solution.

Spark too offers windowing functions and the capability to write to many sinks, Azure Data Lake Storage included.

If you want to read more about Structured Streaming with Spark, please refer to the link in the *Further reading, Structured Streaming with Spark* section.

Security in your streaming solution

Secure access to sources and sinks in your solution is paramount. There are some considerations that you might want to go through when implementing a streaming solution with ASA.

Connecting to sources and sinks

If you examine the *Integrating sources* and *Writing to sinks* sections, you will find the authentication mode in the list of the connectors in almost every one except Event Hubs and IoT Hub, where you would use key and connection strings to connect.

Implementing authentication with either service users and passwords or managed identities will already create very secure access into your sources and sinks. Azure Active Directory implements a multitude of security measures to eliminate the possibility for attackers to break into your solution.

With the use of managed identities, you are implementing a service principal. This is a kind of Azure user that can only be used with Azure services. You can compare them to the on-premises service users that can be configured to only be used with a service and not, for example, for an interactive session.

Please find a link to an overview about Azure managed identities in the *Further reading*, *Managed identities* section.

Understanding ASA clusters

ASA clusters are an additional offering when it comes to the Azure Stream PaaS component.

With ASA clusters, you can create a shared environment for ASA jobs that can be used by several developers from your company.

An ASA cluster can scale to 216 SUs in comparison to the 192 of single ASA jobs. But the more important option of clusters in comparison to single jobs is the capability to connect to **Virtual Networks** (**vNets**) using private endpoints. This means when you are implementing network security in your Azure tenant and plan to "hide" your resources within vNets, you will need to use an ASA cluster as single ASA jobs can't be used with private endpoints.

You will find a link to an introduction to ASA clusters in the *Further reading*, *ASA clusters* section.

Monitoring your streaming solution

As seen in *Figure 8.9*, you can see some information already on the **Overview** page of your ASA job. If you navigate to the **Monitoring** section of your ASA job, you can get further insights into your job.

In the **Logs** section, for example, you are presented with a list of predefined queries that will produce insights into all kinds of errors that can occur when you are running your ASA job:

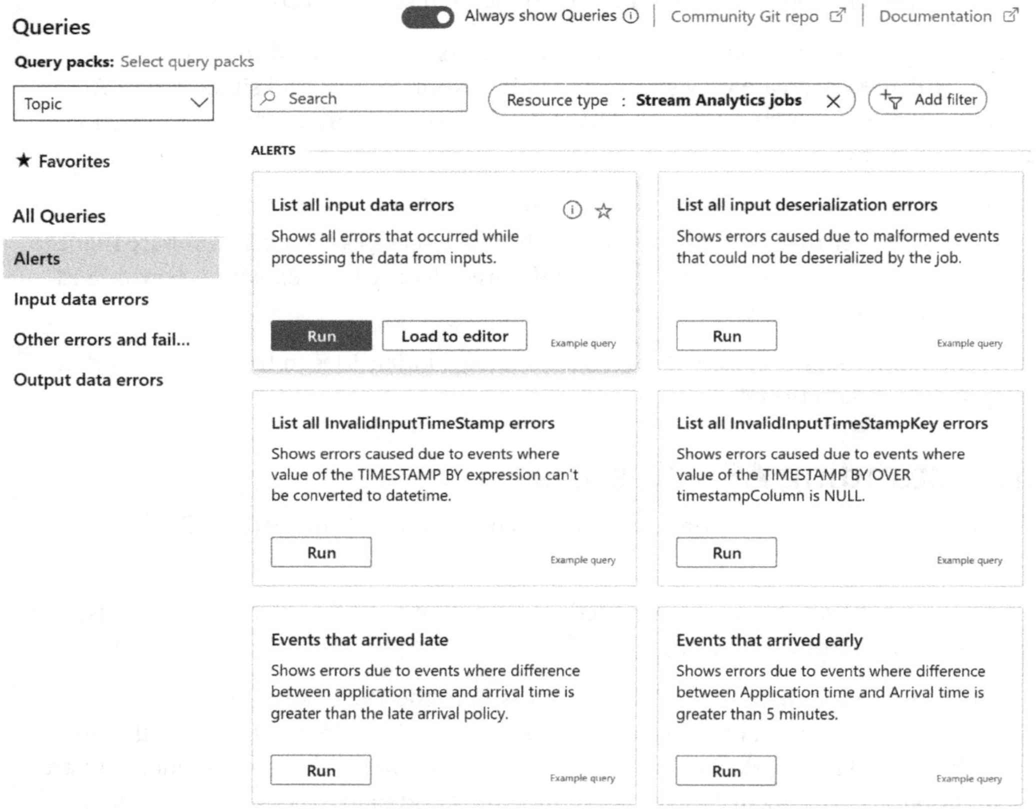

Figure 8.16 – Available error queries in the Logs section

If you proceed to **Metrics**, you are taken to a chart editor where you can select from the available ASA metrics:

Chart Title ✎

⤵ Add metric ⁺▼ Add filter ∷ Apply splitting   ⌁ Line chart ∨ ⧉ Drill

Scope	Metric Namespace	Metric	Aggregation	
cloudscalestreaming	Stream Analytics job sta... ∨	⦿ Select metric ∨	Select aggregation ∨	⊗

```
                                         ∷ Backlogged Input Events
    100                                  ∷ CPU % Utilization (Preview)
                                         ∷ Data Conversion Errors
     90                                  ∷ Early Input Events
                                         ∷ Failed Function Requests
     80                                  ∷ Function Events
                                         ∷ Function Requests
     70                                  ∷
```

60	Select a metric above to see data appear on this chart or learn more below:	
50	⁺▼	▲▲	◯ ▄ⁿ
	Filter + Split ⧉	Plot multiple metrics ⧉	Build custom dashboards ⧉

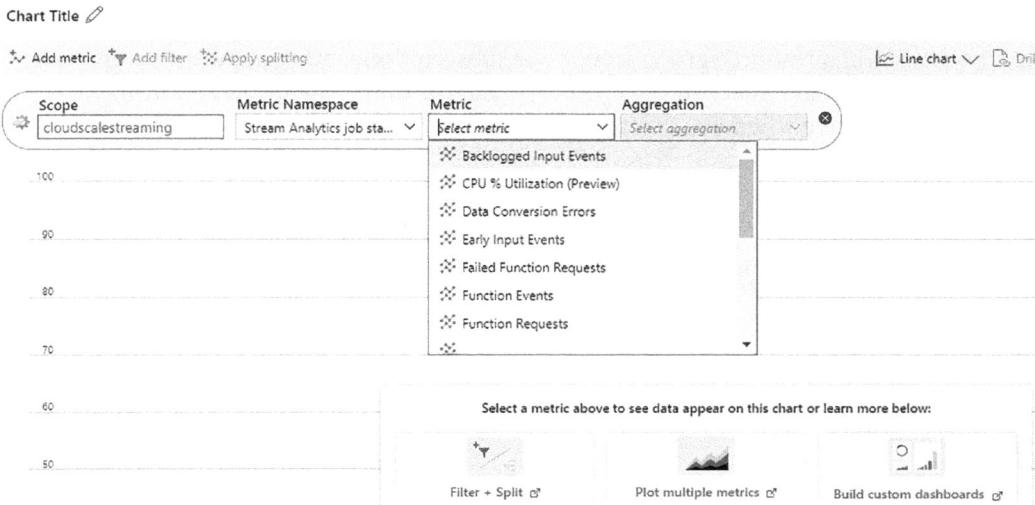

Figure 8.17 – ASA metrics view

You have metrics such as backlogged input events, data conversion errors, early input events, and failed function requests. This section will give you a deep insight into your ASA job.

If you want to set up alerts for your ASA job, such as the SU percentage utilization, for example (remember the *Understanding streaming units* section), this is the place to do so.

Let's implement an alert to be sent when the SU percentage utilization hits 70%:

1. In the **Alert rules** section, hit **+ New alert rule**. In the following blade, you will find your ASA job already selected in the **Resource** section.

2. In the **Condition** section, please click **Add condition**, and in the following dialog, select **SU % utilization** as the signal name.

3. In the **Configure signal logic** dialog, scroll to the bottom and enter 70 in the **Threshold value** field, and hit **Done**. The condition will show in the list.

4. Next, you can set up an action group in the **Actions** section. Please check **Application Insights Smart Detection** and hit **Select**. The action group name will show.

Finally, name your alert in the **Alert rule details** section. You can set a name and a description, select a resource group from your subscription, and configure a severity for your rule. Once you have checked the box below **Enable alert rule upon creation**, your rule will become active when you create it:

Figure 8.18 – Creating an alert rule

You will need to click **Manage Alert rules** to see a list of your implemented rules. From there, you can control your rules and manage them.

Using Azure Monitor

ASA, like all the other PaaS components on Azure, integrates with Azure Monitor. You can configure ASA to deliver telemetry to Azure Monitor. By doing so, you will be able to put your streaming analysis telemetry into correlation with other components of your solution. You can therefore produce insights in a wider focus.

For example, you might want to correlate your ASA insights with the logs of your Event Hubs and Azure Data Lake Storage, which may act as input and output for your ASA job.

In the **Monitoring** section, you can do the correlation in the diagnostic settings. If you enable them, you can send log details from **Execution, Authoring,** and **All Metrics**.

You can select to **send the details to your Log Analytics workspace**, **archive to a storage account**, and/or **stream to an event hub**.

Please find information about Azure Monitor in the *Further reading, Azure Monitor* section.

Summary

In this chapter, you have learned how to provision an ASA job. You have seen how to connect to sources and sinks and how to use them as inputs and outputs. You have also learned about ASA SQL and its windowing functions.

Furthermore, you have seen that ASA SQL queries can route data from the input to different outputs, creating different granularities. You have examined the capabilities to add reference data to your queries and how to add further functionality such as user-defined functions and machine learning using functions.

Finally, we have talked about SUs, the performance metrics of ASA, and how partitioning will help you to improve performance. You have examined security questions and have learned about monitoring. If all the features of ASA do not deliver on your requirements, there are additional technologies available on Azure, such as Spark clusters in Synapse or Databricks that can be used to implement streaming.

We have touched on the topic of machine learning now already several times. If you are interested in the available capabilities on Azure, please proceed to *Chapter 9, Integrating Azure Cognitive Services and Machine Learning*.

Further reading

- *Event Hubs*: https://docs.microsoft.com/en-us/azure/event-hubs/event-hubs-about

- *IoT Hub*: https://docs.microsoft.com/en-us/azure/iot-hub/about-iot-hub

- *Service Bus*: https://docs.microsoft.com/en-us/service-bus-messaging/service-bus-queues-topics-subscriptions

- *Sinks*: https://docs.microsoft.com/en-us/azure/stream-analytics/stream-analytics-define-outputs

- *Optimizing Azure Stream Analytics*: https://docs.microsoft.com/en-us/azure/stream-analytics/stream-analytics-parallelization

- *Window functions*: https://docs.microsoft.com/en-us/stream-analytics-query/windows-azure-stream-analytics

- *Job recovery*: https://docs.microsoft.com/en-us/azure/stream-analytics/stream-analytics-concepts-checkpoint-replay

- *Reference data*: https://docs.microsoft.com/en-us/stream-analytics-query/reference-data-join-azure-stream-analytics

- *Joins*: https://docs.microsoft.com/en-us/stream-analytics-query/join-azure-stream-analytics

- *MATCH_RECOGNIZE*: https://docs.microsoft.com/en-us/stream-analytics-query/match-recognize-stream-analytics

- *Typical query usage*: https://docs.microsoft.com/en-us/azure/stream-analytics/stream-analytics-stream-analytics-query-patterns

- *Functions*:

  Functions overview: https://docs.microsoft.com/en-us/azure/stream-analytics/functions-overview

  C# user-defined functions: https://docs.microsoft.com/en-us/azure/stream-analytics/stream-analytics-edge-csharp-udf-methods

- *ASA clusters*: https://docs.microsoft.com/en-us/azure/stream-analytics/cluster-overview

- *Managed identities*: https://docs.microsoft.com/en-us/azure/active-directory/managed-identities-azure-resources/

- *Structured Streaming with Spark*: https://docs.microsoft.com/en-us/
  azure/architecture/reference-architectures/data/stream-
  processing-databricks

- *Azure Monitor*: https://docs.microsoft.com/en-us/azure/azure-
  monitor/overview

9
Integrating Azure Cognitive Services and Machine Learning

Advanced intelligence efforts are gathering speed and the potential for integrating machine learning models with data processing, for example, to detect fraud, predict device failure, or simply to warn a company before a customer moves on, is promising. The seamless integration of ML models into your data pipelines, and with this the shortest possible time to production, is therefore key.

In this chapter, you are going to examine predefined machine learning models that you can provision and instantly consume as services on Azure. You will learn how to implement them in a Synapse Spark notebook to use them with your data.

In the second part, we will have a look at the Azure Machine Learning service and the available features there. You will examine the graphical interface of Azure ML and learn how to implement your own ML model and expose it to your modern data warehouse environment.

This chapter will not give you a deep dive into machine learning and AI implementation. This would go beyond the scope of this book as it fills multiple publications already.

The topics covered in this chapter are the following:

- Understanding Azure Cognitive Services
- Using Cognitive Services with your data
- Examining Azure Machine Learning
- Using Azure Machine Learning with your modern data warehouse

Technical requirements

You will again provision Azure services and combine them. Therefore, you will need the following:

- An Azure subscription where you have at least contributor rights, or you are the owner
- The right to provision Azure Cognitive Services
- The right to provision the Azure Machine Learning service
- The right to register an app in Azure Active Directory
- The right to provision the Azure Key Vault service
- Your data lake from *Chapter 3*, *Understanding the Data Lake Storage Layer*
- Your Synapse Analytics workspace from *Chapter 4*, *Understanding Synapse SQL Pools and SQL Options*, and *Chapter 6*, *Using Synapse Spark Pools*

Understanding Azure Cognitive Services

With Azure Cognitive Services, Microsoft delivers a wide offering of predefined and ready-to-use machine learning models. Some of these services can even be adjusted to individual customer data, for example, to recognize customer vocabulary in written or spoken text or custom images. All of them come with easy-to-use APIs that you can embed into your applications.

As these services are provisioned just like any other on Azure, you will be able to reduce the amount of time it takes to implement AI capabilities in your modern data warehouse environment massively.

These services come pre-trained, but many of them can be adjusted or retrained with your company- or project-specific data to reflect your requirements.

Examining available Cognitive Services

Let's browse through the available Cognitive Services and understand the collections the purpose of each one to locate possible candidates for your usage.

You will find the following collections of Cognitive Services on Azure:

Understanding the Decision collection

We will check out the Decision collection here:

- **Anomaly Detector**: This can be used to recognize potential significances in your datasets. Anomaly Detector works on time series datasets and adjusts itself to use the most appropriate model. The sensitivity level can be customized to tune the model.

- **Content Moderator**: This supports you in detecting offensive and undesirable content in your application. You can use Content Moderator with videos, images, and text, and even add in a human review application to combine machine learning components with human interaction to fine-tune prediction confidence.

- **Metrics Advisor**: This builds on Anomaly Detector and helps you to analyze time series datasets to do root cause analysis by identifying significances and sending alerts, and helps you to diagnose and monitor unusual correlations in your data. It also has an easy-to-use interface for interaction with your data.

- **Personalizer**: This adds reinforcement learning-based methods to your application to personalize your content and offer options to your users. Not just a recommender engine, this service offers the so-called **Apprentice** mode. This will enable Personalizer to run parallel to the original application and learn behavior by observing.

Understanding the Language collection

We will check out the Language collection here:

- **Immersive Reader**: This adds reading support to your application. Immersive Reader will read text aloud. You can choose to translate to other languages and even highlight the text read to the user.

- **Language Understanding**: This offers you a visual interface to create a Language Understanding model to be used with your applications. The Language Understanding cognitive service, for example, will help to extract the intent of a text to be used in a bot conversation. It can be customized to your needs to enable your application to extract information from conversations with your users and react accordingly.

- **QnA Maker**: This supports you with a quickly trainable question and answer model to be added to a conversational environment. You can upload questions and answers from semi-structured files and instantly use them as a knowledge base. The service will be able to recognize similar questions and find the answers accordingly.

- **Text Analytics**: This delivers **Natural Language Processing (NLP)** capabilities to your application. Text Analytics can help to identify key phrases, entities such as people, places, or organizations, and extract common topics and trends. Additionally, you can use the service to classify terminology, for example, in a medical environment. Another feature of the service is easy sentiment analysis of phrases and texts. All the functions support a wide range of languages.

- **Translator**: This service helps you to detect languages and translate text in real time between a wide range of languages.

Understanding the Speech collection

We will check out the Speech collection here:

- **Speech to Text**: Use this service to transcribe spoken words into indexable and searchable text. A wide number of languages are supported by this offering and you can additionally add your own vocabulary to reflect your company's products and language specifics.

- **Text to Speech**: This service will support you in creating applications that can "talk" to your users. The service offers a lot of different voices and a wide range of languages that can be used to *talk* to your users. This offering can even use emotional variants for the output.

- **Speech Translation**: This service enables real-time translation from audio into a wide range of different languages. You can even add your company's vocabulary for specific output according to the words and terms used in your company.

- **Speaker Recognition**: You can use this service to recognize speakers according to their individual voices. This service can even recognize individuals within a group.

Understanding the Vision collection

We will check out the Vision collection here:

- **Computer Vision**: When you need to recognize and analyze the content of images and videos, or you want to perform text recognition in your images or videos, you can use Computer Vision. From the labeling of the content in the source material up to the recognition of the physical movement in a 3D area, this service can support you in these analytics.

- **Custom Vision**: This service is used to implement image recognition. You can use your own images and tags to train the model.

- **Face**: Further specialized in the recognition of faces and emotions, this service can be trained to even recognize persons from your own collection of images.

- **Form Recognizer**: This is an interesting one as well. Form Recognizer supports you in extracting information from images of forms and documents and can help you transform your documents into table-oriented data for further analysis.

- **Video Indexer**: This is another option for detecting and extracting metadata from videos. Video Indexer will support you in transcribing spoken words into searchable text. It can be trained to recognize people and will return emotions presented in a video. Even displayed text can be extracted and returned with the additional delivery of the time index within the film by Video Indexer.

Getting in touch with Cognitive Services

Let's get familiar with Cognitive Services and give it a try. Navigate to `https://azure.microsoft.com/en-us/services/cognitive-services/speech-translation/`. Scroll down the page just a little to this view:

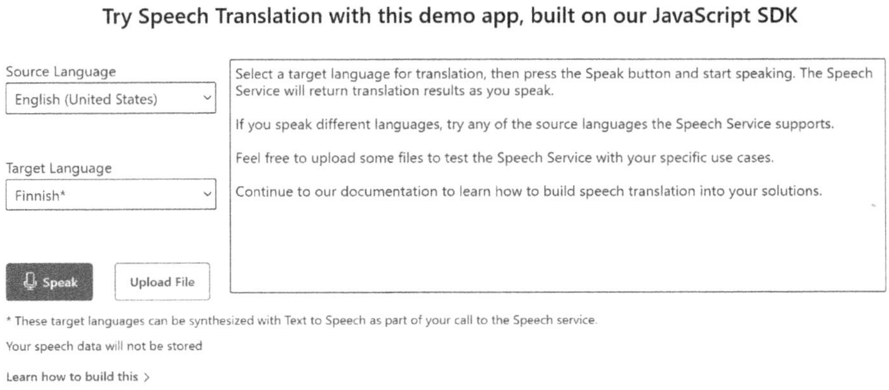

Figure 9.1 – Speech Translation

First, select your mother tongue in the **Source Language** field, and then select a target language. Now you can hit the **Speak** button and say some words that you want the service to translate. The translated text will appear in the textbox. What do you think?

This is just the tip of the iceberg of the capabilities of the Speech Translation service. For example, you can also load a sample `.wav` file in this demo.

The service itself can be extended to recognize individual protocols. You can add and train your company-specific vocabulary to support you in your international communication.

In the upcoming section, *Using Cognitive Services with your data*, let's examine another cognitive service and use it with a Synapse Spark notebook to analyze incoming text and detect its sentiment.

Using Cognitive Services with your data

In this section, you will learn how to provision a cognitive service in your subscription and then use it with a Synapse Spark notebook on incoming data. Let's do this with the Text Analytics cognitive service:

1. First, proceed to the Azure portal and hit + **Create a resource**. Search for `Text Analytics` and from the quick results, select **Text Analytics**.

2. On the starting blade, hit **Create** and start the provisioning of your cognitive service.

3. On the **Basics** blade, in the **Project details** section, select the right subscription and either select an existing resource group or create a new one. In the **Instance Details** section, select a region where you want to provision your Text Analytics service. You'll want to have it in the same region as the Synapse Analytics workspace where you will create and run your notebook.

4. Add a name for your cognitive service and finally select a pricing tier. To take care of your budget, you want to use the Free F0 (5K transactions every 30 days) **Stock Keeping Unit (SKU)**. Either hit **Next: Tags** or jump directly to **Review + Create**.

5. On the **Tags** blade, optionally enter any tag you need, for example, to report on your services. Finally, you come to **Review + Create** for one last overview before you provide the service.

6. When everything is as you intended, hit **Create** and initiate provisioning.

7. Once the service is available, you can navigate to it by clicking **Go to resource**.

On the **Quick start** blade of your Text Analytics cognitive service, you will find some useful links, for example, the API key (section 1 of the **Quick start** blade) of your service. This one is very important when you want to connect to the service from another one. You will need one of the two keys later in this section. So, it is a good idea to copy it now and paste it into a text file.

In section 2 of the **Quick start** blade, you will find links to API consoles. Maybe you want to instantly start playing around with your service? Then this is your place to understand request structures and find code examples in different languages (C#, Java, JavaScript, ObjC, PHP, Python, and Ruby) of how to implement a call.

In the *Implementing the call to your Text Analytics cognitive service with Spark* section, you will develop a routine to use this service with a Spark notebook.

Important note

These services will generate costs in terms of your Azure subscription. Remember to either pause, scale down, or even remove provisioned artifacts from this chapter from your Azure subscription when you are done with your work as you will no longer need them.

Understanding the Azure Text Analytics cognitive service

Now, first let's examine what this service is capable of.

The Text Analytics cognitive service can support you in the following tasks:

- **Language Detection**: For a given input, the service will return a language identifier and a score that the service sets to give you an idea of the confidence in how reliable the analysis result is. The value of the score will range between 0 and 1, where 1 is the best possible score.

    Nowadays, languages can be analyzed in a very detailed way, and you will be able to detect not only the language itself, but also dialects and variants of the language, including regional specifics.

    Refer to the *Further reading, Text Analytics Language Detection*, section for details on the implementation of language detection.

- **Entity Extraction**: This service will support you in extracting and classifying entities in your text. This may be, for example, persons and the type of person (a project manager, a wife, a daughter, and so on) or locations, events and event types, dates or time ranges, and many more. This is exposed in a separate endpoint for general entities to distinguish it from other functionalities.

  Additionally, `Named Entity recognition` will expose another endpoint to support you in detecting **Personally Identifiable Information** (**PII**) in your data. This can be birthdates or phone numbers, URLs, and so on.

  `Entity Linking` is the third endpoint that you can use. It uses Wikipedia as its knowledge base in the background and helps you to resolve entities found in your text data and understand ambiguous data. For example, if your text talks about a bank, the Text Analytics cognitive service will help you to understand whether the text refers to a piece of furniture or a financial institute. It will do so by linking the entity to the context it is used in.

  Refer to the *Further reading, Text Analytics Named Entity Recognition*, section for an overview of the usable endpoints and details regarding implementation.

- **Key Phrase Extraction**: The Text Analytics Key Phrase endpoint can be used to extract main phrases from your text such as *trip, beautiful weather*, and *great guide* from "*We had a trip in beautiful weather and had a great guide*." This part can help you to get a quick insight into a text of your choice.

  Refer to the *Further reading, Text Analytics Key Phrases*, section to dive into the details of this endpoint.

- **Sentiment Analysis**: You can use the Sentiment Analysis endpoint to throw text phrases at the service and let it return its sentiment value. For example, "*I feel bad today*" would certainly result in negative sentiment, whereas "*What a wonderful day today*" would create positive sentiment. With the sentiment, the service will return a confidence score to enable you to decide on the reliability of the analysis.

  You can set a flag (`opinionMining=true`) in your request to enable additional **Opinion Mining** with your Sentiment Analysis. This will lead to a more detailed result regarding parts of the given phrase, whereas without the flag, you will only get one sentiment value for the whole phrase.

  Again, you will find a link to more details in the *Further reading* section.

- **Text Analytics for health**: At the time of writing this book, the Text Analytics for health version of the Text Analytics cognitive service is still in preview. This version of the service targets medical named entities that can be recognized, extracted, and linked. Additionally, this version performs negation detection on the phrases sent to it.

> **Note**
>
> Sentiment Analysis works best on smaller phrases compared to Key Phrase Extraction. This function works better on higher volumes of data. You can improve the performance of your analysis when you arrange your input accordingly.

We will use the Sentiment Analysis endpoint in the next section, *Implementing the call to your Text Analytics cognitive service with Spark*.

Implementing the call to your Text Analytics cognitive service with Spark

In this section, you will use your newly created Text Analytics cognitive service to analyze the sentiment of an input file and output it to a new target.

Let's create an input file for this first. Write down some phrases for your analysis and put them into a text file or a CSV, perhaps some phrases such as these:

```
We had a trip in beautiful weather and had a great guide.
The sights were amazing but the prices were far too high.
Then all of a sudden the weather got bad and the mood turned
bad too.
We drove from the west of the island to the east.
When we arrived in the east the wind and the rain stopped again
and the trip was saved.
When we returned home the memories that stayed were so amazing.
```

Maybe you want to create two new folders (a source and target for your analysis) in your data lake and put this file (with a name such as `textAnalyticsSource.csv`) into the new source folder.

Here is what you are going to implement in this section:

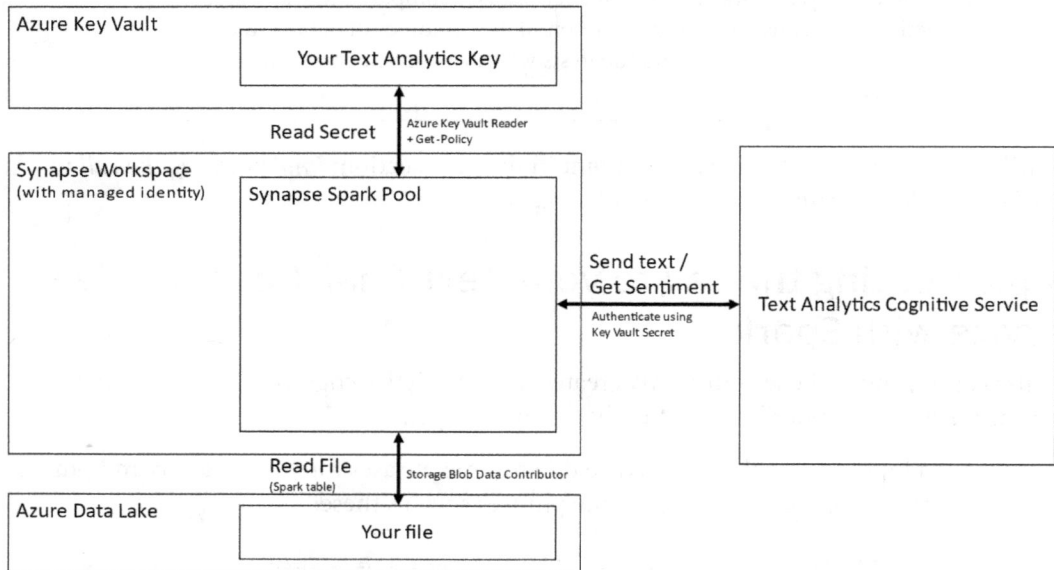

Figure 9.2 – Text Analytics with Synapse Spark

Let's first start implementing your Azure Key Vault in the following section, *Setting up Azure Key Vault*.

Setting up Azure Key Vault

Azure Key Vault is the way to go to store your secrets and keys securely in Azure. In this case, the objects that you will create further on in the book require Key Vault in order for the secrets to be used.

In the following steps, we will store the API key of your Text Analytics service as a secret in Azure Key Vault to use it securely in your Synapse Spark environment. Therefore, first create an Azure Key Vault service in your subscription:

1. Navigate to the Azure portal and hit **+ Create a resource**. Search for Key Vault. From the quick results, select **Key Vault** and, on the **Description** blade, hit **Create** to start the provisioning.

2. On the **Basics** blade, select your subscription and either select an existing resource group or create a new one. Further down in the **Instance details** field, complete the **Key vault name** field and then select the region where your service will be created. Finally, select a pricing tier. Then, hit **Next: Access policy >**.

3. On the **Access policy** blade, you can control how the secrets in your key vault can be accessed. Leave everything with the default settings here for the time being. Then, hit **Next: Networking >**.

4. Here, as in other services, you can control the **Network** settings and how your service should be "visible" to different networks. Also leave this setting as it is now. Now, either hit **Next: Tags >** or go directly to **Review + Create**.

5. Finally, you can start the provisioning again with **Create**.

6. When the service is ready, you can hit **Go to resource** and start using it.

Now, switch back to your Text Analytics cognitive service and navigate to the **Keys and Endpoint** section in the navigator. Next to the **Keys and Endpoint** entries, you will find buttons to copy the content from there. Copy Key 1 and put it aside for later use.

With Key 1, navigate to your newly created key vault and start creating a secret:

1. In the navigator of your key vault service, select **Secrets**. On the **Details** blade, click **+ Generate/Import**.

2. On the **Create a secret** blade, leave **Upload options** as **Manual**, and enter a name for your new secret. Copy Key 1 from your Text Analytics cognitive service into the **Value** field (the one that you set aside at the outset).

3. Leave the other options as they are (you could, of course, use the **Set activation date** and **Set expiration date** options).

4. Finally, hit **Create** and store your new secret in your new key vault.

5. Following creation of the secret, add your Synapse workspace managed identity as a reader to your key vault. To do so, navigate to **Access Control (IAM)** and hit **+Add** > **Add role-assignment**. From the drop-down list in the dialog on the right, select **Key Vault Reader** as the role. In the **Users** field, start typing the name of your Synapse workspace and hit *Enter*. Select the user that will be displayed as the Synapse managed identity and click **Save**:

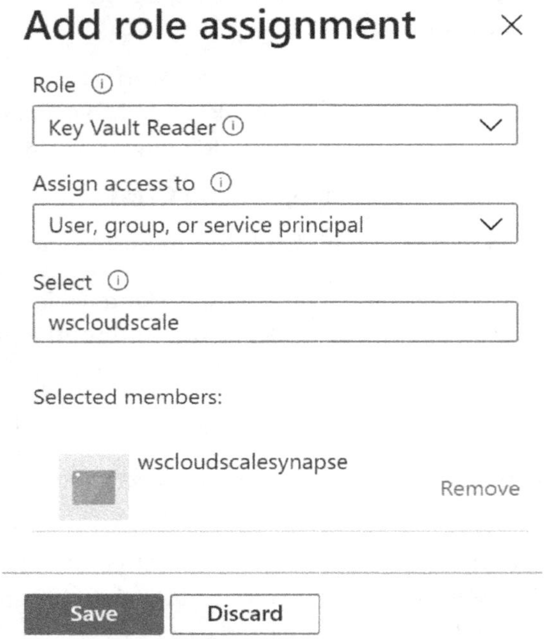

Add role assignment   ✕

Role ⓘ

Key Vault Reader ⓘ   ⌄

Assign access to ⓘ

User, group, or service principal   ⌄

Select ⓘ

wscloudscale

Selected members:

wscloudscalesynapse   Remove

Save	Discard

Figure 9.3 – Key Vault Add role assignment

6. Additionally, you will need to set an access policy within your key vault. In the navigator blade of your key vault, navigate to **Settings** > **Access Policies** and click the link **+ Add Access Policy** in the middle of the **Details** blade. In the following dialog, open the drop-down field next to **Secret permissions** and check **Get**. Then, next to **Select principal**, click on the link, and again select your Synapse workspace managed identity. Click **Select** on the dialog on the right and then Add on the main blade. You are then taken back to the **Access policies** blade. Finally, click **Save** to store your changes.

Note

The step of saving the policies is sometimes forgotten. When you later run into errors with your notebook, you'll want to check whether you saved your policies before you left the Key Vault configuration.

Implementing your Text Analytics cognitive service in a Spark notebook

In the last sequence, you will now create a Spark table from your file in your Data Lake and use the wizard (as I told you, it's Microsoft: they give you wizards) to create a notebook that uses the Text Analytics cognitive service to process the data in the file and return the sentiment for each row that you have entered in your file.

Proceed to the Synapse workspace that you created in *Chapter 4*, *Understanding Synapse SQL Pools and SQL Options*, and which was used again in *Chapter 6*, *Using Synapse Spark Pools*. You can start by entering `https://web.azuresynapse.net` in your browser's address field. On the following screen, you can select your subscription and workspace. When you have logged in successfully, you can proceed with the following steps:

1. First, add your Key Vault service as a linked service to your Synapse workspace. From **Management Hub**, navigate to **Linked Services** and click + **New**. You can search for `Key Vault` or select `Azure` and then select **Key Vault** from the results. Go through the configuration and select the one created above.

2. If you haven't yet done so, navigate to your data lake and create `Source` and `Target` folders for this analysis. During this step, upload the aforementioned file (something like `textAnalyticsSource.csv`) to the newly created source folder.

   As soon as your file is uploaded, right-click it and select **New notebook** > **New Spark table**:

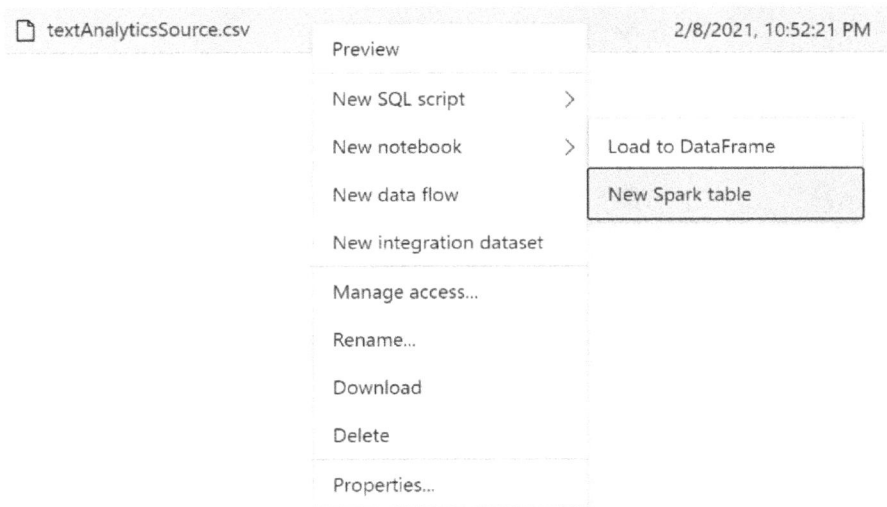

Figure 9.4 – Creating a Spark table from a file

3. In the notebook that then opens, replace [YourTableName] in the last line with a table name of your choice, perhaps something like textAnalyticsSource, and then click the **Run** button on the left side of the cell. (You will be prompted to attach your notebook to a Spark cluster. Maybe you still have the one from *Chapter 6, Using Synapse Spark Pools*, left over. Why not use that one?):

Figure 9.5 – Creating a Spark table from a file

4. Now, navigate to your Spark pool in the **Workspace** section in the **Data** hub:

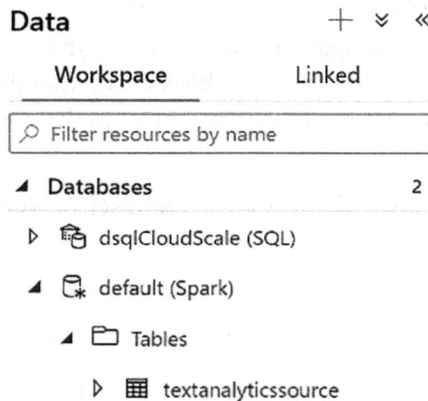

Figure 9.6 – Finding your Spark table

5. Now, when you right-click the Spark table, you will find the menu item **Machine Learning**. Select **Machine Learning > Enrich with existing model**:

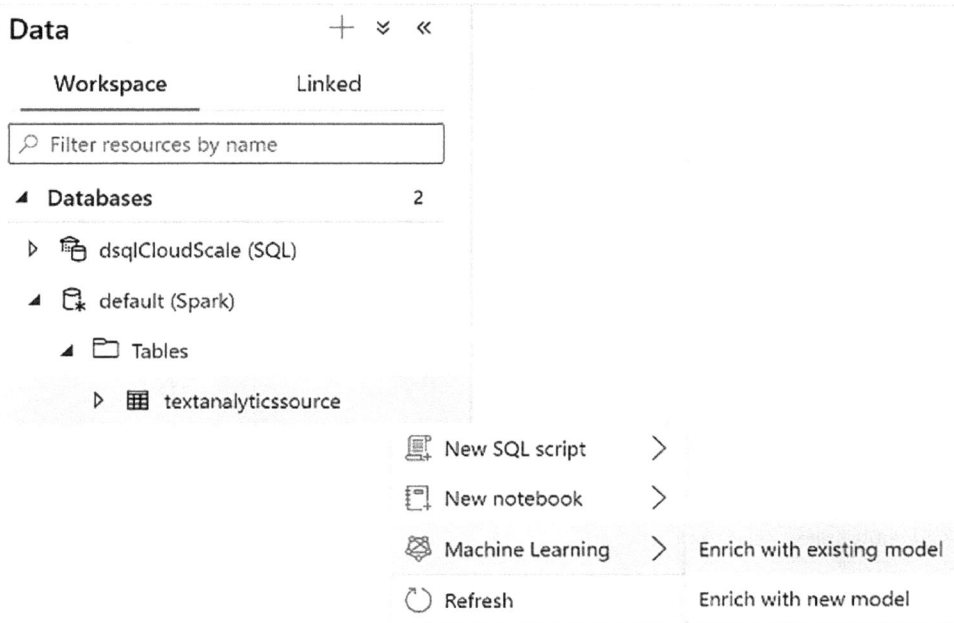

Figure 9.7 – Enriching data with a machine learning model

6. A dialog blade will appear from the right. Select **Text Analytics - Sentiment Analysis** followed by **Continue**:

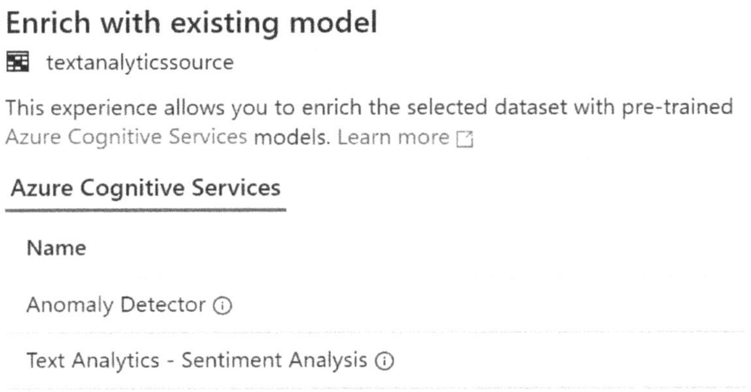

Figure 9.8 – Selecting the Azure cognitive service

7. In the following dialog, select your subscription, your Text Analytics cognitive service, the Azure Key Vault linked service created above in your Synapse workspace, and the secret name that you created in your key vault:

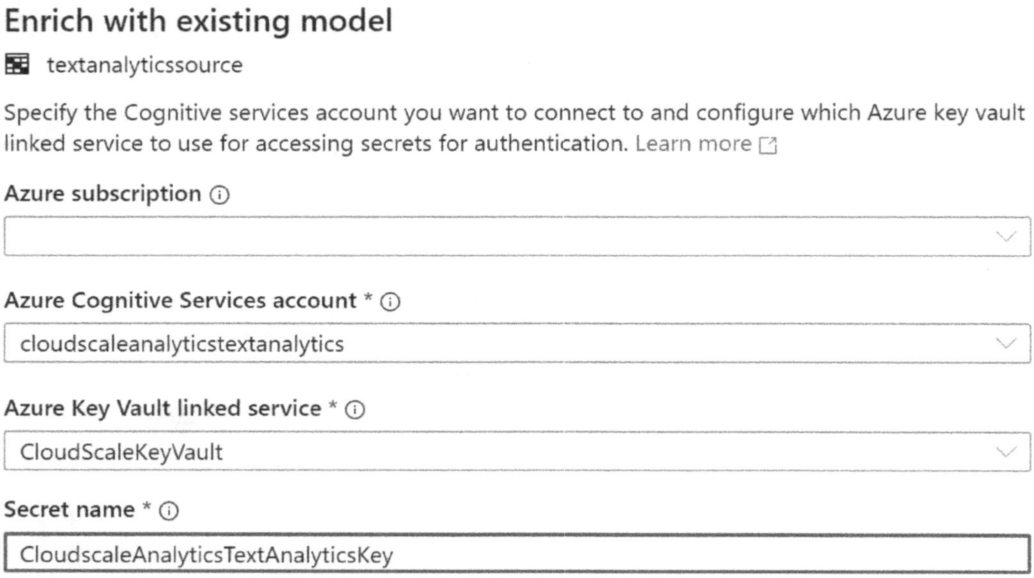

Enrich with existing model

⊞ textanalyticssource

Specify the Cognitive services account you want to connect to and configure which Azure key vault linked service to use for accessing secrets for authentication. Learn more ☐

Azure subscription ⓘ

Azure Cognitive Services account * ⓘ

cloudscaleanalyticstextanalytics

Azure Key Vault linked service * ⓘ

CloudScaleKeyVault

Secret name * ⓘ

CloudscaleAnalyticsTextAnalyticsKey

Figure 9.9 – Configuration of the necessary values

8. In the final dialog, select the language of the text that will be sent to the Text Analytics service and then select the **Text column** field from the file that you want to analyze. Finally, click **Open Notebook**:

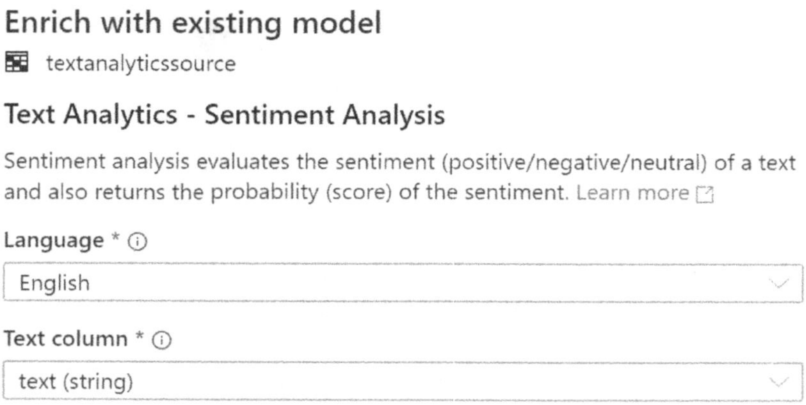

Enrich with existing model

⊞ textanalyticssource

Text Analytics - Sentiment Analysis

Sentiment analysis evaluates the sentiment (positive/negative/neutral) of a text and also returns the probability (score) of the sentiment. Learn more ☐

Language * ⓘ

English

Text column * ⓘ

text (string)

Figure 9.10 – Language and column settings

9. To kick off the analysis, you will now need to attach the notebook to your Spark pool and then click **Run all**.

10. Maybe you want to visualize the results now and show the count of the sentiments in a bar chart? Above the output in the result cell, you will find the **View** switch to toggle the output between the **Table** and **Chart** views. Give it a try and change the display to show bars for **Count of the Sentiment**. Hint: The chart can be manipulated using the small button on the far-right side of the output area.

11. Maybe as a final step, you might want to add a `write` command to your notebook to store the results of your analysis in your data lake:

```
results.write.parquet('abfss://[YOURFILESYSTEM]@
[YOURDATALAKEACCOUNT].dfs.core.windows.net/[YOURPATH]/
textAnalyticsResult.parquet')
```

This sequence now uses the wizard that Microsoft provides with Synapse to consume the Text Analytics cognitive service.

When you examine the **Quick start** blade of your Text Analytics cognitive service, you will find a lot of samples and documentation regarding the service. And as the wizards now only support Sentiment Analysis and Anomaly Detector, you will need to create REST API calls with your input data.

> **Note**
> When you start to try building your REST API calls, the **API Console** links on the **Quick start** blade are a great source of examples.

Now we have examined predefined services on Azure that can already take you very far when you want to implement Advanced Analytics using Cognitive Services in your modern data warehouse. You can analyze images and videos, translate and understand the sentiment and tone of a given text, recognize voices, faces, and speakers, and much more besides.

But what are your options when Azure Cognitive Services is not delivering the exact functionality that you are searching for? You will need to create a suitable machine learning model for your requirement yourself. This will, of course, require you to understand some more details about machine learning and its implementation. In any case, Microsoft tries to help you in your approach with the Azure Machine Learning service and Machine Learning studio. In the next section, *Examining Azure Machine Learning*, you will be provided with an overview of this service and make yourself familiar with its options.

Examining Azure Machine Learning

Azure Machine Learning will offer you a wide collection of capabilities to develop, train, and deploy machine learning models. Additionally, the environment supports you with the automation and management of your models. This includes the versioning and tracking not only of the models but also the training data that you use to build and retrain them.

Browsing the different Azure ML tools

As mentioned, Azure Machine Learning comes with a collection of tools and functionalities to support you in any aspect of the machine learning life cycle:

- **Azure ML Designer**: A graphical interface to build machine learning models on a point-and-click basis.

- **Jupyter Notebooks**: A programming interface where you can build your own ML models using Python.

- **R Scripts/Notebooks**: Programming interface where you can build your own ML models using R.

- **Many Models Solution Accelerator**: Offers you capabilities to work with thousands of ML models if needed.

- **ML Extension for Visual Studio Code**: Integrates Azure ML with Visual Studio Code to enable local development.

- **Machine Learning CLI**: A command-line interface for the most important tasks, such as running experiments, registering ML models, and controlling the life cycle of models.

- **Open-Source Frameworks**: Use your favorite frameworks with Azure ML, for example, TensorFlow, scikit-learn, and others.

- **Reinforcement Learning with Ray RLib**

Additionally, Azure ML will support you in your MLOps activities and help you to deploy your model to your target environment of choice, even on different compute resources. You can modularize your models and restart selected steps if needed or create pipelines that can be run in batch mode on bigger datasets.

But where do you train and later run your models after you have developed the code or their graphical representation? With Machine Learning Studio, you can define compute clusters of your choice to do so. You can choose from a wide variety of VM sizes, from small 2 core versions with 7 GB RAM general-purpose ones up to the 128 core versions with 3.8 TB RAM memory-optimized ones. Additionally, there are VMs available that will add GPU support to your clusters with access to several different GPU versions and sizes.

When it comes to performance and cost, we might even reach some limits in Azure ML Studio. But there is another option to deploy your machine learning model in a productive environment: containers. During provisioning, you saw the option to add an Azure container registry. Azure ML will even support you in deploying your machine learning model to the container registry and make it available in a containerized environment.

Another very attractive option is to use **Open Neural Network Exchange (ONNX)** models in your Synapse dedicated SQL pools. With Azure Machine Learning, you can store an ML model in ONNX format.

This model can be imported into a dedicated SQL Pool table and used within the database. By doing so, you can directly use the model with the data in the database without moving data back and forth from the data lake. Refer to the *Understanding further options to integrate Azure ML with your modern data warehouse* section for more information.

When it comes to datastores and datasets, Machine Learning studio will support you in accessing different storage components on Azure. Azure Data Lake Gen2 is one of them and we are going to use it in the following example.

Examining Azure Machine Learning Studio

Let's go and create an Azure Machine Learning service to browse ML Studio and examine the capabilities described in the previous section:

> **Important note**
> These services will generate costs in terms of your Azure subscription. Remember to either pause, scale down, or even remove provisioned artifacts from this chapter from your Azure subscription when you are done with your work as you will no longer need them.

1. Proceed to the Azure portal and click **+ Create a resource**. Search for Machine Learning and, from the quick results, use **Machine Learning**. On the **Description** blade, select **Create**.

2.   You already know the drill by now – fill in the required fields: **Subscription,** **Resource Group,** and **Name,** and select a region. You might create a new storage account or select an available one, key vault, and so on. For the time being, you won't require a container registry, so select **None** there.

3.   On the **Networking** blade, select **Public endpoint** on this occasion.

4.   On the **Advanced** blade, you can leave the settings as they are.

5.   On the **Tags** blade, well, you know what you can do there already.

6.   After revisiting all the settings on the **Review + create** blade, click **Create** and start provisioning. When this is done, you will again see the **Go To Resource** button.

7.   When you can navigate to the service after provisioning, add your Synapse workspace managed identity as a contributor in **Access Control** to your Azure Machine Learning workspace.

8.   Finally, you will need to create an app registration to have a service principal in your Azure Active Directory (refer to the *Further reading, Creating a service principal* section), add it to your data lake storage, and give it the STORAGE BLOB DATA CONTRIBUTOR role. During creation, copy the Client ID and the created Client secret to Notepad or similar. You will need it shortly. Additionally, add this one to your Azure Machine Learning service with the Contributor role (**Access Control** in the Machine Learning service). Certainly, you will need to change this later for fewer security settings.

Now, without further ado, let's move straight on to the first tool in Azure ML Studio in the following section, *Understanding the ML designer*.

Understanding the ML designer

ML studio offers you a graphical interface to build your models on a low-code/no-code basis. It is called the designer and offers many predefined machine learning algorithms, together with a lot of transformational artifacts that you can put together in a drag-and-drop experience.

For the example and to start with, we are going to examine this area. There are far more sophisticated options for data scientists, but to get a quick start, the designer gives us all we need.

There is a multitude of transformations for all kinds of actions performed on your data, grouped into the following topics:

- Input and Output
- Data Transformation
- Feature Selection
- Statistical Functions
- ML Algorithms
- Model Training
- Model Scoring and Evaluation
- Python Language
- R Language
- Text Analytics
- Computer Vision
- Recommendation
- Anomaly Detection
- Web Service

The `ML Algorithms` group, for example, already offers 19 different algorithms that you can use to train a model. They are grouped into regression functions, clustering, and classification functions.

When you check the `Data Transformation` group, you will find all kinds of functions to manipulate your dataset, including `Apply SQL Transformation`, for example, or `Group Data into Bins`, and many more.

You will find a link to an overview of all the available modules in the ML designer in the *Further reading, Azure ML designer available modules*, section.

Creating a linear regression model with the designer

Let's proceed and create a simple model that we will later use in a Synapse pipeline to perform batch scoring:

1. Navigate to your newly created Azure ML service in the Azure portal. On the **Overview** blade, find the **Launch Studio** button. This will take you to Azure ML Studio.

2. You will first need to create a datastore as a connector to your data lake. Search for `Datastores` and click **+ New datastore**.

3. In the dialog on the right, enter a name and select **Azure Data Lake Storage Gen2** for **Datastore type**.

4. Select your subscription, storage account, and filesystem in the fields.

5. In the **Authentication type** field, leave **Service Principal**. In **Tenant ID**, the tenant of your subscription should already be selected. In the **Client ID** and **Client Secret** fields, paste the values that you copied when you created your app registration in the *Examining Azure Machine Learning studio* section when you prepared the environment.

6. Now we can go and create the machine learning model. Find the **Designer** entry in the navigator and click it. Here, you can, of course, now examine the available samples. When you are ready to proceed, click **New pipeline**. The graphical designer interface will open:

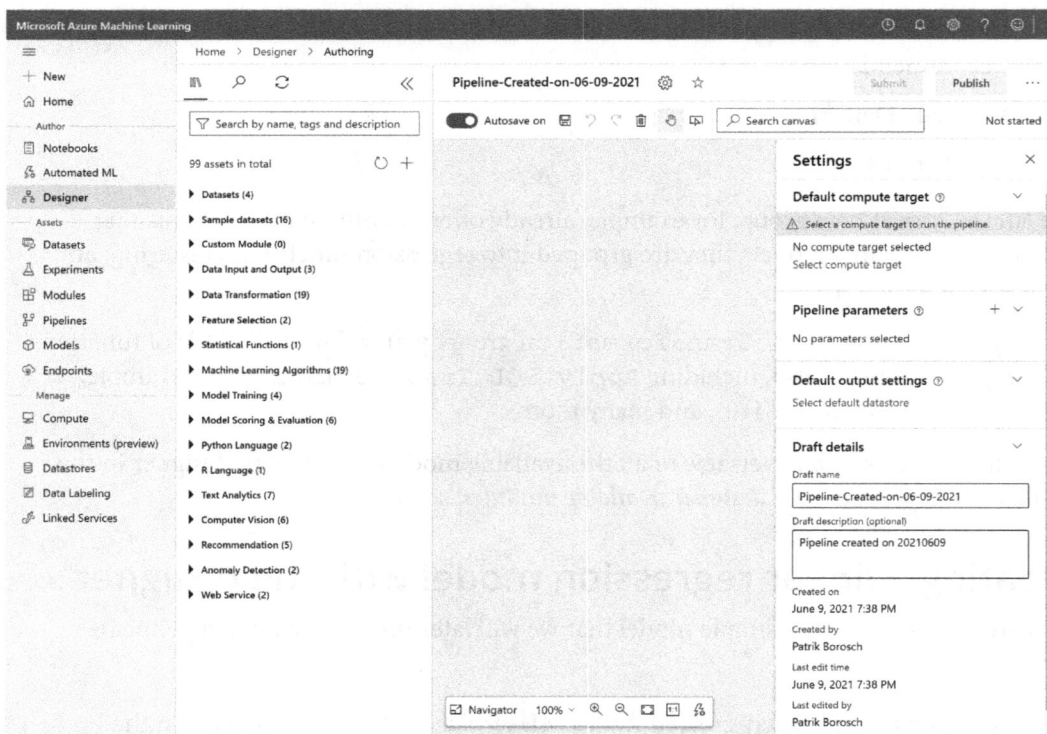

Figure 9.11 – Empty Designer canvas

7. In the upper-right area, **Default compute target** is highlighted in red. Click **Select compute target** and create a new one.

8. First, type `Import` into the search field of the **Asset navigator** field. Drag the **Import Data** asset onto the canvas. In the **Configuration** blade on the right, select the newly created datastore from *Step 2*.

9. In the **Path** field, type in the path to your source and the filename you want to use. Maybe use the `airdelays.csv` file from earlier exercises. Enter the path of your data lake without the data lake storage account name and the filesystem name of your data lake. It should look like this:

   `[YOURTOPLEVELFOLDER]/[YOURPATH]/.../source/airdelays.csv`

10. You can click on **Preview schema** to check the structure of the file.

11. As there are columns in the file that are completely empty, we can remove them from the set. Search for **Select Columns in Dataset** and select the following columns from your file:

    ```
    YEAR,MONTH,DAY_OF_MONTH,DAY_OF_WEEK,FL_DATE,UNIQUE_
    CARRIER,FL_NUM,ORIGIN_AIRPORT_ID,ORIGIN,ORIGIN_STATE_
    ABR,DEST_AIRPORT_ID,DEST,DEST_STATE_ABR,CRS_DEP_TIME,DEP_
    TIME,DEP_DELAY,DEP_DELAY_NEW,DEP_DEL15,DEP_DELAY_
    GROUP,CRS_ARR_TIME,ARR_TIME,ARR_DELAY,ARR_DELAY_NEW,ARR_
    DEL15,ARR_DELAY_GROUP,CANCELLED,DIVERTED,CRS_ELAPSED_
    TIME,ACTUAL_ELAPSED_TIME,FLIGHTS,DISTANCE,DISTANCE_GROUP
    ```

12. Now, search for the word `clean` and select **Clean Missing Data** on the canvas. Connect the output node from **Import Data** to the input node of **Clean Missing Data**.

13. In the **Configuration** tab on the right, in the **Columns to be cleaned** field, select **All columns** and then adjust **Cleaning mode** to **Remove entire row**.

14. In the next step, search for and add the **Split Data** asset to the pipeline. Connect the left output node from **Clean Missing Data** to the **Split Data** asset.

15. In the **Configuration** blade of the **Split Data** asset, set the **Fraction of rows** property to 0.7. This will split the incoming dataset into 70% and 30% portions. We will use 70% to train your model and the other 30% to validate it.

16. Now, let's get the machine learning algorithm that we want to use in this model. Search for `linear` and drag the **Linear Regression** asset onto the canvas.

17. Typically, we would train the model next. Search for train and add the **Train Model** asset to the canvas. From the **Linear Regression** algorithm, connect the output node to the left input node of the **Train Model** asset. From the **Split Data** asset, use the left output node and connect it to the right input node of the **Train Model** asset. On the **Configuration** blade on the right, you will need to select the column that you want the model to predict. Enter ARR_DELAY.

18. In the next step, we will score the trained model against the 30% data portion that the algorithm didn't yet see.

19. Search for the **Score Model** asset and add it to your pipeline.

20. Now, connect the output node of the Train Model asset to the left input node of the Score Model asset, and use the right output node of the Split Data asset to connect it to the right input of the Score Model asset.

21. The penultimate step involves an Evaluate Model asset. Search this one and add it to the canvas. The output of the Score Model asset goes to the left input of the Evaluate Model asset.

22. Finally, we will add an Export Data asset to the pipeline. Connect the output of the Score Model asset to the Export Data asset.

23. In the Configuration blade of the export data, select Azure Data Lake Storage Gen2 for the Datastore type. In the Datastore field, select the datastore that you created earlier and type in the path to your target. The path follows the same rules as in the Import Data asset from *Step 4*. For **File format**, select the format that you want. Perhaps use CSV as you can easily check this one. In *Figure 9.13*, you see the model as it is supposed to look.

24. As **Autosave** is switched on by default, the pipeline is already stored in your repository.

25. You can now click **Submit** to start the training of your machine learning model. In the dialog Set up pipeline run, you can click **Create new** and name it. Now, on the dialog, click **Submit** again and initiate the training. This will take a short while. You can watch the model as it is processed step by step:

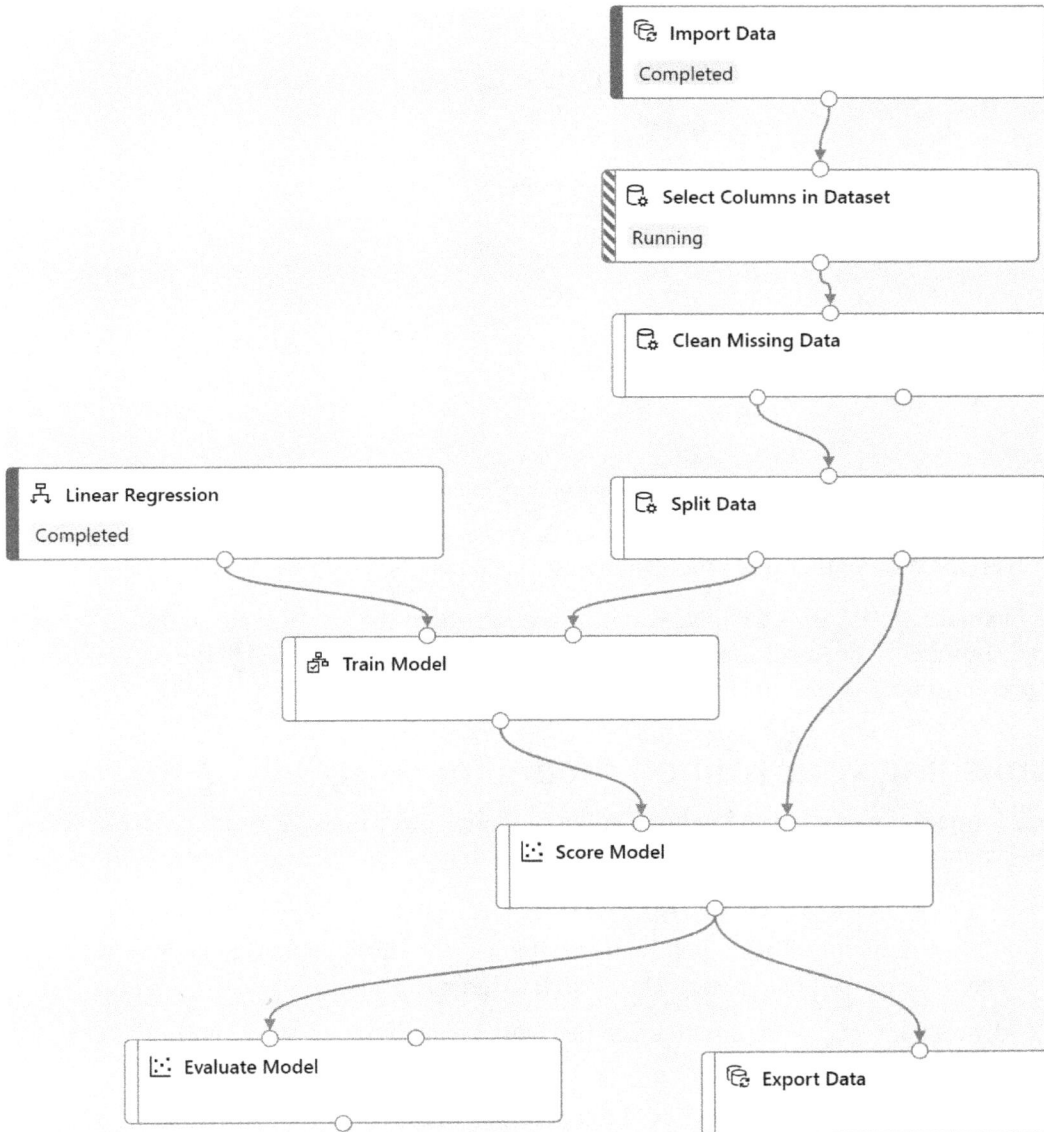

Figure 9.12 – Azure Machine Learning model created in the designer

As the pipeline finishes, you can navigate to your data lake (Synapse or Storage Explorer) and check whether the target file was created.

When you examine the assets and check the **Configuration** blade on the right, you can always examine the details and even the data that was handled by the single asset during the process:

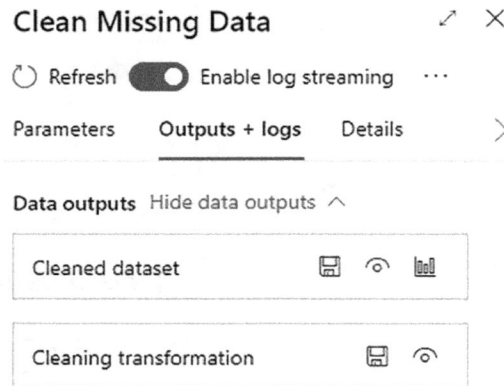

Figure 9.13 – Options in the Configuration blade

When you click the *eye* symbol in the artifacts, you can examine additional information and even the dataset that was produced by the particular asset.

The **Evaluate Model** asset will give you information about the metrics of the model. These are the numbers (the lower the better) a data scientist will look at to judge the accuracy of the prediction of the model.

Publishing your trained model for usage

There is one more step needed to publish the model in a way that you can then use it with your Synapse pipelines; for example, you will need to create an inference pipeline from your already trained model:

1. Click **Create inference pipeline** in the upper-right area of your **Designer** canvas. From the drop-down menu, select **Batch inference pipeline**.

2. A new version of your pipeline will be created in a second window. From there, you can click **Publish**.

3. In a dialog, you will be prompted to either use an existing or create a new pipeline endpoint. Create a new one and click **Publish**:

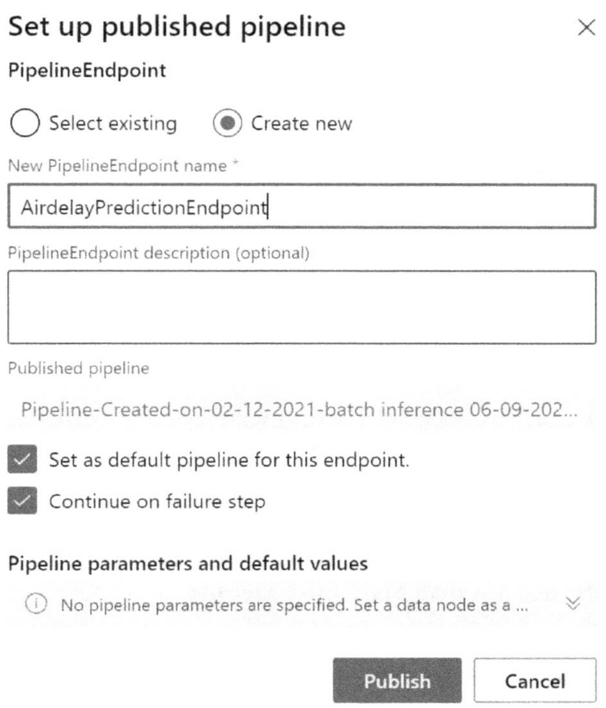

Figure 9.14 – Setting up an endpoint for a published pipeline

4. Once the pipeline has been published successfully, you will see a link in the upper area of the Designer canvas. Either click on that or go to Endpoints and select the newly created one. Note the endpoint name and the pipeline endpoint ID for later.

At this point, you have trained a machine learning model and published it with an endpoint so that you can use it from other services such as Synapse pipelines, for example. In the next section, *Using Azure Machine Learning with your modern data warehouse*, you will do exactly this. You will create a data pipeline and trigger the Azure ML model that you have created.

Using Azure Machine Learning with your modern data warehouse

Machine learning models can help you in many situations to improve business processes. Customer churn, fraud detection, and machine failure predictions are examples of where machine learning can support you in finding answers to tricky questions in a way that you would not, or only with excessive effort, be able to find otherwise.

However, a machine learning model that is not integrated into your daily business routine or one that will only be processed by a specialist on an on-demand basis will not perform with the full efficiency that might be possible.

One of the advantages of Synapse pipelines (and, of course, the Azure Data Factory standalone version as well) is the tight integration with other Azure services. Azure Machine Learning is one of them. Let's use our model from above and integrate it with an Azure pipeline. This will enable you to integrate Azure ML with all the data that you land in your data lake.

Connecting the services

To be able to use Azure Machine Learning within your Synapse workspace, you will first need to add your Azure ML service as a linked service to your Synapse workspace:

1. On the **Management** hub in Synapse Studio, go to **Linked Services** and click **+ New**.

2. **For New linked service**, select **Compute** from the tabs above the **Connectors** list and, in the results, select **Azure Machine Learning**.

3. Name your linked service and select your subscription and the **Azure_Machine Learning Workspace** name from previously.

4. This is now the place where you need the service principal ID and the secret key that you generated in the *Examining Azure Machine Learning studio* section. Click **Create** to store your new linked service.

Now you are ready to integrate your machine learning model with a Synapse pipeline. Let's move on and navigate to the **Integration** hub in your Synapse workspace and start to create a new pipeline:

1. In the **Assets** blade, click the + symbol next to the **Integrate** header and select **Pipeline**. On the **Properties** blade on the right, you can name your new pipeline (maybe something like `BatchscoringAzureML`).

2. From **Activities**, open the **Machine Learning** node at the bottom of the list and drag a **Machine Learning Execute Pipeline** activity to the canvas. The **Settings** blade will open below the canvas and you can start configuring it.

3. Give the activity a name that will tell you what happens here. Then, switch to the **Settings** tab.

4. In the Azure Machine Learning linked service, select the Azure ML service that you created as you worked through this chapter.

5. In the Machine Learning pipeline ID type, you want to select the pipeline endpoint ID, and below that, for the machine learning pipeline endpoint name, select the endpoint that you created earlier. Then, select the highest pipeline version and set the Experiment name that you chose during model training.

6. Then, click **Publish all** and store your work.

This is the minimal setup that you need to batch score data (`airdelays.csv`) in your source folder and harvest the result in your target folder. You could now choose to add **Copy Data** or other activities that would land the data in your source folder or get the results from the target folder and do something with it.

Perhaps you want to delete the result file in the target folder from the test run when you have trained your model. And then, finally, let's trigger the pipeline now and see what happens. Above the pipeline, click **Add trigger** and select **Trigger now**. On the following dialog, just click **OK**. The pipeline will start running now and you can monitor it on the **Monitor** hub:

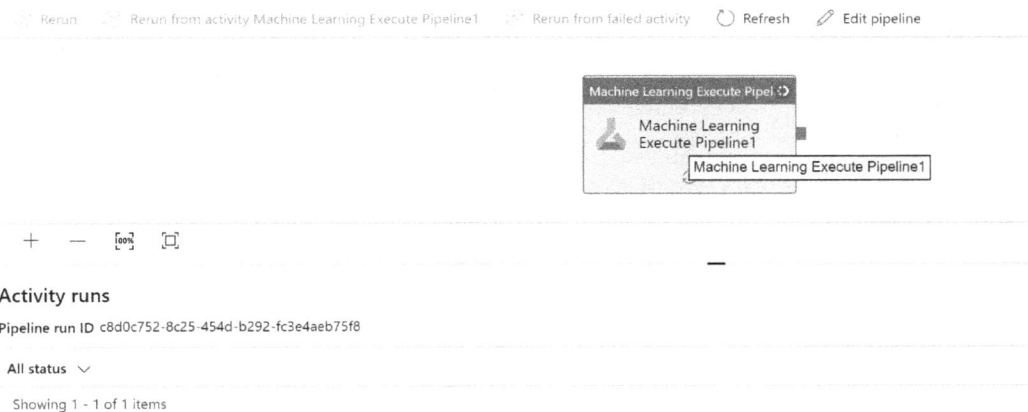

Figure 9.15 – Monitoring the pipeline run

Congratulations! You have created and trained a machine learning model in Azure Machine Learning and you have successfully integrated it with Synapse Studio.

Understanding further options to integrate Azure ML with your modern data warehouse

You have now seen how to integrate an Azure Machine Learning model with Synapse pipelines (Azure Data Factory will work accordingly). However, other options are available for integrating Azure ML into your modern data warehouse.

Embedding native prediction in Synapse dedicated SQL pools

When you have created Azure Machine Learning models and you can provide your trained model as a so-called **Open Neural Network Exchange (ONNX)** model, you have the option to import the model into a table in your Synapse dedicated SQL pool.

With an Azure ML linked service available in Synapse Studio, you will find an additional entry in the context menu when you right-click a table in your dedicated SQL pool: **Machine Learning** (you used it with the Spark table in the *Implementing your Text Analytics cognitive service in a Spark notebook* section of this chapter.)

In this case, with the dedicated SQL pool, you will need to have a table ready, which the model will be imported into. Once the ONNX model is present in your table, you will be able to use the T-SQL command `PREDICT()` with this model and integrate it, for example, with stored procedures in your database. You'll find more information on `PREDICT()` in the *Further reading, Native prediction in T-SQL*, section.

Using Azure Machine Learning compute from Synapse Spark

In your Spark notebooks, you have the option to connect to your Azure ML workspace. Libraries such as `azureml.core` and `azureml.train` will expose functions to interact with the workspace, its datastores and datasets, and, of course, models. Compute is also available and can be integrated into your model training.

One very interesting use case would be to create and train Auto ML models from within Synapse using Azure ML Auto ML capabilities. Refer to the *Further reading, Auto ML*, section for more information.

Summary

In this chapter, you have learned about several predefined cognitive services on Azure and how to use them in your modern data warehouse. You have created a Spark notebook and analyzed the sentiment of a given text with the Text Analytics cognitive service.

In the second part of the chapter, you examined the Azure Machine Learning service. This can be seen as one of the central services on Azure when it comes to the implementation of Machine Learning and AI.

You have learned about the different options that a data scientist finds in Azure ML and have implemented your own machine learning model using the graphical designer of Azure ML.

Finally, you connected Azure ML to your Synapse workspace and integrated the ML model with a Synapse pipeline.

Additionally, we discussed other options for integrating Azure ML with your modern data warehouse.

In the upcoming chapter, *Implementing the Presentation Layer with Synapse Analytics*, we will examine how to import, model, and deliver data to all the consuming tools, such as dashboarding and reporting tools or APIs.

Further reading

You will find additional material in the following section.

- **Text Analytics Language Detection**: `https://docs.microsoft.com/en-us/azure/cognitive-services/text-analytics/how-tos/text-analytics-how-to-language-detection`

- **Text Analytics Named Entity Recognition**: `https://docs.microsoft.com/en-us/azure/cognitive-services/text-analytics/how-tos/text-analytics-how-to-entity-linking?tabs=version-3-preview`

- **Text Analytics Key Phrases**: `https://docs.microsoft.com/en-us/azure/cognitive-services/text-analytics/how-tos/text-analytics-how-to-keyword-extraction`

- **Azure ML designer available modules**: `https://docs.microsoft.com/en-us/azure/machine-learning/algorithm-module-reference/module-reference`

- **Creating a service principal**: `https://docs.microsoft.com/en-us/azure/active-directory/develop/howto-create-service-principal-portal`

- **Native prediction in T-SQL**: `https://docs.microsoft.com/en-us/azure/synapse-analytics/machine-learning/tutorial-sql-pool-model-scoring-wizard`

- **Auto ML**: `https://docs.microsoft.com/en-us/azure/synapse-analytics/spark/apache-spark-azure-machine-learning-tutorial`

10
Loading the Presentation Layer

Now that you have seen the many modules that you have available on Azure to build your modern data warehouse architecture, let's dive into the Presentation layer in this chapter and the one following, *Chapter 11, Developing and Maintaining the Presentation Layer*.

In this chapter, you are going to learn how to load data into your Presentation Layer using either PolyBase, the COPY command, or, alternatively, Synapse pipelines.

You will see how you can include Synapse serverless SQL pools and examine the benefits of using this tool, implementing SQL directly in your data lake.

Additionally, we will investigate how you can leverage Spark to extend the functionality of your data loads beyond the SQL language.

Finally, we will examine the options in terms of Synapse regarding metadata exchange between the different compute components and how this will increase developer efficiency.

The topics covered in this chapter are as follows:

- Understanding data loading with Synapse-dedicated SQL pools
- Loading data into Synapse-dedicated SQL pools
- Using Synapse serverless SQL pools

- Integrating data with Synapse Spark pools
- Exchanging metadata between computes

Technical requirements

If you want to build the artifacts discussed in this chapter, you will need the following:

- A Synapse workspace with a `System Administrator` RBAC role (see *Chapter 4, Understanding SQL Pools and SQL Options*)
- Azure Data Lake with a `STORAGE_BLOB_DATA_CONTRIBUTOR` role for you and your Synapse workspace managed identity (see *Chapter 3, Understanding the Data Lake Storage Layer*)

Understanding the loading strategy with Synapse-dedicated SQL pools

The different options that you have available for the table design of a dedicated SQL pool, distributed or replicated tables, and the decision regarding the use of column stores or heaps and partitioning on top will influence the way in which you load data into it.

Certainly, loading into a hash-distributed table can be quite a quick process. But when you consider the additional compute step to calculate the hash keys to distribute the incoming rows to their target distribution and compare it to a round-robin-distributed table, where this step is not required, you can imagine that loading data into the latter will be faster.

Another consideration for a staging table in a dedicated SQL pool would be to use heap tables instead of column store ones. Again, you can avoid additional compute overhead for the column store and load data quickly.

In the end, it all comes down to performance. Therefore, following the arguments presented previously, you will first want to bring your data into a round-robin-distributed heap table. This will guarantee quick loading into the database from outside.

If you are working on smaller datasets, you might even stop here and use round-robin-distributed heap tables throughout your model. They will always ensure sound performance.

When you are working on massive datasets, however, you may need to use hash-distributed tables to gain query performance. This will then add another step to your loading sequence.

From your round-robin heap staging table to a hash-distributed table, dedicated SQL pools offer an additional language element: CREATE TABLE AS SELECT. This command uses all the available worker nodes that the database offers at the time of issuing the command. With this, you will benefit from all the parallel power that the database can offer and again achieve optimal performance.

In this step, you will then have another option in terms of working on your data warehouse model. You can take care of historization and versioning, for example, and establish your Kimball model or data vault:

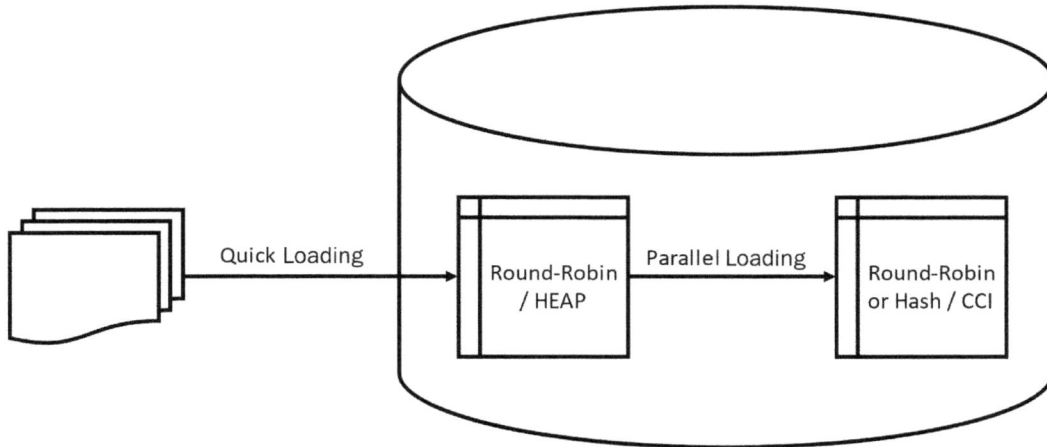

Figure 10.1 – Loading strategy for dedicated SQL pools

In the next section, *Loading data into Synapse-dedicated SQL pools*, we will examine how to physically get data into your round-robin heap staging table.

Loading data into Synapse-dedicated SQL pools

With a rich pedigree established by the Microsoft Parallel Data Warehouse appliance, which was the grandfather of Azure SQL Data Warehouse, Synapse-dedicated SQL pools benefit from a mature **massively parallel processing** (**MPP**) technology that implements some very useful pieces of technology.

For example, there is PolyBase, which will enable your dedicated SQL pool to either load data from your data lake into the SQL pool or query the data from where it is stored in your data lake without the need to transport it into the pool.

Another option is to use the COPY statement. This SQL feature was added to speed up loading data from your data lake into your SQL pool.

The third option for bringing data into your dedicated SQL pool is, of course, to use either Synapse pipelines or Azure Data Factory (see *Chapter 5, Integrating Data in Your Modern Data Warehouse*).

Examining PolyBase

PolyBase is a feature that allows the reading and writing of data from and to file stores using all the parallelism of the worker nodes that the service has configured at the time of the connection.

This engine comes with a **Data Movement service** that is available on every compute node that is configured in the dedicated SQL pool. This allows parallel loading from sources that PolyBase can connect to. The data is loaded, bypassing the control node, and directly routed over the worker nodes into the storage, which makes the architecture very fast for data loading.:

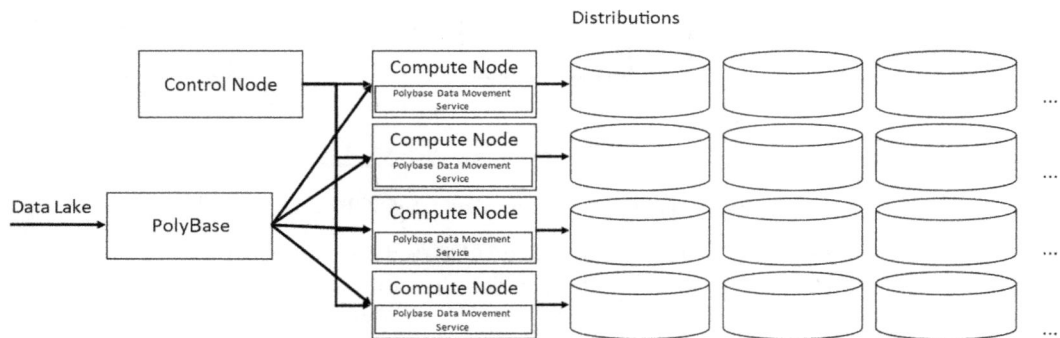

Figure 10.2 – PolyBase architecture schema

You can use PolyBase in different use cases:

- Loading data from your data lake into your dedicated SQL pool

- Reading and joining data from your data lake with data from your dedicated SQL pool using SQL while leaving it in your data lake

But before we start with your first PolyBase experience, let's examine the rights that you need to connect to data in your data lake.

Understanding rights setup for PolyBase

When you, as an administrator, want to set up a PolyBase external table, you will need the following rights configuration to succeed:

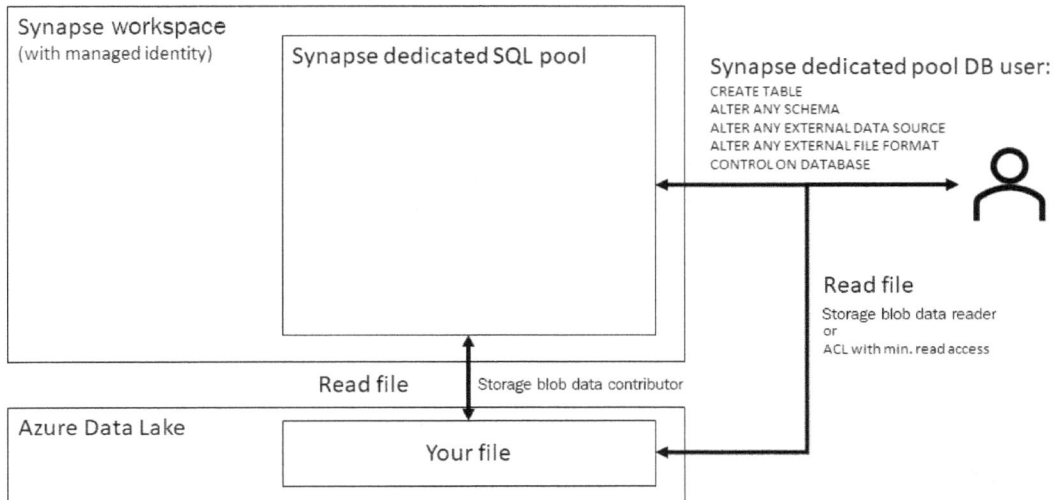

Figure 10.3 – Rights configuration; creating an external table

Now, as you have an initial idea of how security works, let's get connected to your data lake in the *Creating a PolyBase external table* section.

Creating a PolyBase external table

To access your data in your data lake with PolyBase, you will use the concept of an **external table** in your dedicated SQL pool.

The external table represents either a file or a folder in your data lake. You will be able to query it from within your dedicated SQL pool just as if it was a SQL table. You can join it to other tables or views, you can write to it, and you can create views on it. You can even set up row-level security based on the data in the data lake.

The construct of a PolyBase external table comes with some additional objects that you will require to set it up:

- **DATABASE SCOPED CREDENTIAL:** This object will be used to access the data lake. You will need to create a master key if one doesn't yet exist in your database. That will encrypt the credential. After that, you will create your database-scoped credential.

- **EXTERNAL DATA SOURCE**: This object will add the location of your data to your system. Here, you will set your data lake and the type of connection, and you will use the database-scoped credential from the preceding bullet point.

- **EXTERNAL FILE FORMAT**: Using EXTERNAL FILE FORMAT will add information about the files to be read. You will name the format type, for example, DELIMITEDTEXT, RC (Hive RC), ORC (Hive ORC), Parquet, or JSON, the compression type (Gzip or default), and format options such as FIELD_TERMINATOR and STRING_DELIMITER.

- **EXTERNAL TABLE**: This object combines the preceding objects into a selectable table object. As is the case with other tables, you will need to name the column structure and the data types for the columns.

Creating the database-scoped credential

To start the sequence, you will first need to create a master key in your dedicated SQL pool. In your Synapse workspace, please navigate to the **Development** hub. Here, you can start a new SQL script and enter the following:

```
CREATE MASTER KEY;
```

> **Note**
> Dedicated SQL pools don't require a password for the master key. But if you do enter one, please ensure that you put it somewhere safe and that you remember it.

The same command with a password would look like this:

```
CREATE MASTER KEY ENCRYPTION BY PASSWORD = '[YOURPASSWORD]'
```

This key will now be used to encrypt your credentials, which you will create with the following command:

```
CREATE DATABASE SCOPED CREDENTIAL [YOURCREDENTIALNAME]
WITH IDENTITY = 'Managed Service Identity';
```

As Synapse will use its managed service identity and will pass through the credentials of the user who is querying the data lake, you need to set Managed Service Identity in the database-scoped credential.

Creating an external data source

In the next step, you will create an external data source. You will provide a type (with your data lake, this will always be "Hadoop") and a location using the **Azure Blob File System (ABFS)** driver. This one is explicitly optimized for use with Data Lake Storage Gen2.

> **Note**
>
> Please don't forget to replace the parts in brackets, [ ], in all the statements in this chapter with your own paths or names and so on.

The command to create your external data source looks like this:

```
CREATE EXTERNAL DATA SOURCE [YOUREXTERNALDATASOURCENAME]
WITH
(
    TYPE = hadoop
    , LOCATION = 'abfss://[YOURFILESYSTEMNAME]@
[YOURDATALAKEACCOUNT].dfs.core.windows.net'
    , CREDENTIAL = [YOURCREDENTIALNAME!
);
```

For all the possible options when creating an **external data source**, please check the *Further reading*, *EXTERNAL DATA SOURCE* section.

Creating an external file format

Now we are already at the point where we are teaching the database how the file(s) that we want to query is/are formatted.

In this case, let's load the .csv file that we created when we predicted the air delays in the chapter on machine learning (see *Chapter 9*, *Integrating Azure Cognitive Services and Machine Learning*).

Now, the command to create your **external file format** looks like this:

```
CREATE EXTERNAL FILE FORMAT [YOUREXTERNALFILEFORMATNAME]
WITH
(
    FORMAT_TYPE=DELIMITEDTEXT
    , FORMAT_OPTIONS (
        FIELD_TERMINATOR = ','
```

```
        , USE_TYPE_DEFAULT = TRUE
    )
);
```

Please review the *Further reading*, *EXTERNAL FILE FORMAT* section for the other possible options to influence the file formats when you need to read different files from your data lake.

Creating an external table

Now, as we have created all the necessary objects, we can start to create the **external table**. As mentioned previously, this object will act as a pointer to your file in your data lake.

You will need to name the columns and their data types to make them available for the database engine. Additionally, you will use the artifacts created previously to marry the structure, data format, and location into one object to make it available for querying. The basic command should look like this:

```
CREATE EXTERNAL TABLE [YOUREXTERNALTABLENAME]
(
    [YEAR]                VARCHAR(20)
    , [MONTH]             VARCHAR(20)
    , DAY_OF_MONTH        VARCHAR(20)
    , DAY_OF_WEEK         VARCHAR(20)
    , FL_DATE             VARCHAR(30)
    , UNIQUE_CARRIER      VARCHAR(20)
    , FL_NUM              VARCHAR(20)
    , ORIGIN_AIRPORT_ID   VARCHAR(20)
    , ORIGIN              VARCHAR(20)
    , ORIGIN_STATE_ABR    VARCHAR(20)
    , DEST_AIRPORT_ID     VARCHAR(20)
    , DEST                VARCHAR(20)
    , DEST_STATE_ABR      VARCHAR(20)
    , CRS_DEP_TIME        VARCHAR(30)
    , DEP_TIME            VARCHAR(30)
    , DEP_DELAY           VARCHAR(20)
    , DEP_DELAY_NEW       VARCHAR(20)
    , DEP_DEL15           VARCHAR(20)
    , DEP_DELAY_GROUP     VARCHAR(20)
```

```
    , CRS_ARR_TIME        VARCHAR(20)
    , ARR_TIME            VARCHAR(20)
    , ARR_DELAY           VARCHAR(20)
    , ARR_DELAY_NEW       VARCHAR(20)
    , ARR_DEL15           VARCHAR(20)
    , ARR_DELAY_GROUP     VARCHAR(20)
    , CANCELLED           VARCHAR(20)
    , DIVERTED            VARCHAR(20)
    , CRS_ELAPSED_TIME    VARCHAR(20)
    , ACTUAL_ELAPSED_TIME VARCHAR(30)
    , FLIGHTS             VARCHAR(20)
    , DISTANCE            VARCHAR(20)
    , DISTANCE_GROUP      VARCHAR(20)
    , Scored_Labels       VARCHAR(50)
)
WITH
(
    LOCATION = '/[YOURDATALAKEPATH]/airdelayspredict.csv'
    , DATA_SOURCE = [YOUREXTERNALDATASOURCENAME]
    , FILE_FORMAT = [YOUREXTERNALFILEFORMATNAME]
);
```

Now you should be able to query your external table:

```
select top 100 *
from [YOUREXTERNALTABLENAME]
```

Did you get a result set back?

> **Note**
>
> You may have recognized that the data types in the EXTERNAL TABLE
> statement are all set to VARCHAR(xx). This is a method to ensure that all
> rows can be loaded into your database. There, you can handle them according
> to your requirements. You may want to choose other data types according to
> the data that comes from your files. In this case, you want to add REJECT_
> ROW_LOCATION and other REJECT_ options to your EXTERNAL TABLE
> statement. These settings will handle rows that don't fit into the column
> structure during queries and will write them to an additional file where you can
> analyze them and maybe even reload them after correction.

To find more information on the REJECT_ options of the CREATE EXTERNAL TABLE statement, please review the *Further reading, EXTERNAL TABLE* section.

> **Note**
>
> When you examine the CREATE EXTERNAL TABLE statement and check the LOCATION option, you see the explicit file that we want to connect named there. The LOCATION option can also point to a folder. In this case, you need to take care that the files stored in that folder are all the same structure. This is a handy feature when you need to merge a set of files into one table.

Now that you have established a connection to your data in your data lake, you can decide how you want to use the data in your MDWH setup. You can leave the data where it is and query your external table together with other tables in your dedicated SQL pool. Keep in mind that this might come with a performance debt with the roundtrip to the data lake.

You can otherwise decide to load the data from your data lake into your dedicated SQL pool. You can do this, for example, by using the SQL INSERT INTO command. This will perform quite well on smaller datasets and delta loads. For initial loads and bigger volumes, dedicated SQL pools offer the CREATE TABLE AS SELECT (**CTAS**) statement.

Using CREATE TABLE AS SELECT (CTAS)

The CTAS statement was introduced with the MPP technology to make use of the parallelism that dedicated SQL pools offer with their worker nodes. As it is a minimally logged operation, it will not only gain performance by leveraging the power of all available worker nodes but will also speed up by avoiding excessive logging. This, on the other hand, makes it an all-or-nothing operation.

CTAS will always create a new table. This is another fact that you need to remember when you plan your loading sequence. You will need to create your new table using the CTAS statement and, after that, rename the old table and then the newly created one. These operations are quite quick as they are metadata operations (refer to the *Renaming a table in a Synapse-dedicated SQL pool* section).

> **Important note**
>
> When loading data using PolyBase external tables, there is a 1 MB limit on the row size that you can load into your dedicated SQL pool. Additionally, the number of files that you can load in a single attempt is 1,000,000. When you try to load from more than 1,000,000 files, you will get the following error: `Operation failed as split count exceeding upper bound of 1000000.`

Now, let's finalize our example and bring in the data from the external table that you created in the *Creating an* external table section. The statement looks like this:

```
CREATE TABLE [YOURSTAGINGTABLE]
WITH (
    DISTRIBUTION=ROUND_ROBIN
    , HEAP
)
AS
    SELECT *
    FROM [YOUREXTERNALTABLENAME];
```

Now that was easy. And when you now query the newly created table, you will query the data from within your database:

```
select *
from [YOURSTAGINGTABLE];
```

Renaming a table in a Synapse-dedicated SQL pool

As mentioned previously (in the *Using CREATE TABLE AS SELECT (CTAS)* section), you will create a new table using the CTAS statement. But when you are loading in a running system, the target table will most probably be there already. In this case, your loading sequence will be as follows:

1. CTAS your data into a new table, [YOURSTAGINGTABLE_new].
2. Rename your old table [YOURSTAGINGTABLE_old].
3. Rename your new table with the target name [YOURSTAGINGTABLE].
4. Get rid of your old table.

The RENAME statement works as follows:

```
RENAME OBJECT [YOURSCHEMANAME].[YOURTABLENAME_
new] TO [YOURTABLENAME];
```

> **Note**
>
> Don't use the schema.table notation for the target name in the RENAME OBJECT statement. This will result in an incorrect target name, such as dbo.dbo.mytablename. The target name comes without any additional information. The database and schema will be derived from the source table in the RENAME OBJECT statement.

You'll find a complete example of a load using a PolyBase external table, including renaming, in the GitHub repository that belongs to this book.

Loading data into a dedicated SQL pool using COPY

Unlike PolyBase external tables, the **COPY** statement will not establish a permanent connection to a file or a folder in your data lake.

COPY will simply load data from your data lake file or folder into a table in your dedicated SQL pool. It does not require additional objects, such as EXTERNAL DATA SOURCE or EXTERNAL FILE FORMAT, as PolyBase does, and another big advantage is the lower privileges needed to run a COPY statement with your dedicated SQL pool.

The COPY statement even has some more advantages. One of them is that you can name different files and wildcards in your location paths. Others include the fact that you can use custom row terminators, for example, or use your own default values for columns in your target and specify the source-to-target column mappings for your loads.

The COPY statement supports the following file types:

- CSV
- Parquet
- ORC

Understanding rights setup for COPY INTO

With the COPY statement, you are in another rights and security setup compared to PolyBase. You can load data from the following sources:

- Azure Blob storage
- Azure Data Lake Storage Gen2

Depending on the storage type and the file type, you will need different authentication types and different drivers to connect to the storage and load the files:

Storage Type	File Format	Authentication
Blob storage	CSV	Service Principal Managed Service Identity Shared Access Signature Storage Account Key Azure Active Directory
	Parquet	Shared Access Signature Storage Account Key
	ORC	Shared Access Signature Storage Account Key
Data Lake Gen2	CSV	Service Principal Managed Service Identity Shared Access Signature Storage Account Key Azure Active Directory
	Parquet	Service Principal Managed Service Identity Shared Access Signature Storage Account Key Azure Active Directory
	ORC	Service Principal Managed Service Identity Shared Access Signature Storage Account Key Azure Active Directory

Figure 10.4 – Storage types and authentication methods

When you are reading Parquet or ORC files from your Data Lake Gen2 account, please note the following:

- Shared acces signature authentication needs to use the `.blob.core.windows.net` endpoint.

- Managed service identity authentication needs to use the `.dfs.core.windows.net` endpoint.

> **Note**
>
> Performance-wise, the `.blob` endpoint turns out to be the faster alternative. So, if you don't require the `.dfs` one because of authentication limitations, you could consider the `.blob` one.

When you check the syntax of the COPY statement (please refer to the *Further reading*, *COPY INTO* section for a full overview of the syntax, arguments, and more), you will find an argument (**CREDENTIAL (Identity = ", SECRET = ")**) whose setting depends on the authentication method that you choose:

Authentication Method	Identity	SECRET	RBAC or permission required
Shared Access Signature	Shared Access Signature (constant)	The Shared Access Signature	READ + LIST
Service Principal	OAuth 2.0 Token Endpoint	The Application Service Principal Key (AAD)	Storage Blob Data Contributor Or Storage Blob Data Owner Or Storage Blob Data Reader
Storage Account Key	Storage Account Key (constant)	The Storage Account Key	
Managed Identity	Managed Identity (constant)		Storage Blob Data Contributor Or Storage Blob Data Owner
Azure Active Directory			Storage Blob Data Contributor Or Storage Blob Data Owner

Figure 10.5 – Authentication methods and required inputs

Now, let's get going and load data into your dedicated SQL pool using a COPY statement. First, we need to have a table to load the data into:

```
CREATE TABLE [YOURCOPYSTAGINGTABLE]
(
    [YEAR]                 VARCHAR(20)
,   [MONTH]                VARCHAR(20)
,   DAY_OF_MONTH           VARCHAR(20)
,   DAY_OF_WEEK            VARCHAR(20)
,   FL_DATE                VARCHAR(30)
,   UNIQUE_CARRIER         VARCHAR(20)
,   FL_NUM                 VARCHAR(20)
,   ORIGIN_AIRPORT_ID      VARCHAR(20)
,   ORIGIN                 VARCHAR(20)
,   ORIGIN_STATE_ABR       VARCHAR(20)
,   DEST_AIRPORT_ID        VARCHAR(20)
,   DEST                   VARCHAR(20)
,   DEST_STATE_ABR         VARCHAR(20)
,   CRS_DEP_TIME           VARCHAR(30)
,   DEP_TIME               VARCHAR(30)
,   DEP_DELAY              VARCHAR(20)
,   DEP_DELAY_NEW          VARCHAR(20)
,   DEP_DEL15              VARCHAR(20)
,   DEP_DELAY_GROUP        VARCHAR(20)
,   CRS_ARR_TIME           VARCHAR(20)
,   ARR_TIME               VARCHAR(20)
,   ARR_DELAY              VARCHAR(20)
,   ARR_DELAY_NEW          VARCHAR(20)
,   ARR_DEL15              VARCHAR(20)
,   ARR_DELAY_GROUP        VARCHAR(20)
,   CANCELLED              VARCHAR(20)
,   DIVERTED               VARCHAR(20)
,   CRS_ELAPSED_TIME       VARCHAR(20)
,   ACTUAL_ELAPSED_TIME    VARCHAR(30)
,   FLIGHTS                VARCHAR(20)
,   DISTANCE               VARCHAR(20)
,   DISTANCE_GROUP         VARCHAR(20)
```

```
    , Scored_Labels        VARCHAR(50)
)
WITH
(
    DISTRIBUTION = ROUND_ROBIN
    , HEAP
);
```

And then we issue the COPY INTO statement:

```
COPY INTO [YOURCOPYSTAGINGTABLE]
FROM 'https://[YOURDATALAKEACCOUNT].dfs.core.windows.net/
[YOURPATH]/ airdelayspredict.csv'
WITH
(
    FILE_TYPE = 'CSV'
    ,MAXERRORS = 0
    ,FIELDTERMINATOR = ','
    ,FIRSTROW = 2
);
```

Please note that we didn't use the CREDENTIALS argument here. This statement worked using credential passthrough and pushes your Azure Active Directory credentials to your Data Lake account. See *Figure 10.5, Authentication methods and required inputs.*

And again, you can now select the content of the table that has just been loaded like this:

```
select *
from [YOURCOPYSTAGINGTABLE];
```

As mentioned previously, one of the advantages of the COPY statement is the option to read from multiple folders or files in one statement. You just need to name the different locations, separated by a comma:

```
...
FROM 'https://[YOURDATALAKEACCOUNT].dfs.core.windows.net/
[YOURPATH]/airdelayspredict.csv',
'https://[YOURDATALAKEACCOUNT].dfs.core.windows.net/
[YOURALTERNATIVEPATH]/airdelayspredict2.csv',
'https://[YOURDATALAKEACCOUNT].dfs.core.windows.net/
[YOURADDITTIONALPATH]/*'
...
```

Another advantage of the COPY statement, unlike the CTAS statement that we used previously, in the *Using CREATE TABLE AS SELECT (CTAS)* section, is the fact that we didn't need to perform the rename actions. The `COPY INTO` statement appends data into your table, whereas the CTAS statement always creates a new table.

But what if you don't want to code SQL to load data into your dedicated SQL pool? Let's examine the next section, *Adding data with Synapse pipelines/Data Factory*, where we use a pipeline with a copy activity to land data in our dedicated SQL pool.

Adding data with Synapse pipelines/Data Factory

There is another way of landing data into dedicated SQL pools instead of coding the elements and the load itself.

In *Chapter 5*, *Integrating Data in Your Modern Data Warehouse*, we examined the capabilities of Synapse pipelines and Azure Data Factory. Perhaps you remember the copy activity and the sink options that it offers. And yes, dedicated SQL pools are one of them.

Let's rebuild the previous case with a Synapse pipeline and a copy activity:

1. Please navigate to your Synapse workspace and to the integration hub.

2. Now, please click the + button and select **Copy Data tool**:

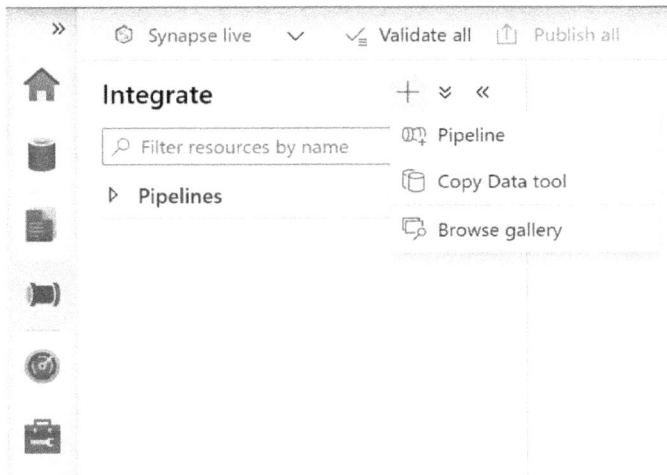

Figure 10.6 – Starting a copy pipeline

3. On the first blade, select a **task name** and leave the **Task cadence/schedule** field at **Run once now**. Then, click **Next >** to go to the next page of the wizard.

4. On the **Source data store** page of the wizard, please either select your Data Lake connection if it is there already or create a new one. Please refer to *Chapter 5, Integrating Data in Your Modern Data Warehouse*, if you don't remember how to create it. Then, click **Next >** to proceed to select the source file.

5. You are then taken to the **Choose the input file or folder** page, where you can click **Browse** next to the **File or folder** input field to start your selection. As we are only using one file, you can also uncheck the **Recursively** option:

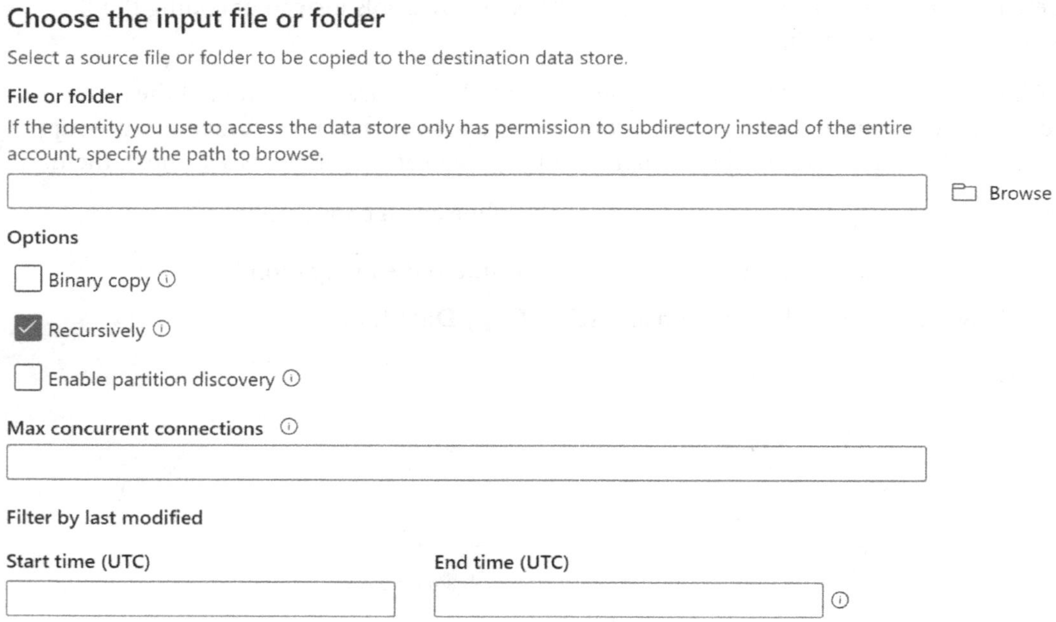

Choose the input file or folder

Select a source file or folder to be copied to the destination data store.

File or folder

If the identity you use to access the data store only has permission to subdirectory instead of the entire account, specify the path to browse.

[   ]   📁 Browse

Options

☐ Binary copy ⓘ

☑ Recursively ⓘ

☐ Enable partition discovery ⓘ

Max concurrent connections ⓘ

[   ]

Filter by last modified

Start time (UTC)   **End time (UTC)**

[   ]   [   ] ⓘ

Figure 10.7 – Choosing the input file or folder

6. When you click **Next >**, you are taken to **File format settings**. You'll see a data preview in the lower section of the page and file options above. If the content is not displayed correctly in the preview, try and adjust the format settings here. Then, click the **Next >** button once again.

7. You are then taken to the **Destination data store** selection page. In *Chapter 5, Integrating Data in Your Modern Data Warehouse*, you already created a connection to your dedicated SQL pool. Let's reuse that one now and then click **Next >**.

8. On the **Table mapping** page, you can now either type in a new name for a table to be created in your dedicated SQL pool or you can click the **Use existing table** link below the input field and change the view to a selector, where you can choose an existing table from your dedicated pool. Let's create a new one and name it dbo . airdelayprediction, as in *Figure 10.8*:

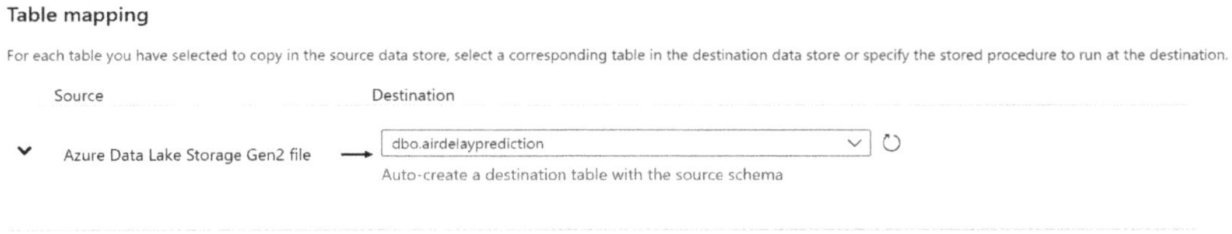

Table mapping

For each table you have selected to copy in the source data store, select a corresponding table in the destination data store or specify the stored procedure to run at the destination.

Source	Destination
∨ Azure Data Lake Storage Gen2 file ⟶	dbo.airdelayprediction ∨ ↻ Auto-create a destination table with the source schema

Figure 10.8 – Table mapping: creating a new table

9. Go to **Column mapping** by clicking **Next >**. You are now presented with the interface to map columns from your source to your target. When you create a new table, this is the place where you can influence the column names for your new table.

Below the column mapping, you'll find additional options, including **Pre-copy script**, which will be fired before the table gets loaded. Further on in the **Advanced** section, you have inputs for **Write batch timeout**, **Write batch size**, and **Max concurrent connections**. These are options that come in handy when you have several sessions writing in parallel to your target and resources are limited. You can proceed by clicking the **Next >** button:

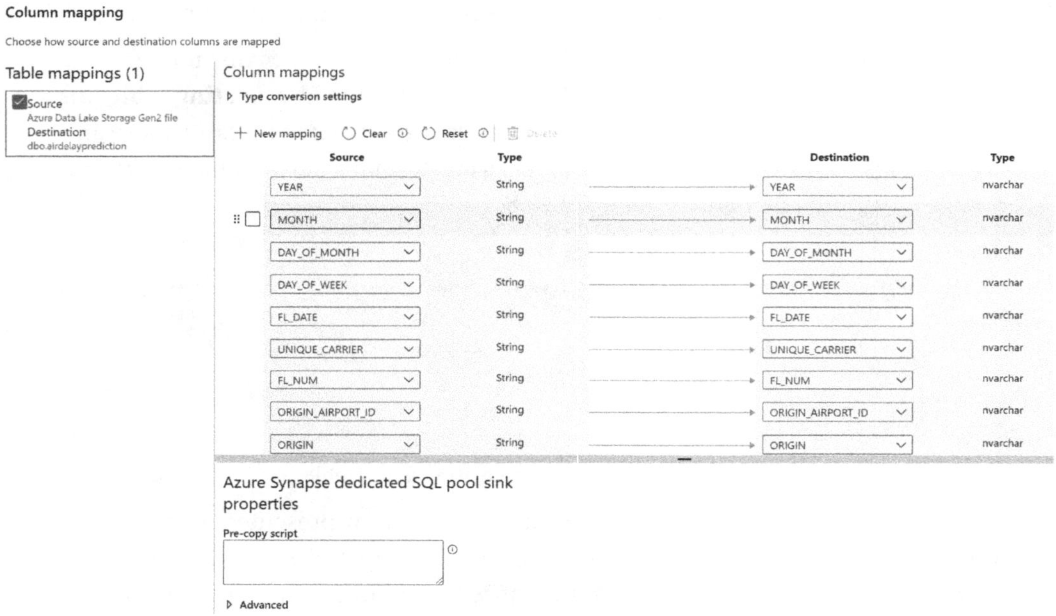

Figure 10.9 – Column mapping

10. On the penultimate page of the wizard, you can adjust the **Settings** options for your load. This is the page where you are going to configure the usage of **PolyBase**, for example:

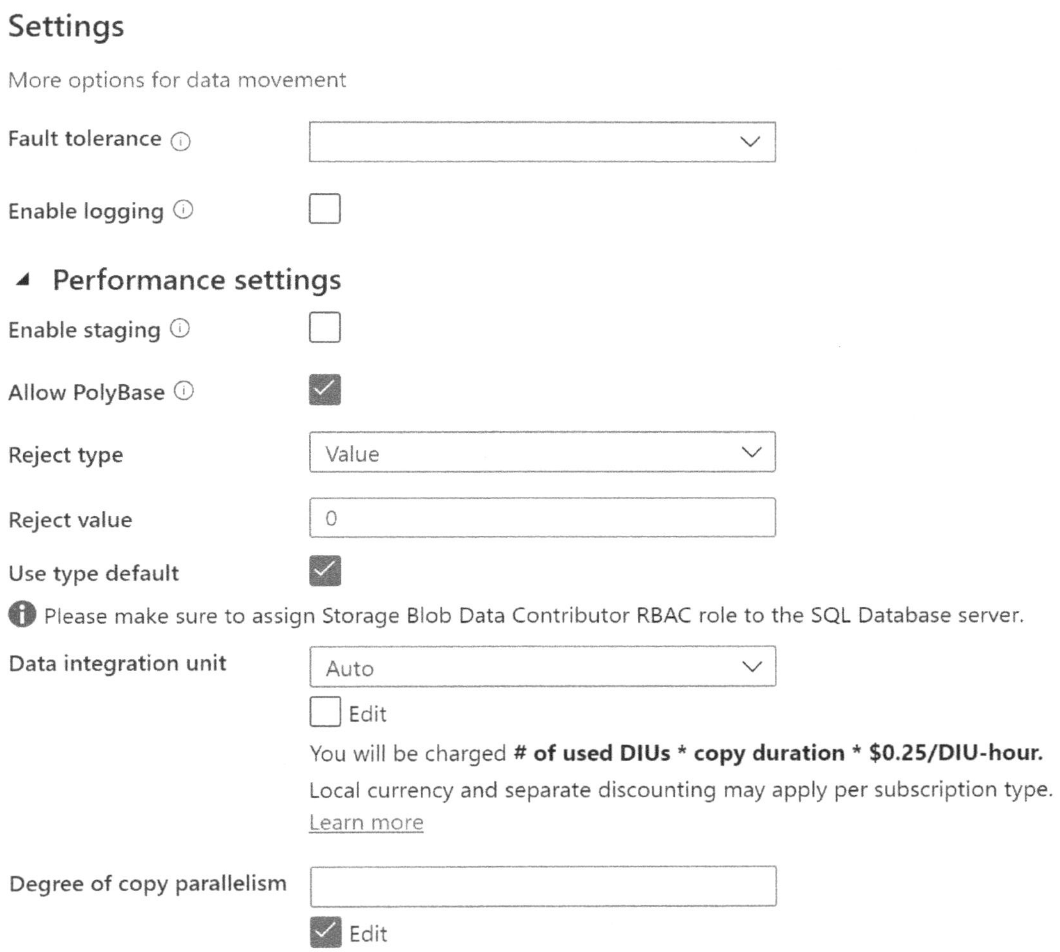

Settings

More options for data movement

Fault tolerance ⓘ   [⌄]

Enable logging ⓘ   [ ]

◢ **Performance settings**

Enable staging ⓘ   [ ]

Allow PolyBase ⓘ   [✓]

Reject type   [Value   ⌄]

Reject value   [0   ]

Use type default   [✓]

ⓘ Please make sure to assign Storage Blob Data Contributor RBAC role to the SQL Database server.

Data integration unit   [Auto   ⌄]

                            [ ] Edit

You will be charged **# of used DIUs * copy duration * $0.25/DIU-hour.**
Local currency and separate discounting may apply per subscription type.
Learn more

Degree of copy parallelism [   ]

                            [✓] Edit

Figure 10.10 – The Settings page

The **Fault tolerance** drop-down field, for example, will offer you options to **Select all rows** or to **Skip incompatible rows** but keep the copy job running when there are errors.

Enable logging will extend your log with the filenames that are copied as well as the rows and error rows that occur during your load.

Performance settings shows **Enable staging** for the **PolyBase** option. Checking **Enable staging** will display another selector where you will need to name a Data Lake folder or a storage account that will be used to stage files when you are reading from other systems.

> **Important note**
>
> The staging option comes in handy when you are reading data from behind a firewall and you can't reach your dedicated SQL pool directly.
>
> Be aware that text columns from your source that may contain commas within the text will cause errors as they will be interpreted as column delimiters.
>
> At this point in time, there is no option to influence the delimiter for the staged file.
>
> In this case, you will need to build a dual-step pipeline where you first unload your source to your data lake into a file where you can control the delimiter. From there, you can load your data with another copy activity to the database.

The **Settings** page additionally contains settings for **Reject type** and **Reject value**, giving you configuration options to control the number of failed rows before your copy job will fail.

Use type default will control what the job will do in the case of missing values. If you check the box, numeric columns will be set to 0, string columns will receive an empty string (""), and date columns will be set to Jan 1, 1900.

The **Data integration unit** and **Degree of copy parallelism** options give you access to further performance settings of the pipeline/Data Factory engine. Use these settings to control resources for jobs executed in parallel.

Finally, click **Next >** to jump to the **Summary** page.

> **Important note**
>
> As outlined on the **Settings** page, don't forget to assign the STORAGE_ BLOB_DATA_CONTRIBUTOR RBAC role to the SQL database server. This is the managed identity of your Synapse workspace. Otherwise, the load will fail.

11. On the **Summary** page, you can review all the settings before you finally start the job. If everything is OK, click **Next >**.

12. The job will be deployed and started. You can wait for the start and directly switch to the monitor when the job is running:

Azure Data Lake Storage Gen2 ⟶ Azure Synapse dedicated SQL pool

Deploying ..

▷ Validate copy runtime environment ✅

Deployment step	Status
⟩ Creating datasets	In progress 🔁
⟩ Creating pipelines	Pending 🔁
⟩ Running pipelines	Pending 🔁

Figure 10.11 – Job deployment

13. The monitor will give you insights into your job:

Details   ↻ Refresh

Learn more on copy performance details from here.

Activity run id: c2dcadfd-ef5c-4ed3-8db5-3acace4a133b

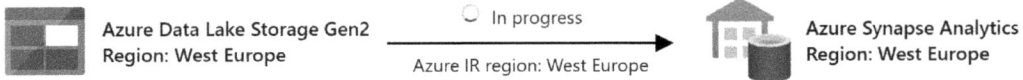

Azure Data Lake Storage Gen2
Region: West Europe

○ In progress
Azure IR region: West Europe

Azure Synapse Analytics
Region: West Europe

Data read: ⓘ	56.019 MB	Data written: ⓘ	43.474 MB
Files read: ⓘ	1	Rows written: ⓘ	230,000
Rows read: ⓘ	444,780	Peak connections: ⓘ	2
Peak connections: ⓘ	10		

Copy duration	00:01:02
Throughput: ⓘ	925.212 KB/s

◢ Azure Data Lake Storage Gen2 → Azure Synapse Analytics

Start time	6/9/21, 8:22:28 PM
Used DIUs ⓘ	4
Used parallel copies ⓘ	1
◢ Duration	00:01:02

Details	Working duration	Total duration
⊘ Queue ⓘ		00:00:02
◉ Transfer ⓘ	Listing source ⓘ 00:00:00	00:01:00
	Reading from source ⓘ 00:00:00	
	Writing to sink ⓘ 00:00:47	

Data consistency verification ⓘ   Not verified

Figure 10.12 – Monitoring details

Let's examine the copy activity that you have created. Once the pipeline has finished running, please switch to the **Integration** hub and select it. Click the pipeline that you have created and select it in the details area. The copy activity settings are displayed below. Please select **Sink** and examine the settings there:

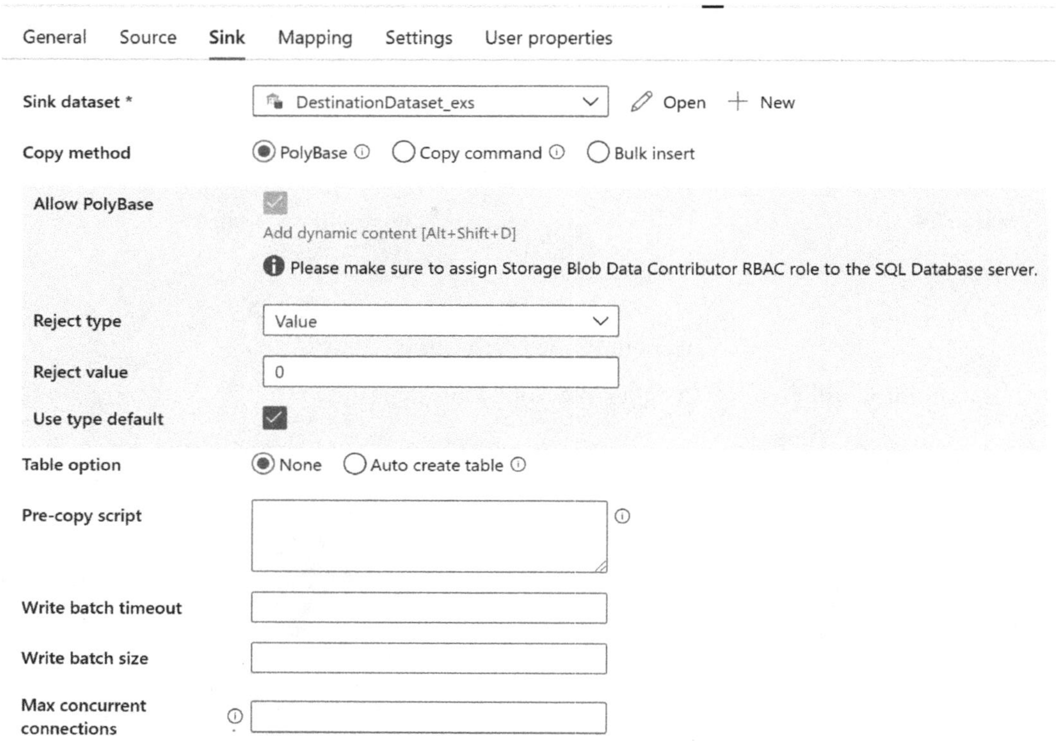

General   Source   **Sink**   Mapping   Settings   User properties

Sink dataset *	DestinationDataset_exs ⌄ ✎ Open + New
Copy method	⦿ PolyBase ⓘ ◯ Copy command ⓘ ◯ Bulk insert
Allow PolyBase	☑

Add dynamic content [Alt+Shift+D]

ⓘ Please make sure to assign Storage Blob Data Contributor RBAC role to the SQL Database server.

Reject type	Value ⌄
Reject value	0
Use type default	☑
Table option	⦿ None ◯ Auto create table ⓘ
Pre-copy script	ⓘ
Write batch timeout	
Write batch size	
Max concurrent connections	ⓘ

Figure 10.13 – Copy activity Sink settings

Below the dataset name, you will find **Copy method**. This is the place where you can select **PolyBase**, **Copy command** , or **Bulk insert**. Now, PolyBase is selected. You might switch to **Copy** or to **Bulk insert**.

Maybe give it a try and compare the performance. Perhaps you want to compare the performance between the pipeline activity and the preceding SQL construct in the *Loading data into a dedicated SQL pool using COPY* section, too. Which one is the fastest?

The PolyBase statement will be the fastest. Do you remember why? It is minimally logged. And although we're using PolyBase in the pipeline too, the pipeline logs more information during its run.

This is acceptable in many cases as the graphical interface, on the other hand, offers a lot of advantages. But if you really need the punch and performance is the absolute key to your job, you may want to think about implementing a stored procedure that uses the PolyBase construct in your dedicated SQL pool, one that does the staging, the CTAS, and the renaming for you. And you can still orchestrate the stored procedure using Synapse pipelines' stored procedure activities.

However, Synapse does offer more SQL options than the dedicated SQL pools. In the next section, *Using Synapse serverless SQL pools*, you will examine how to leverage your Data Lake Storage Gen2 with **T-SQL (Transact SQL)**. You'll see how to create views based on files in your data lake and consume them with Power BI, and you will learn how to wrangle data in your data lake and even create new files with SQL.

Using Synapse serverless SQL pools

Serverless SQL pools are an offering that was introduced with the Synapse workspace. It is a SQL engine that works in a distributed manner in the background. It can therefore offer a high fault tolerance even when you run it against bigger volumes of data in your data lake.

The serverless SQL engine comes in handy for different types of workloads:

- Ad hoc data analysis using SQL

- Transforming, cleansing, and preparing your data in the data lake using SQL

- Virtual data warehousing on multiple files and folders without transporting your data any further from your data lake

As the name suggests, the serverless SQL pool does not cause upfront resource provisioning in your Synapse workspace. It is charged on a volume basis (TB queried per month) based on competitive rates (see the *Further reading, Synapse pricing* section). This means that you will not generate any costs when you aren't running queries on the serverless SQL pool.

Serverless pools don't offer the database storage concept. They have a metadata store for objects such as views or stored procedures. However, you don't store data in tables in a serverless SQL pool. The only data storage that is available for this SQL engine is your data lake or a Blob storage (Azure Storage) account.

Browsing data ad hoc

Let's go and run a query against a file in your data lake using the serverless SQL pool. Therefore, please navigate to your Synapse workspace and to the data hub. Please navigate to your source folder from *Chapter 5, Integrating Data in Your Modern Data Warehouse.*

When you locate the `airdelays.csv` file, right-click it and select **New SQL script** and then **Select TOP 100 rows**:

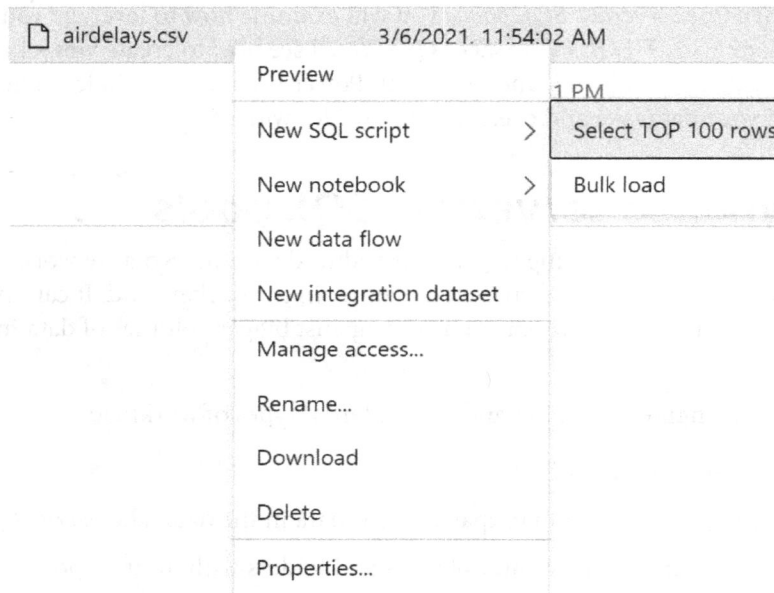

Figure 10.14 – New SQL script

We have done this before. A new SQL script is created and displayed showing a `SELECT` statement using the `OPENROWSET` function of the serverless SQL pool:

```
-- This is auto-generated code
SELECT
    TOP 100 *
FROM
    OPENROWSET(
        BULK 'https://[YOURDATALAKEACCOUNT].dfs.core.windows.
net/[YOURPATH]/airdelays.csv',
        FORMAT = 'CSV',
        PARSER_VERSION='2.0'
    ) AS [result]
```

When you click **Run** in the ribbon above the query (alternatively, you can hit *Shift +
Enter*), the query will be executed, and the result set will be displayed below the query text.

Now, it may be the case that the query result might not instantly show up correctly. The
file is using a semicolon, ; , as the column delimiter, and the serverless pool searches for
a comma, , , by default. Please add the following to your query text below the PARSER_
VERSION entry to adjust the column delimiter to reflect the semicolon:

```
,  FIELDTERMINATOR=';'
```

But there is still an issue. The first line is not recognized as a header. To use the first line of
the file as a header row, you can add the following to your query:

```
,  HEADER_ROW=TRUE
```

For a complete overview of the options of the OPENROWSET() function, please visit the
Further reading, OPENROWSET() section.

Now you can start and use your SQL knowledge on your file. Perhaps you only need
a certain set of columns in your query. After you have taught the parser the header row,
you can now select the columns just as you would do in a "normal" SQL query.

> **Note**
> If you need to display the filenames and/or paths in your query for subsequent
> use, or if you want to filter on certain files (see the *Analyzing multiple
> files* section below) serverless SQL pools offer the filename() and
> filepath() functions.

Mapping columns and data types

In CSV files, serverless SQL will perform **automatic schema discovery**, trying to use the
most suitable data type for the columns read. Using a WITH() section in your query, you
have the option to control data types and even select a subset of the columns read from
the file yourself.

You can name columns and add their data types by adding the ordinal number behind the
column entry in your WITH() section.

In our preceding case, you could, for example, extract the first three columns by adding
the following code fragment to your query:

```
. . .

    )

    WITH
```

```
(
        ad_year              INT 1
      , ad_month             INT 2
      , ad_dayofmonth        INT 3
) as [result]
```

Analyzing multiple files

Let's check the options when you want to analyze several files in your data lake in one go.

An easy option would be to change the preceding statement to read all CSV files in the same folder by replacing the filename with the asterisk wildcard (*):

```
BULK 'https://[YOURDATALAKEACCOUNT].dfs.core.windows.net/
[YOURPATH]/*.csv',
```

Really easy, isn't it?

> **Important note**
> Please ensure that all the CSV files in this folder have the same structure, otherwise, your SQL statement will run into an error.

When you omit the filenames completely after the last / in your path, you are sending a signal to the engine to read the entire folder (the same as /*.* or /*).

You can also consider reading multiple folders in a path by adding the wildcard to the folder names (.../raw*/). Ensure that you add a / as a final entry in the path, otherwise, the engine will search for files with the pattern entered.

Another option would be to traverse a folder structure recursively, meaning you want to read all folders and all files below a certain folder. You can do this by adding two asterisks to the end of the folder where you want to start the reading: .../raw/sales/**.

These options can even be combined. So maybe you want to traverse all the files with a certain name pattern in all the folders that start with the word raw: .../raw*/air*.csv.

Once again, please ensure that all the files that you read in one statement are of the same structure, otherwise, your query will fail.

Please find more information on Parquet and JSON files in the *Further reading*, *Reading multiple Parquet and JSON files with serverless SQL* section. You will find examples of reading partition folders with the Parquet format, for example, or how to successfully parse JSON files with the JSON_VALUE() and OPEN_JSON() functions.

Using a serverless SQL pool to ELT

Extract, load, transform (ELT) is increasingly replacing the classical data integration pattern (ETL), where data is transformed during the transport phase.

In our modern data warehouse or data lakehouse, or whatever you want to call it, we try to land data in our data lake layer as quickly as possible to keep the pressure of complex queries away from the source systems. Another reason for bringing data into your data estate and then transforming it is elastic compute. When we consider Azure, and Synapse, in particular, you can benefit from tremendous scalability. You could say, here you have the power.

When you examine the performance of the serverless SQL pools, you will find that most of your queries are answered in quite surprisingly short periods of time. So why not leverage this engine and the SQL language, which is familiar to most data engineers out there, for big data too?

Serverless SQL pools come with an extension very similar to the CTAS statement of the dedicated SQL pools. In this case, it is called **CETAS**, **CREATE EXTERNAL TABLE AS SELECT**. Pretty much like the dedicated SQL pools, this statement will create new objects. In this case, new files in the data lake as serverless SQL do not have a database storage concept in the same way as dedicated pools.

Understanding CETAS

With the CETAS statement, you can create new files in your data lake or Blob storage. You have the option to either create a CSV or a Parquet file.

When you try and remember the different objects that you had to create in the *Creating a PolyBase external table* section, you will think of the following artifacts:

- External data source
- External file format
- External table

And, surprise, surprise, you will have to create similar structures. So, let's start right away and create the external file format.

Bur first, maybe create another folder where you can land the data that you have transformed using your serverless SQL pool. Perhaps something like `.../target/ SQLServerlessTarget/` or similar?

When you have successfully created your folder, please navigate to the **development** hub, and create a new SQL script. Perhaps name it something along the lines of `CreateExternalTableAsSelect`.

Creating the external data source

First off, you can't create an external data source in either the "master" database of your serverless SQL pool or in the "default" one. To be able to create tables and views and so on, you will need to create a new database where this metadata is stored.

Please create your new database with the following statement, refresh the database picker in the ribbon above the development section, and then select it for all the subsequent statements:

```
CREATE DATABASE [YOURSERVERLESSPOOLDATABASENAME];
```

The statement looks pretty much like what you created in the dedicated SQL pool. Let's try the following:

```
CREATE EXTERNAL DATA SOURCE [YOUREXTERNALDATASOURCENAME]
WITH
(
     LOCATION='https://[YOURDATALAKESTORAGE].dfs.core.windows.
net/[YOURPATH]/SQLServerlessTarget/'
);
```

Did you succeed? Great job!

When you examine the documentation (see the *Further reading, Serverless EXTERNAL DATA SOURCE* section), you will additionally find a **CREDENTIAL** option and its usage in this statement. When you omit the **CREDENTIAL** option, the serverless SQL pool will pass through your Azure Active Directory identity to the Data Lake storage. Convenient, isn't it?

Creating the external file format

As already mentioned, you have the option to either create a CSV or a Parquet file. `EXTERNAL FILE FORMAT` is the place where you nail this. Let's create a Parquet file in the target folder and therefore use the following statement:

```
CREATE EXTERNAL FILE FORMAT [YOUREXTERNALFILEFORMAT]
WITH
(
```

```
FORMAT_TYPE=PARQUET
   , DATA_COMPRESSION='org.apache.hadoop.io.compress.
SnappyCodec'
);
```

As you can see from the statement, we're using Snappy compression for the Parquet file, too. The other option would be Gzip (`org.apache.hadoop.io.compress.GzipCodec`).

When you want to create a CSV file, you would use `DELIMITEDTEXT` as the `FORMAT_TYPE` option. There again, you have all the options that you need to create the file format of choice, such as column and row delimiters and compression. You will find all the settings in the *Further reading, Serverless EXTERNAL FILE FORMAT* section.

Creating the external table AS SELECT

Moving on, we initiate the main statement that we want to achieve. Let's write out some data with `CREATE EXTERNAL TABLE AS SELECT`. You might want to use the `select` statement from previously (*Browsing data ad hoc*) as a basis and count the number of flights performed per day:

```
CREATE EXTERNAL TABLE [YOUREXTERNALTABLENANE]
WITH
(
    LOCATION='/[YOURTARGETPATH]/[YOURPARQUETFILE].parquet'
    , DATA_SOURCE=[YOUREXTERNALDATASOURCENAME]
    , FILE_FORMAT=[YOUREXTERNALFILEFORMAT]
)
AS
SELECT
    ad_year
    , ad_month
    , ad_dayofmonth
    , count(*) ad_noofflights
FROM
    OPENROWSET(
    BULK 'https://cloudscaleanalyticsadls.dfs.core.windows.net/
cloudscaleanalyticsfs/chapter5/source/airdelays.csv',
    FORMAT = CSV',
    PARSER_VERSION='2.0'
```

```
        , FIELDTERMINATOR=';'
        , HEADER_ROW=TRUE
)
WITH
(
        ad_year          INT 1
        , ad_month       INT 2
        , ad_dayofmonth  INT 3
) as [result]
GROUP BY
    ad_year
    , ad_month
    , ad_dayofmonth
```

You can now, of course, use EXTERNAL TABLE in the same manner as you use it in your dedicated SQL pools. Need to query it? It is available from the dictionary in your serverless SQL pool database that you created previously:

```
select *
from [YOUREXTERNALTABLE]
```

This comes in handy too, right?

> **Note**
>
> You can also use the CETAS statement with dedicated SQL pools when you want to output data from your dedicated pool to your data lake and make it available as files.

Again, please visit the *Further reading*, *Serverless CREATE EXTERNAL TABLE AS SELECT* section for more insights into the CETAS statement.

Building a virtual data warehouse layer with Synapse serverless SQL pools

The third purpose of serverless SQL pools that was mentioned previously might be interesting for many architects and users. Isn't it a tempting thought that you can use the Synapse workspace and the serverless SQL pool to create your data warehouse structures directly from the files that you have brought to your data lake?

You can achieve this using databases, schemas, views, and external tables in the serverless SQL pools.

You have already seen the `CREATE DATABASE` statement in the preceding examples (*Creating the external data source*), and if you already have some SQL knowledge, you will also have an idea of how to create schemas and views. But let's go and create a new schema in your new serverless SQL pool database and then create a view.

To create a new schema, you can simply type the following:

```
CREATE SCHEMA [YOURSCHEMANAME];
```

This can now act as a securable container where you can collect related objects and secure them in one go as you grant or revoke access to users and groups to the entire schema.

Views are created just like in any other database, too. Let's try and use the external table that we created previously (*Creating the external table as SELECT*):

```
CREATE VIEW [YOURSCHEMANAME].[YOURVIEWNAME]
AS
    select *
    from [YOUREXTERNALTABLE];
```

And, of course, we select data from it:

```
select *
from [YOURSCHEMANAME].[YOURVIEWNAME]
```

Understanding client tools for Synapse SQL pools

Irrespective of whether you are using a dedicated SQL pool or you're building a virtual data warehouse with your serverless pool database, you will need a client tool to develop the SQL objects. Certainly, Synapse Studio offers several advantages. It integrates seamlessly with the DevOps setup that you may have chosen there, but there are other options, too, in terms of accessing the environment:

- Azure Data Studio: A cross-platform (Windows, Linux, macOS) SQL client
- SQL Server Management Studio: The database development and administration tool for all Microsoft database products
- Visual Studio Code: A cross-platform (Windows, Linux, macOS) code editor
- Visual Studio: A Microsoft application development IDE

You can use these clients (and others that can connect to a SQL endpoint using a **TDS** (**Tabular Data Stream**) connection) to develop against the SQL pools of Synapse.

Find more information and download links in the *Further reading, SQL client tools* section.

You can find the server names for your connections on the overview blade of your Synapse workspace in the Azure portal:

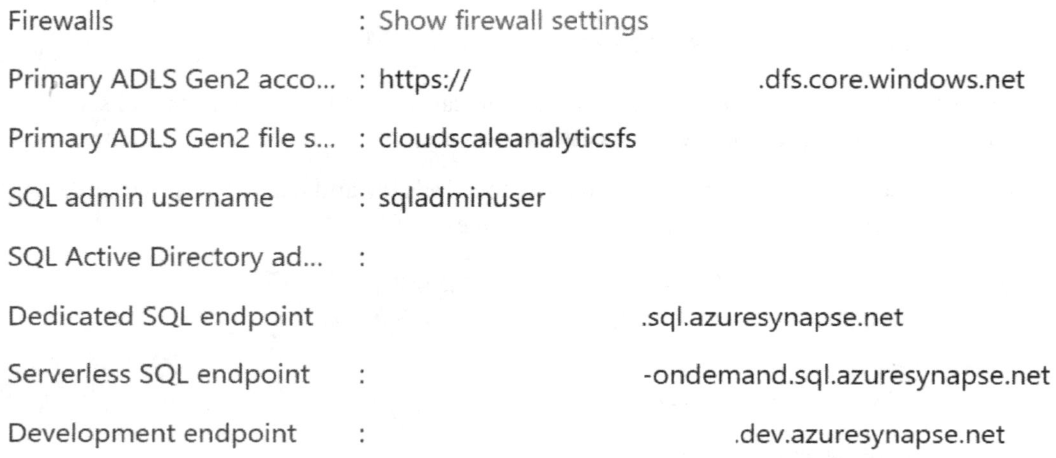

Firewalls   : Show firewall settings

Primary ADLS Gen2 acco...   : https://   .dfs.core.windows.net

Primary ADLS Gen2 file s...   : cloudscaleanalyticsfs

SQL admin username   : sqladminuser

SQL Active Directory ad...   :

Dedicated SQL endpoint   .sql.azuresynapse.net

Serverless SQL endpoint   :   -ondemand.sql.azuresynapse.net

Development endpoint   :   .dev.azuresynapse.net

Figure 10.15 – Synapse server endpoints

The hidden parts in *Figure 10.15* represent the Synapse workspace name that you have used for your environment.

Views and external tables can, of course, be consumed with Power BI and other reporting/dashboarding tools. You can connect them to the serverless SQL pool endpoint and create your analysis using all the artifacts there.

When, for example, you use Power BI in import mode, which means you get the data from your serverless pool and bring it to a Power BI dataset, this is a great example where you don't necessarily need a dedicated SQL pool. Power BI will provide sufficient performance to allow suitable reporting. And the serverless SQL pool will provide enough performance to refresh the Power BI datasets.

> **Note**
> Taking the cost of the serverless SQL pools into consideration, this setup can provide a lot of flexibility and scalability on the data side and great performance on the visualization side at a considerably lower cost.

Integrating data with Synapse Spark pools

If you are a Spark developer and want to use Synapse Spark to wrangle and load your data into your dedicated SQL pools, this is quite an easy thing to accomplish.

JDBC was, and still is, the way to establish the connection and the exchange. There is one caveat regarding the use of JDBC; only interact with the dedicated SQL pools. It will only talk to the control node of your dedicated pool. This is a suboptimal way as both Spark, but also dedicated SQL pools, have a lot of parallelism to offer.

Microsoft adjusted the JDBC driver slightly to benefit from the parallel workers that are part of this game. The JDBC driver will establish a connection between the control node of the dedicated SQL pool and the driver node of the Spark cluster. The Spark engine will issue CETAS statements and send filters and projections over this channel. The data itself will otherwise be exchanged using PolyBase and the Data Lake storage that is attached to your Synapse workspace:

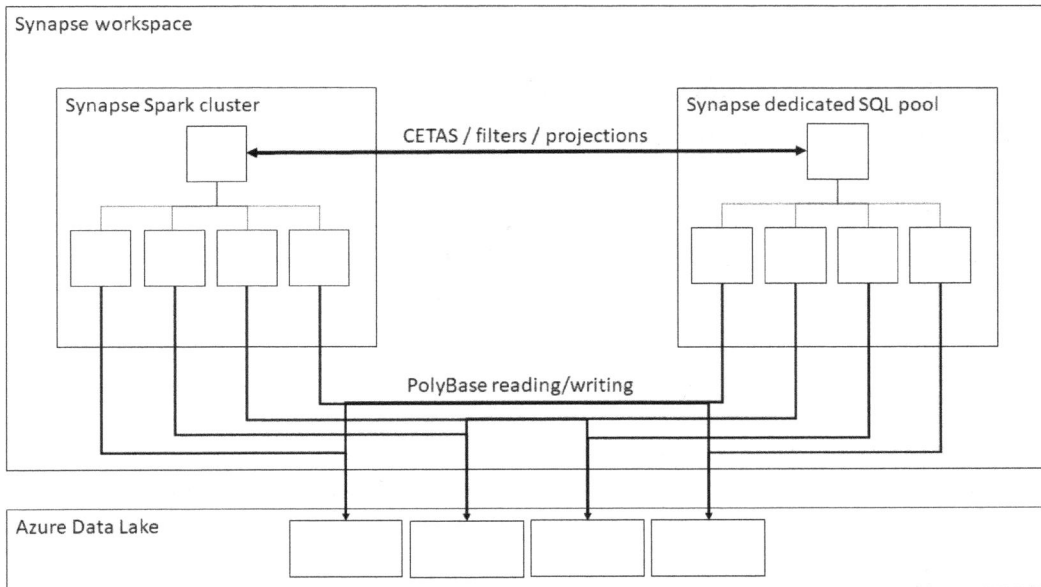

Figure 10.16 – Data integration with Synapse Spark and a dedicated SQL pool

To make use of this construct, you will need to use a Scala notebook. The necessary imports are done for you automatically; you don't need to formulate library imports.

Reading and loading data

The read command to get data from a dedicated SQL pool would appear as follows:

```
val df = spark.read.option(Constants.
SERVER, "[YOURDEDICATEDSQLPOOL].database.windows.net").
synapsesql("[YOURDEDICATEDSQLPOOLDB].[YOURSCHEMANAME].
[YOURTABLENAME]")
```

When you want to write from Spark to your dedicated SQL pool, this looks very similar:

```
df.write(Constants.SERVER, "[YOURDEDICATEDSQLPOOL].database.
windows.net").synapsesql("[YOURDEDICATEDSQLPOOLDB].
[YOURSCHEMANAME].[YOURTABLENAME]", "[TableType])
```

This method is valid between the dedicated SQL pools and the Synapse Spark clusters. When you need to access objects in your serverless SQL pool, please check the next section, *Exchanging metadata between computes*.

Exchanging metadata between computes

The Synapse workspace supports metadata exchange between the participating compute components; for example, if you create a Spark table, this is instantly available within the serverless SQL pool for querying.

Let's check this behavior and try to query the Spark table that you have created in *Chapter 9, Integrating Azure Cognitive Services and Machine Learning*:

```
SELECT *
FROM [default].[dbo].[YOURSPARKTABLE]
```

The Synapse workspace will add more metadata exchange functionality over time. This will allow all the computes to access each other's objects easily and in a frictionless way.

Summary

This chapter was all about Synapse SQL pools. You have seen and examined the different options of loading data into a dedicated SQL pool.

After that, you learned about the serverless SQL pools of Synapse and the benefits that they can add to your modern data warehouse from the perspectives of usability and flexibility, but also from a cost point of view.

Finally, you examined the communication between Synapse Spark pools and the dedicated SQL pools, and the exchange of metadata between the different computes of Synapse workspaces.

In the next chapter, *Chapter 11*, *Developing and Maintaining the Presentation Layer*, you will examine the options to implement your presentation layer and present data optimally to downstream consuming tools and clients.

Further reading

- **EXTERNAL DATA SOURCE**: `https://docs.microsoft.com/en-us/sql/t-sql/statements/create-external-data-source-transact-sql?view=azure-sqldw-latest&tabs=dedicated`

- **EXTERNAL FILE FORMAT**: `https://docs.microsoft.com/en-us/sql/t-sql/statements/create-external-file-format-transact-sql?view=azure-sqldw-latest&tabs=delimited`

- **EXTERNAL TABLE**: `https://docs.microsoft.com/en-us/sql/t-sql/statements/create-external-table-transact-sql?view=azure-sqldw-latest&tabs=dedicated`

- **COPY INTO**: `https://docs.microsoft.com/en-us/sql/t-sql/statements/copy-into-transact-sql?view=azure-sqldw-latest`

- **Synapse Pricing**: `https://azure.microsoft.com/en-us/pricing/details/synapse-analytics/`

- **OPENROWSET()**: `https://docs.microsoft.com/en-us/azure/synapse-analytics/sql/develop-openrowset`

- **Reading multiple PARQUET and JSON files with serverless SQL:**

  a) `https://docs.microsoft.com/en-us/azure/synapse-analytics/sql/query-parquet-files`

  b) `https://docs.microsoft.com/en-us/azure/synapse-analytics/sql/query-json-files`

- **Serverless EXTERNAL DATA SOURCE**: `https://docs.microsoft.com/en-us/azure/synapse-analytics/sql/develop-tables-external-tables?tabs=sql-on-demand#create-external-data-source`

- **Serverless EXTERNAL FILE FORMAT**: `https://docs.microsoft.com/en-us/azure/synapse-analytics/sql/develop-tables-external-tables?tabs=sql-on-demand#create-external-file-format`

- **Serverless CREATE EXTERNAL TABLE AS SELECT**: `https://docs.microsoft.com/en-us/azure/synapse-analytics/sql/develop-tables-cetas`

- **SQL client tools**:

  a) `https://docs.microsoft.com/en-us/sql/azure-data-studio/what-is-azure-data-studio?view=sql-server-ver15`

  b) `https://docs.microsoft.com/en-us/sql/ssms/download-sql-server-management-studio-ssms?view=sql-server-ver15`

  c) `https://visualstudio.microsoft.com/`

  d) `https://code.visualstudio.com/`

Section 4: Data Presentation, Dashboarding, and Distribution

This section explores modern data warehouse practices when it comes to monitoring and analyzing data. We'll learn how to integrate with the machine learning and AI components in Azure and explore various dashboards for data.

This section comprises the following chapters:

- *Chapter 11, Developing and Maintaining the Presentation Layer*
- *Chapter 12, Distributing Data*
- *Chapter 13, Introducing Industry Data Models*
- *Chapter 14, Establishing Data Governance*

11
Developing and Maintaining the Presentation Layer

In the previous chapter, *Chapter 10*, *Loading the Presentation Layer*, and this one, we investigate how to implement and maintain the Presentation Layer of your **Modern Data Warehouse (MDWH)**.

With all the versatile modules of your modern data estate, you have acquired data, transformed it, and even used advanced analytics to predict behavior based on the data that you have collected.

In this chapter, we will examine how to use Azure Synapse, and particularly Synapse Studio, when you implement your Presentation Layer. You will see how to integrate Azure Synapse with Azure DevOps and how you can automate your deployments. In your role as an MDWH developer, you will also enjoy the developer productivity features that Synapse Studio offers.

You will see how to implement backup and **Disaster Recovery (DR)** and how you can monitor your Azure Synapse environment.

Finally, you will learn about the two aspects of security when we dive into the access control and network security of your Azure Synapse environment.

You will find these topics covered in the following sections:

- Developing with Synapse Studio
- Backing up and DR in Azure Synapse
- Monitoring your MDWH
- Understanding security in your MDWH

Developing with Synapse Studio

One of the most important areas for efficient development is the capability to collect all your sources in a central repository, collaborate on artifacts during development, and automate deployments from development to test to production environments.

Synapse Studio is no exception here and integrates with either Azure DevOps or GitHub to enable these features. Other repositories are not yet supported.

Integrating Synapse Studio with Azure DevOps

Let's examine how to integrate your Synapse Studio with Azure DevOps and version your work.

First, you will need to connect your Azure Synapse workspace to your Azure DevOps environment. In *Chapter 5, Integrating Data in Your Modern Data Warehouse*, you saw how to do this with Azure Data Factory. With Azure Synapse, it is pretty much the same. Let's take these steps:

1. Please navigate to your DevOps environment (`https://dev.azure.com/`) and create a new project. Name it something such as `CloudScaleAnalyticsSynapesDevOps`.

2. On the **Create new project** dialog, please select **Private** as you want to have the most control over who can see the project.

    Leave the rest of the settings as the defaults displayed.

3. Now please jump to your Azure Synapse workspace. Here, you have two options to start the configuration of your DevOps environment. You can select **Set up code repository** from the drop-down box in the upper-left corner of the Synapse workspace:

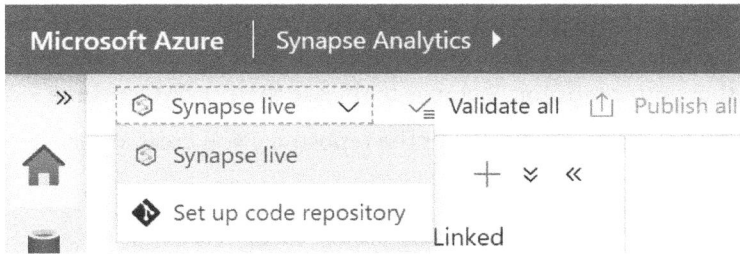

Figure 11.1 – Drop-down box to start

Or you can navigate to the **Management** hub and start the setup from there using the **Git configuration** option in the **Source control** section:

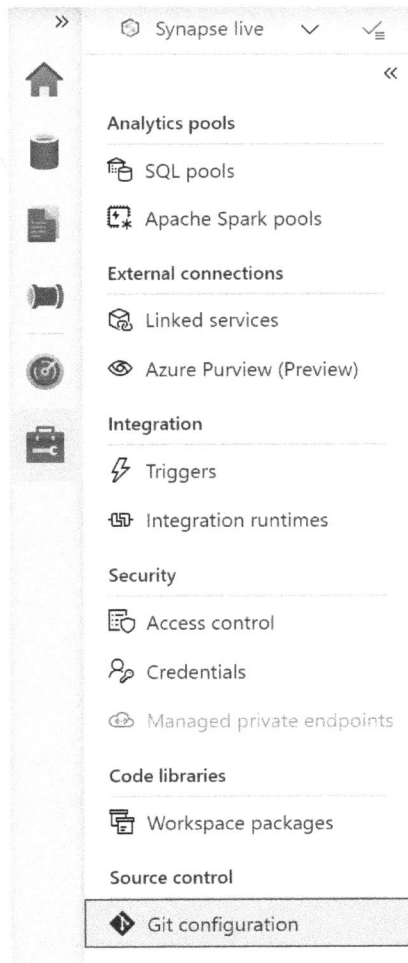

Figure 11.2 – Management hub: Git configuration

4. Once you start the configuration, you are taken to the **Configure a repository** dialog. Please select **Azure DevOps Git** as the repository type and the Azure Active Directory that your repository belongs to. Please then click **Continue**.

5. On the next page, you can now select the repository you created earlier:

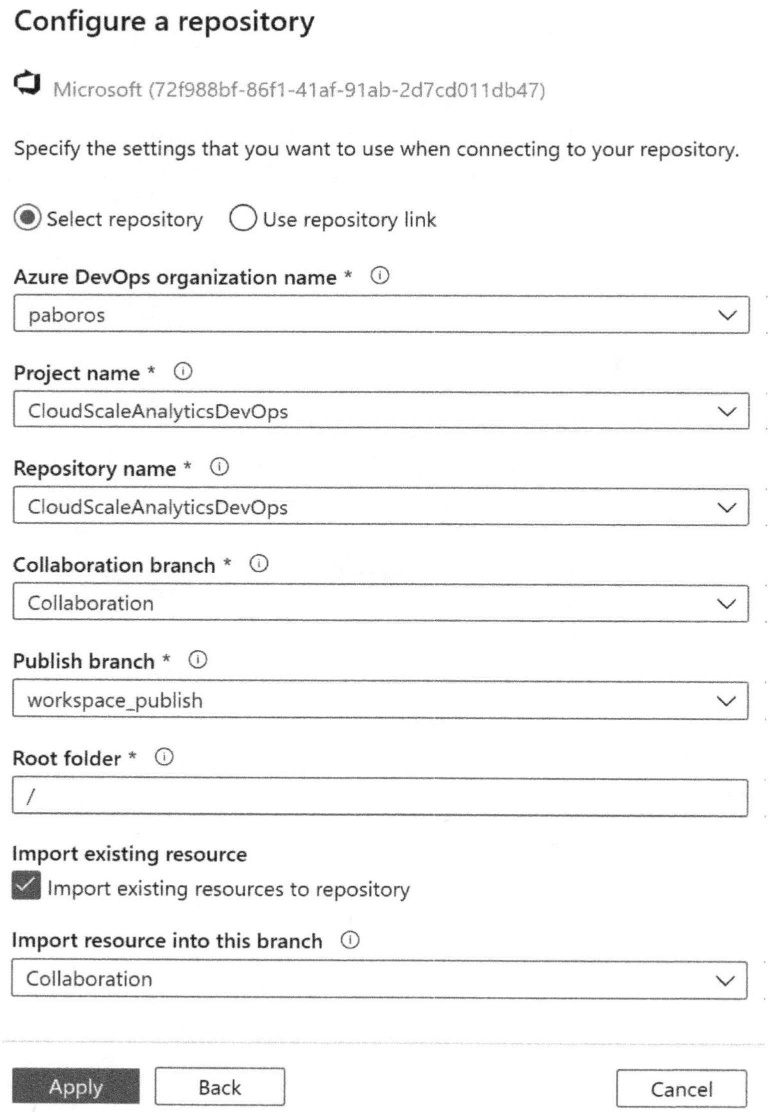

Configure a repository

Microsoft (72f988bf-86f1-41af-91ab-2d7cd011db47)

Specify the settings that you want to use when connecting to your repository.

◉ Select repository   ○ Use repository link

Azure DevOps organization name * ⓘ

paboros	⌄

Project name * ⓘ

CloudScaleAnalyticsDevOps	⌄

Repository name * ⓘ

CloudScaleAnalyticsDevOps	⌄

Collaboration branch * ⓘ

Collaboration	⌄

Publish branch * ⓘ

workspace_publish	⌄

Root folder * ⓘ

/

Import existing resource

☑ Import existing resources to repository

Import resource into this branch ⓘ

Collaboration	⌄

Apply	Back		Cancel

Figure 11.3 – Configure a repository dialog

Leave the **Select repository** radio button selected. With the **Azure DevOps organization name**, **Project name**, and **Repository name** drop-down boxes, you can now select the values from your existing environments. Please create a new collaboration branch. This will act as your master branch in your DevOps repository.

Leave the rest of the settings to the displayed defaults.

Please hit **Apply** to finalize the configuration.

The **Import existing resources to repository** setting will ensure that all resources you have already created in your workspace are imported into the DevOps repository. You may even select the newly created collaboration branch.

6. The workspace will display **Workspace connected** and will instantly display the dialog where you will set the working branch to start your development life cycle.

Your Azure Synapse workspace is now connected to your new Azure DevOps repository and you are set to start your development.

Understanding the development life cycle

You saw a description of the typical development life cycle already in *Chapter 5*, *Integrating Data in Your Modern Data Warehouse*. Let's examine it a little closer:

1. When you start developing a new set of objects or you need to alter an existing script, notebook, or pipeline, the first thing you do is create a working branch from the collaboration branch that you have created during the setup process of your DevOps repository when you connected it to your Synapse workspace:

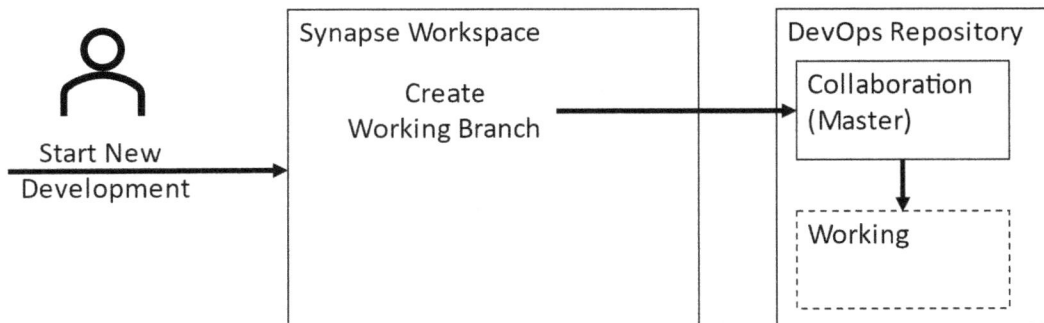

Figure 11.4 – Initial step when starting a new development

You will do this by selecting the drop-down box in the upper-left corner of your Azure Synapse workspace:

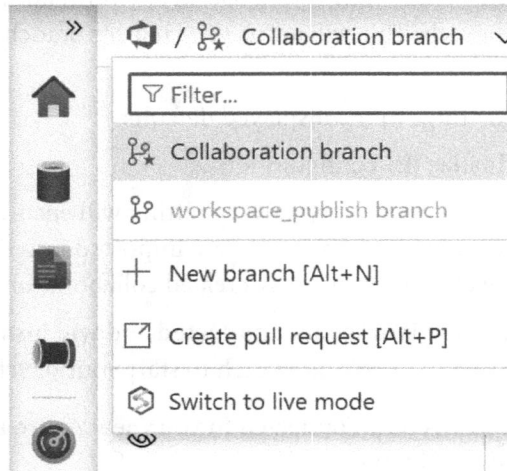

Figure 11.5 – Creating a new working branch

When you hit **+ New branch [Alt + N]**, you will be asked to name it. When you proceed, the new branch name will be displayed instead of the name of the collaboration branch that was there before.

2. You are now in development mode and can start developing your requirements:

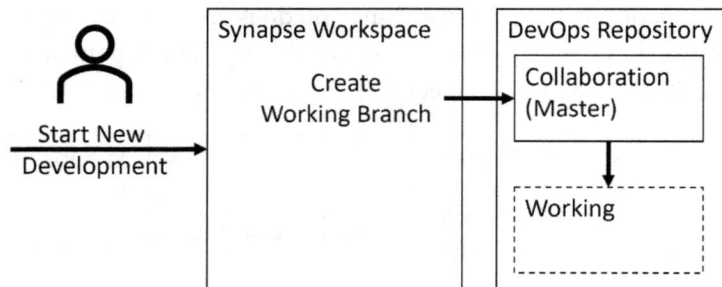

Figure 11.6 – Developing on the working branch

During development, you will need to save your work. You can do this by clicking **Commit all** in the upper ribbon. Unsaved work will be marked with a dot in the tabs next to the names of the objects you have been working on:

Figure 11.7 – Life cycle functions

3. Once you have finished your work, the next step in your development life cycle is to merge the artifacts back from your working branch to your collaboration or master branch:

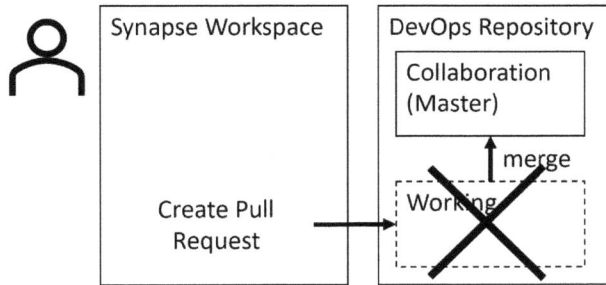

Figure 11.8 – Creating a pull request

You will find the menu item again in the drop-down box below the working branch name:

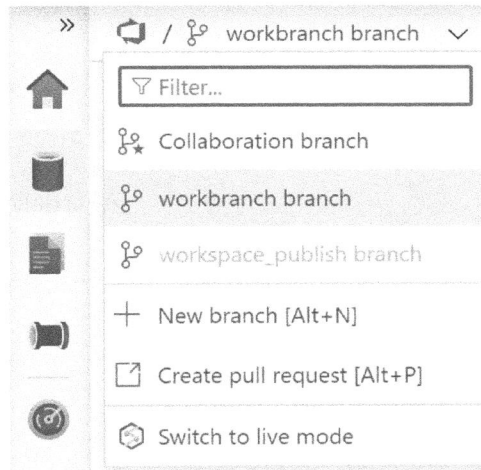

Figure 11.9 – Menu items in the life cycle menu

Please hit **Create pull request [Alt + P]** and start the merge process from your working branch into your collaboration branch. If you didn't implement or change something in *step 2* in your workspace, you will receive a message such as *no changes to merge between the selected branches in the DevOps environment* when you create this pull request.

You will be transferred to your DevOps environment in a pre-populated pull request:

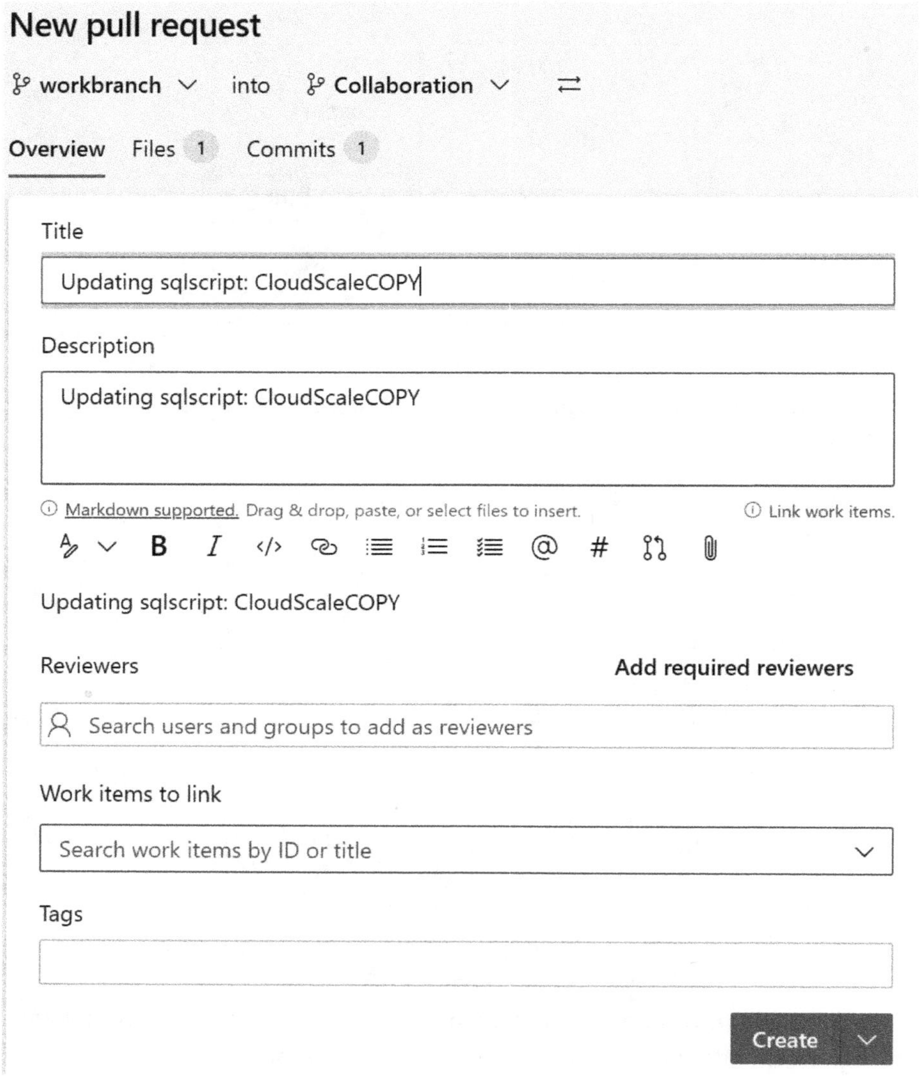

New pull request

⑂ **workbranch** ∨   into   ⑂ **Collaboration** ∨   ⇄

Overview   Files **1**   Commits **1**

Title

Updating sqlscript: CloudScaleCOPY|

Description

Updating sqlscript: CloudScaleCOPY

ⓘ <u>Markdown supported.</u> Drag & drop, paste, or select files to insert.   ⓘ Link work items.

𝒜 ∨   **B**   *I*   </>   ⦾   ≡   ≔   ≣   @   #   ⥂   ⫮

Updating sqlscript: CloudScaleCOPY

Reviewers   **Add required reviewers**

  ⧔ Search users and groups to add as reviewers

Work items to link

  Search work items by ID or title   ∨

Tags

Create ∨

Figure 11.10 – New pull request dialog in Azure DevOps

In the upper area, you can check whether the right branches are selected in the right direction. If everything is correct and you have commented your work as necessary, please hit **Create** and start the merging:

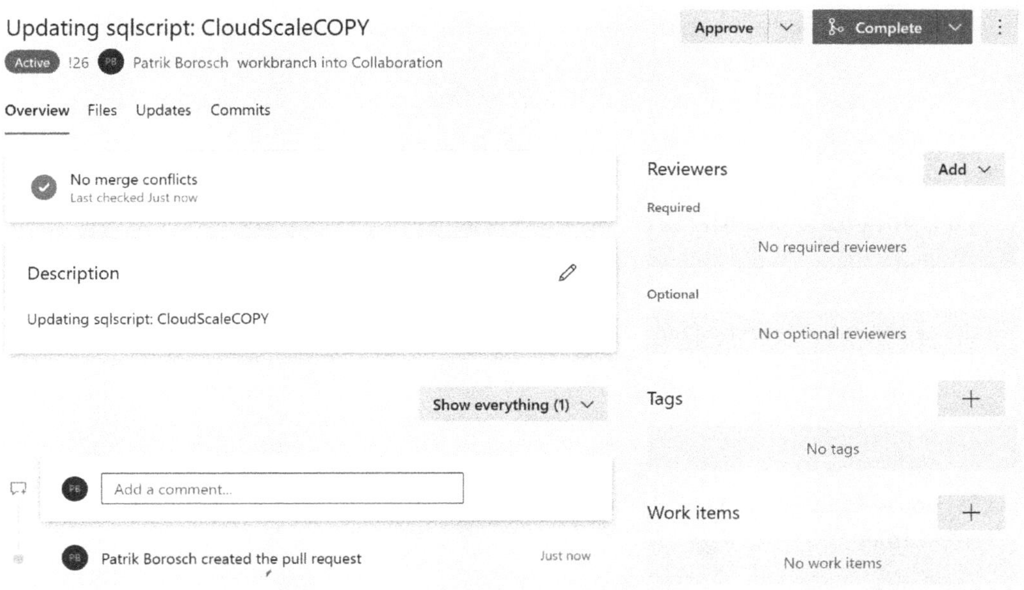

Updating sqlscript: CloudScaleCOPY

Active !26 PB Patrik Borosch workbranch into Collaboration

Approve ∨   ⌾ Complete ∨   ⋮

Overview Files Updates Commits

✓ No merge conflicts
Last checked Just now

Reviewers   Add ∨

Required

No required reviewers

Description   ✎

Optional

Updating sqlscript: CloudScaleCOPY

No optional reviewers

Show everything (1) ∨

Tags   ＋

No tags

PB Add a comment...

Work items   ＋

PB Patrik Borosch created the pull request   Just now

No work items

Figure 11.11 – Complete pull request

When you now hit **Complete**, you can finally select the merge type and select or de-select **Complete associated work items after merging** and **Delete workbranch after merging**.

The Complete associated work items after merging option can be used to automatically update the status of work items in the project management area of the DevOps repository. Work items can be for example user stories, issues, bugs, features, and epics. They are used in the development life cycle to formulate and track the details of your project.

The **Delete** workbranch after merging option will do exactly this. It will purge the working branch from where the artifacts are merged from the repository to keep it clean.

Hitting **Complete merge** will finish the process, and you will have merged your work into the collaboration (master) branch.

> **Note**
> You can configure DevOps, for example, to block developers from approving their own work or to stipulate that work items need to go through a certain workflow before they can be deployed to another environment, among many other options. Please make yourself familiar with these policies as they will help you to implement an extended development life cycle and improve the quality of your work.

4. The update of the collaboration branch does not automatically change the Azure Synapse development environment. This means that up to now, for example, newly created Synapse pipelines that were merged into the collaboration branch do not automatically take effect in the Azure Synapse environment and for other developers.

    You now need to publish the changes to the Synapse service. Once you are back in Synapse Studio, on the collaboration branch, the **Publish** option in the ribbon above the development window will be highlighted. After clicking it, your changes will take effect in the Azure Synapse service:

Figure 11.12 – Publishing changes to the Azure Synapse service

Only now will all your new artifacts be completely available in the development environment:

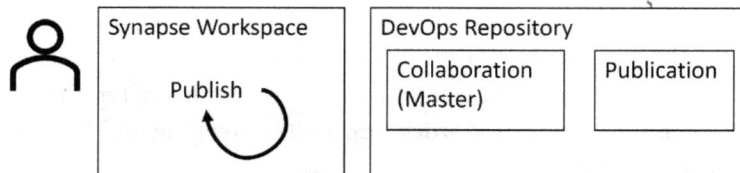

Figure 11.13 – Azure Synapse internal publish

This step will then additionally create your publish branch if it doesn't exist yet. The default name if you haven't changed it during the setup will be `workspace_publish`. But be aware that your work will not yet be transported to the publish branch. You will need to do this in the next step.

5. When a work unit, such as a sprint, for example, is finished, you want to capture the state of your work in another branch from where the objects can further be deployed into your testing environment.

    You will need to have all the necessary work merged into your collaboration branch at the code freeze time. Every developer will need to issue their own pull request, and the merge will need to be carried out in your DevOps environment. After you have made sure that all the work is there, you can now issue another pull request (see *step 3* in this sequence).

    This time, you will issue it from your collaboration branch to your publish branch that was created in *step 4*. By default, it is named `workspace_publish`:

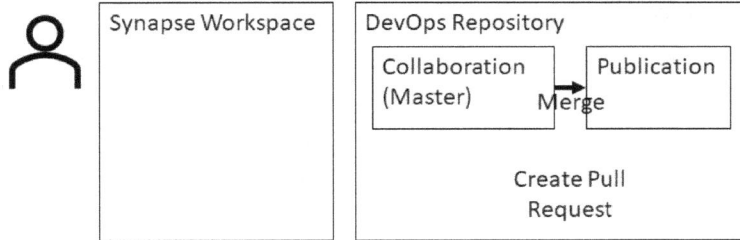

Figure 11.14 – Merging the collaboration branch to the publication branch

This will now bring your work into the publication branch, from where it will be deployed to a testing environment, for example. Your repository content might look like this:

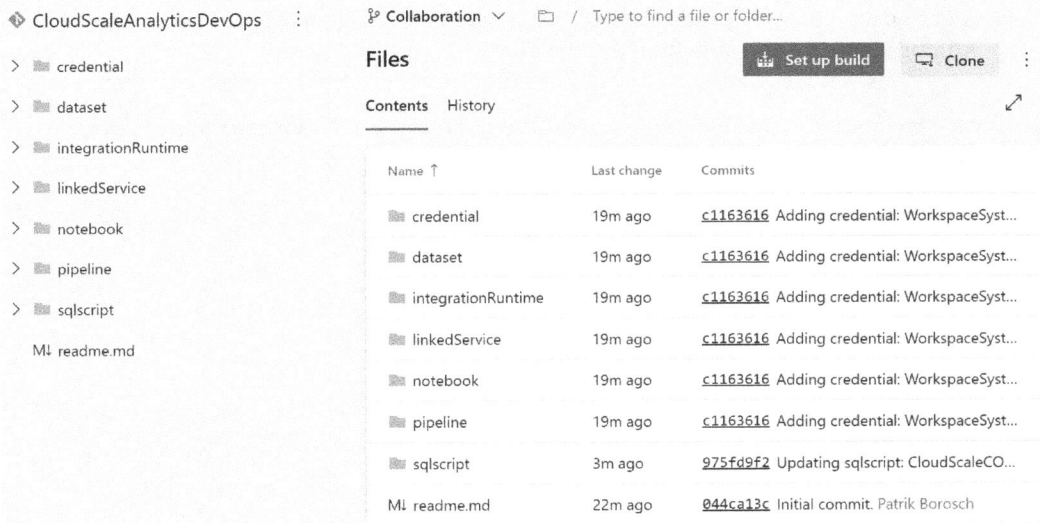

Figure 11.15 – Content in the publish branch

Please see the next section, *Automating deployments*, for the next steps on how to deploy your work.

Automating deployments

In the last section, *Understanding the development life cycle*, you learned how to integrate Synapse Studio with Azure DevOps, and you delivered your work to the publication branch of your development environment.

From here, you need to bring your work to the next stage in your development life cycle, for example, your testing environment. You might want to automate this and do this with a minimal number of manual steps.

Creating a release pipeline

First, we need a DevOps pipeline that will deploy the content of the publish branch to your target Azure Synapse workspace.

But even before that, please create another empty Azure Synapse workspace in another resource group as a target that you will deploy to. If you don't remember how, please check *Chapter 4, Understanding Synapse SQL Pools and SQL Options*:

> **Important note**
>
> Please add your DevOps service principal to the Contributor and User Access Administrator RBAC roles of your new target Synapse workspace in the Azure portal.
>
> Additionally, you need to add the DevOps service principal to the Synapse Administrator RBAC role within the Synapse Studio environment.

1. Please navigate to your DevOps environment, and in the **Navigator** pane, search for `Release`. Please create a new release pipeline.

    The **New release pipeline** wizard will be displayed:

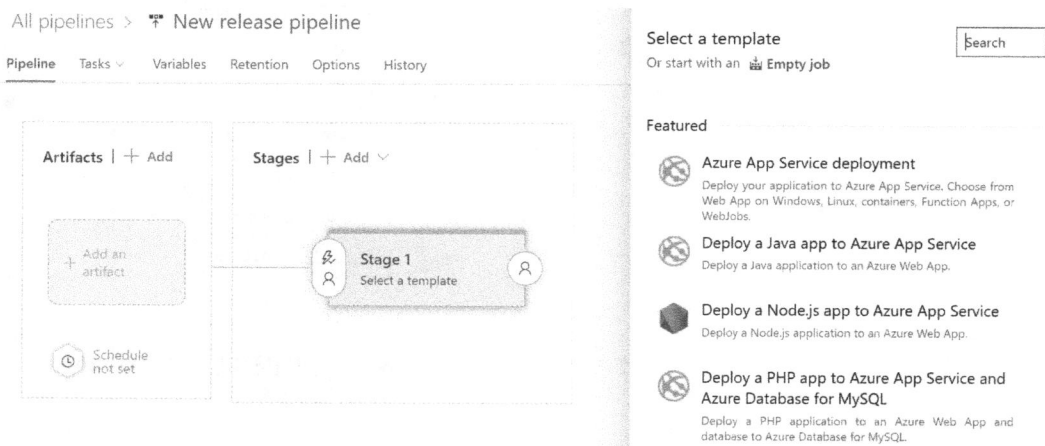

Figure 11.16 – New release pipeline wizard

2. Please click **Empty job** at the top of the **Select a template** dialog that comes in from the right.

3. In the left area, **Artifacts**, please click **Add an artifact**. In the following dialog, you need to connect to your DevOps environment where your repository is stored:

Add an artifact

Source type

5 more artifact types ∨

Project * ⓘ

| CloudScaleAnalyticsDevOps | ∨ |

Source (repository) * ⓘ

| CloudScaleAnalyticsDevOps | ∨ |

Default branch * ⓘ

| Collaboration | ∨ |

Default version * ⓘ

| Latest from the default branch | ∨ |

☐ Checkout submodules ⓘ

☐ Checkout files from LFS ⓘ

Shallow fetch depth ⓘ

| |

Source alias * ⓘ

| _CloudScaleAnalyticsDevOps |

Add

Figure 11.17 – Adding an artifact

Please select your DevOps project, then for **Source (repository)**, select the publish branch (the default is `workspace_publish`), and leave the rest as the default settings.

Please click **Add**.

4. In the **Stages** dialog on the right, you can name your job stage something such as
    `SynapseDeployment`:

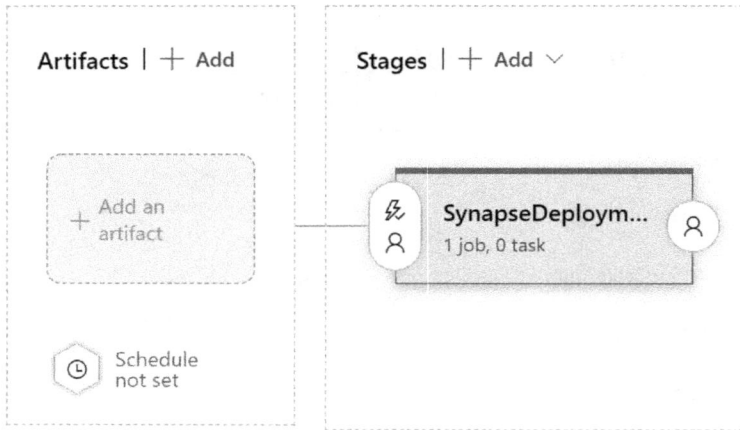

Figure 11.18 – Stage without tasks

5. In the stage, please click on the **1 job, 0 task** link to create the task to move your
    Synapse artifacts. The empty tasks view is displayed:

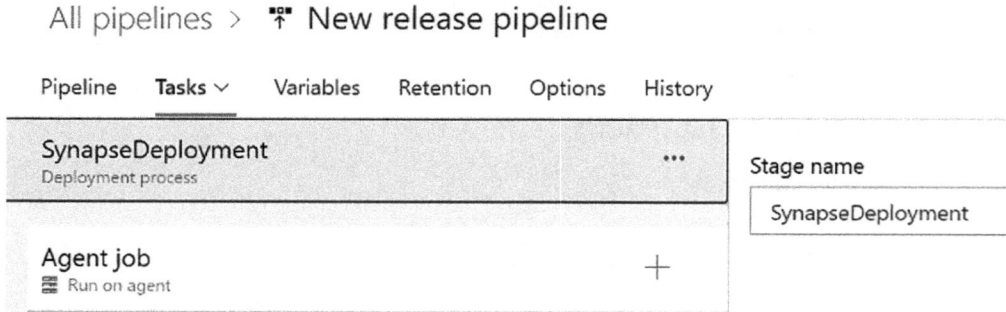

Figure 11.19 – Empty tasks view

6. Please hit the + symbol next to the **Agent job** entry. On the right, you will see all
    kinds of deployment tasks displayed for your selection. Please type `Synapse` into
    the search field, select the displayed **Synapse workspace deployment** task from the
    list, and click **Add**.

> **Note**
> The first time you do this DevOps will ask you to install the Synapse workspace
> deployment task. This is pretty straightforward. You just need to follow the
> wizard.

7. When you now select the newly added task in the list, the options to configure it are displayed on the right:

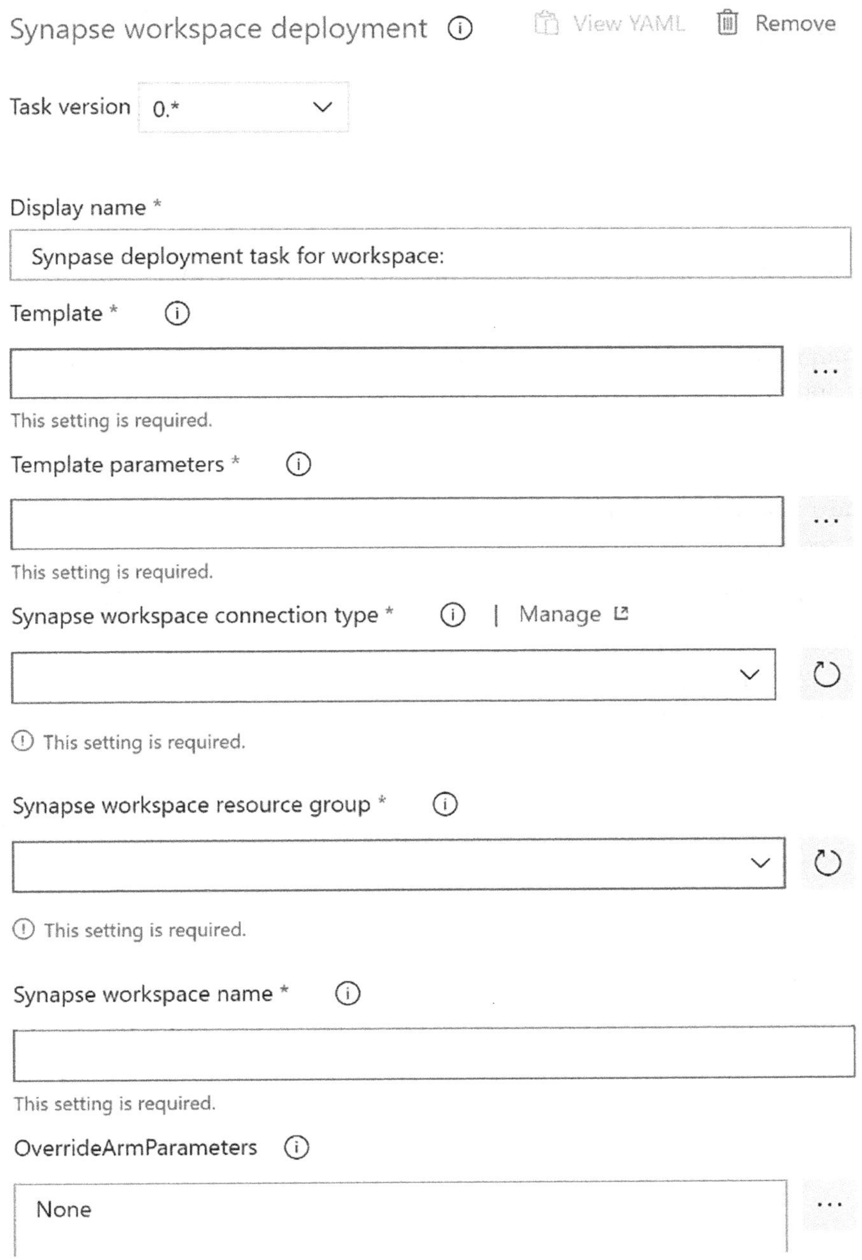

Synapse workspace deployment ⓘ   📋 View YAML   🗑 Remove

Task version   0.*   ⌄

Display name *

Synpase deployment task for workspace:

Template *   ⓘ

...

This setting is required.

Template parameters *   ⓘ

...

This setting is required.

Synapse workspace connection type *   ⓘ | Manage ↗

⌄   ⟳

ⓘ This setting is required.

Synapse workspace resource group *   ⓘ

⌄   ⟳

ⓘ This setting is required.

Synapse workspace name *   ⓘ

This setting is required.

OverrideArmParameters   ⓘ

None   ...

Figure 11.20 – Configuring an Azure Synapse workspace deployment

In the **Display name** field, you can name your task.

In the **Template** field, you will now select the **Azure Resource Manager (ARM)** template that describes your workspace. Please click on the three dots to the right of the field and navigate to the folder that was named after your Synapse workspace. Here, you will find two files: `TemplateForWorkspace.json` and `TemplateParametersForWorkspace.json`. Please select the first one, `TemplateForWorkspace.json`.

In the **Template parameters** field, please select the `TemplateParametersForWorkspace.json` file again by clicking on the three dots and navigating to the folder that was named after your Azure Synapse workspace.

In **Synapse workspace connection type**, you will first need to connect to the subscription where your target workspace was created. If nothing is displayed in the drop-down list, you might need to refresh the list with the button to the right of the field.

The next field you need to select is **Synapse workspace resource group**. This is the target resource group where the target Synapse workspace resides.

For **Synapse workspace name**, of course, you will need to set the target Synapse workspace.

Finally, you will need to set some variables in the **OverrideARMParameters** field. These correspond to the `secureString` parameters in your ARM template. Save your work for a moment and then go back to the ARM template in the workspace folder of the publish branch (`TemplateForWorkspace.json`). Let's examine the file and look for the `secureString` type parameters. In my example, this is the following:

```
1  {
2      "$schema": "http://schema.management.azure.com/schemas/2015-01-01/deploymentTemplate.json#",
3      "contentVersion": "1.0.0.0",
4      "parameters": {
5          "workspaceName": {
6              "type": "string",
7              "metadata": "Workspace name",
8              "defaultValue":
9          },
10         "PatsqlAzureML_servicePrincipalKey": {
11             "type": "secureString",
12             "metadata": "Secure string for 'servicePrincipalKey' of
13         },
14         "wscloudscalesynapse-WorkspaceDefaultSqlServer_connectionString": {
15             "type": "secureString",
16             "metadata": "Secure string for 'connectionString' of          -WorkspaceDefaultSqlServer'"
17         },
```

Figure 11.21 – TemplateForWorkspace.json secureString parameters

Please copy the parameter names and paste them into a Notepad file or something similar.

But where can you find the values that you will need for those parameters? Please navigate back to your development Azure Synapse workspace, to the **Management** hub. The parameters are pointing to linked services that you have created there, and you need to pick up some settings from there. Please go to the `...-WorkspaceDefaultSqlServer` linked service and click the button with the curly brackets, {}. A JSON file will be displayed. Search for the connection string and copy it to your Notepad file next to the name of the parameter:

```
1  {
2      "name":                    -WorkspaceDefaultSqlServer",
3      "type": "Microsoft.Synapse/workspaces/linkedservices",
4      "properties": {
5          "typeProperties": {
6              "connectionString": "Data Source=tcp:       .sql.azuresynapse.net,1433;Initial Catalog=@{linkedService().DBName}"
```

Figure 11.22 – Linked service JSON file

> **Note**
>
> Please **only** get the string without the double quotes. It will cause an error if you add the double quotes to the string.

For the second `secureString`, you will need the service principal key for your Azure Machine Learning service. Remember how you created an app registration in your Azure Active Directory, and you needed to create a new secret in *Chapter 9, Integrating Azure Cognitive Services and Machine Learning*, in the *Examining Azure Machine Learning* section?

Hopefully, you still have the key somewhere because it is needed now for the second parameter that is in the template. If you didn't keep it, don't worry. You can create another secret in Azure Active Directory for the app registration of your Azure Machine Learning app. Please check the description in *Chapter 9, Integrating Azure Cognitive Services and Machine Learning*, in the *Further reading, Creating a service principal* section.

Now let's use these collected values in your release pipeline. Please navigate back to your DevOps environment to the pre-started release pipeline from previously and navigate to the **Variables** section:

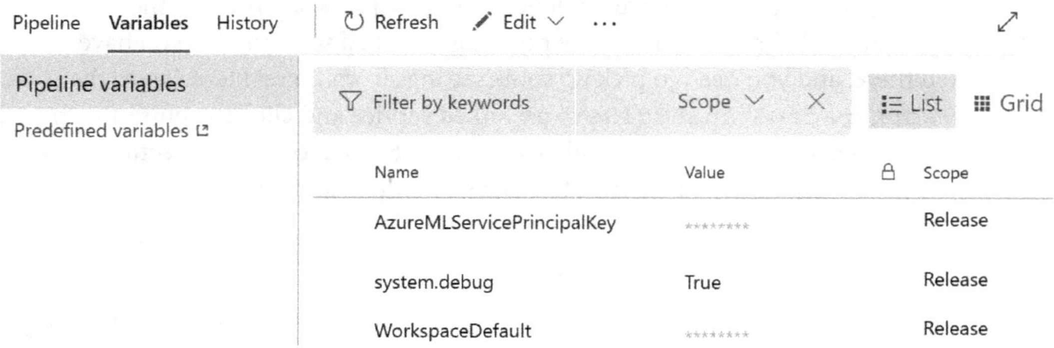

Figure 11.23 – Pipeline variables section

Please click **+ Add**, enter a variable name for your `WorkspaceDefault` SQL server, and paste the connection string that you have copied into the **Value** field. As this should be a `secureString` value, please click the lock symbol next to the **Value** field. The connection string will disappear, and the value will only be displayed as asterisks.

Repeat this step for `AzureMLServicePrincipalKey` and lock this one too.

If you want to receive additional information about your release pipeline, you may add `system.debug` as an additional variable and set it to `True`.

Now comes the moment where you will finish your pipeline configuration. Please go back to the **Tasks** section and to **Synapse deployment task for workspace:[YOURWORKSPACE]**. The last missing field is now **OverrideAParameters**. Please click the three dots and start mapping the parameter names from the template to the variables in your release pipeline:

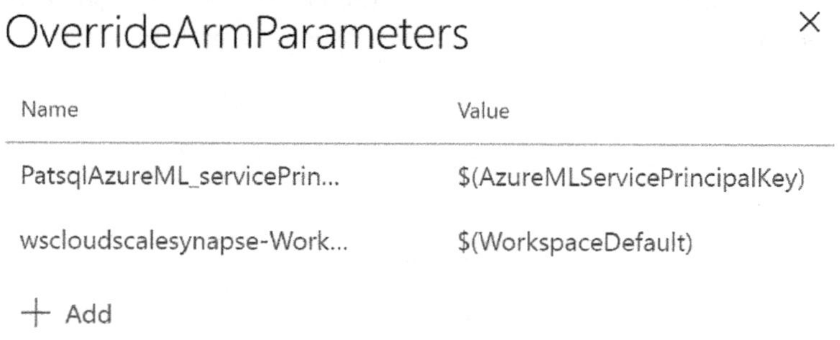

Figure 11.24 – Pipeline variables mapping to ARM template parameters

> **Note**
> The pipeline variables need to be wrapped in $ ( ) .

Finally, click **Save** again and store your work.

8. This next step will show whether you have configured everything correctly. Please look to the upper-right area of your DevOps window. You will find a **Create release** button. When you click this, you will manually trigger your release pipeline to deliver the content of your repository to the target environment.

   If you are ready to proceed, please click **Create** in the dialog and start.

9. To watch the progress of the release, you can click **View releases**. You will find a logging window that will present you with the progress of your deployment.

Up to this point, you have created a release pipeline and have manually triggered it to deploy your artifacts to the target environment.

The cherry on top of this section will now be to automate the release pipeline. We're going to do this based on changes in the publish branch when, for example, the project lead issues a pull request from the collaboration branch after finishing a sprint:

1. Please navigate to your newly created release pipeline and check the **Artifacts** section in the graphical representation. You can get there by editing the pipeline:

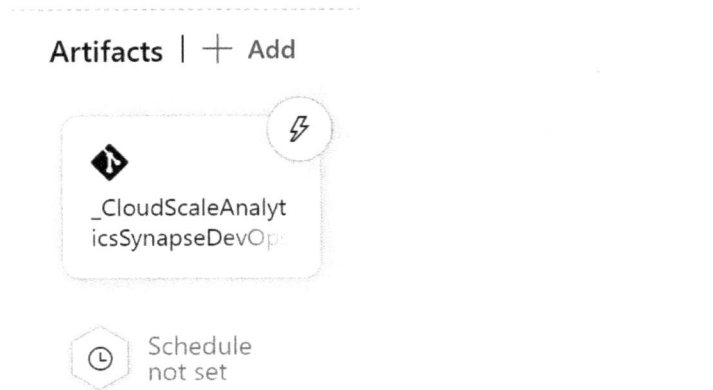

Figure 11.25 – The Artifacts section in the release pipeline

You could now initiate a schedule. But this is not what we want in this case.

The lightning button in the upper-right corner will bring you to the **Continuous deployment trigger/Pull request trigger** dialog. We will use **Continuous deployment trigger** as this will trigger a deployment based on the change of a selected repository.

2. Please enable **Continuous deployment trigger**.

3. You can now add a new branch filter by clicking **+ Add** and configuring **Type** to the **Include** filter for the `workspace_publish` branch:

Continuous deployment trigger

Git: _CloudScaleAnalyticsSynapseDevOps

 Enabled

Creates a release every time a Git push occurs in the selected repository.

Branch filters ⓘ

Type	Branch
Include ⌄	⅛ workspace_publish ⌄ 🗑

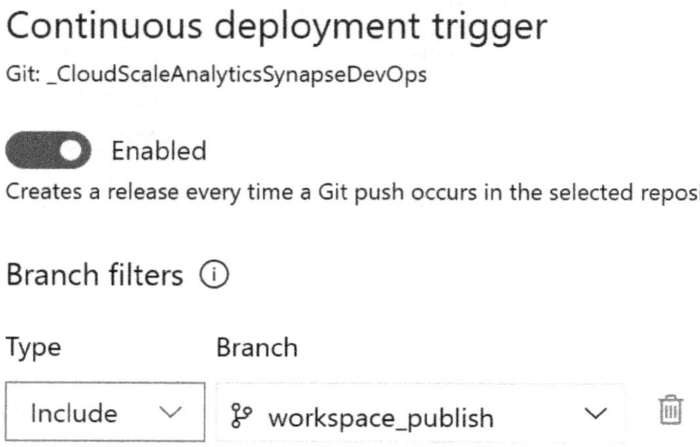

Figure 11.26 – The Include branch filter

4. Now, please click **Save** and store your work.

5. To test this configuration, now you will need to go through the typical life cycle as described in the *Understanding the development life cycle* section. Check out the repository into a `working` branch, change something in the SQL script, for example, and create a pull request into the `collaboration` branch. Once the pull request is created, hit **Publish** in the Azure Synapse workspace, and then go to the DevOps environment. There, please issue another pull request from `collaboration` into your `workspace_publish` branch. Once this pull request has finished, please go to your release pipeline, and check the releases. There you should either find a release that is actively running, or it might be queued and will run soon.

When you check the stage, you might find something like this:

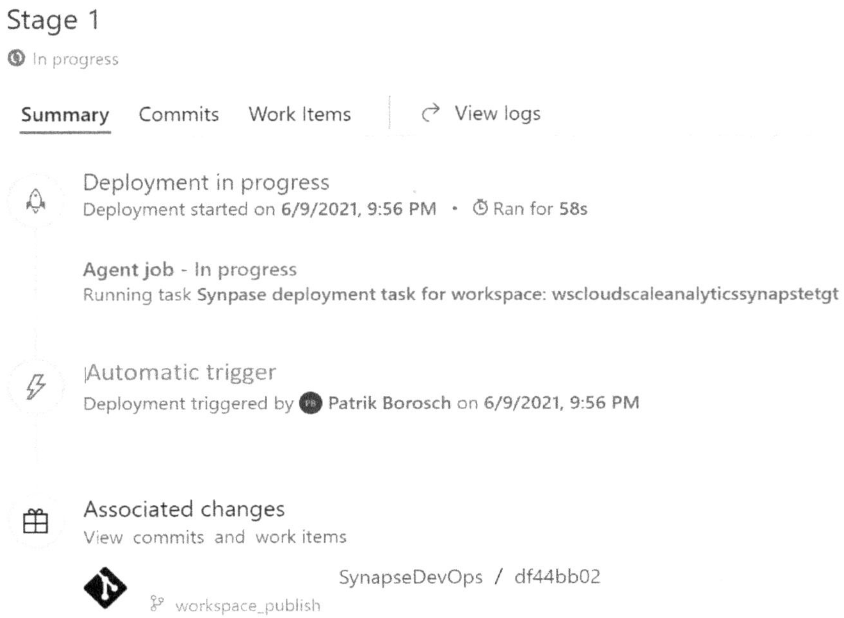

Figure 11.27 – Automatically triggered deployment

If you run into errors, for example, relating to authentication, please check the different areas where you set up authentication. For example, check whether the service principal for DevOps is configured as Contributor for your Azure Synapse workspace at the Azure portal level and as Synapse Administrator in the Synapse workspace itself.

If you encounter more and other errors, please visit the *Further reading, Continuous integration* section.

Understanding developer productivity with Synapse Studio

Synapse Studio offers you a truckload of wizards and supporting functions to make your life as a developer easier. You have already seen some of them and this section will give you a brief overview.

Using the Copy Data Wizard

In *Chapter 5, Integrating Data in Your Modern Data Warehouse*, you saw a hint about the Copy Data Wizard. It is available in Synapse pipelines and in Azure Data Factory too.

It is not just a good decision to examine and learn how the Copy activity can be used. You can also use it to quickly load data into your environment.

In Synapse Studio, you will find the wizard on the home screen when you look for **Ingest**:

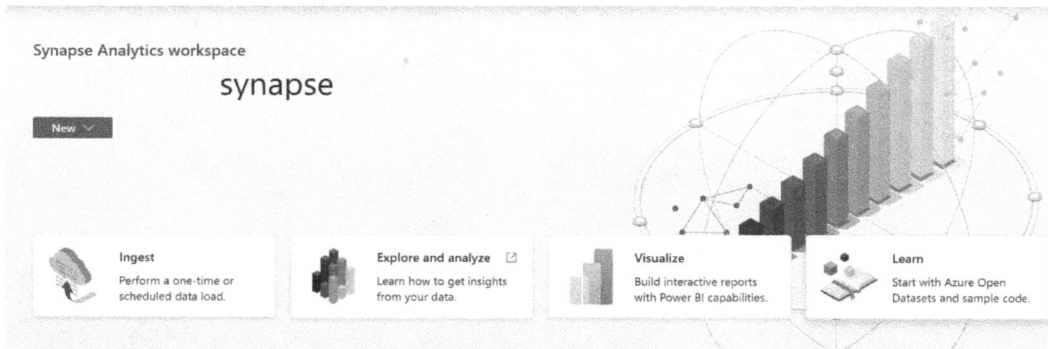

Figure 11.28 – Copy Data Wizard in Synapse Studio

> **Note**
>
> When you are using Azure Data Factory in standalone mode, you will find it on the home screen too, and it is named **Copy data**.

The wizard will take you through several steps:

1. Select the task type for your Copy data pipeline. You have two options here: using the built-in copy task or the newly added create a metadata-driven copy task. The latter will add additional tables to a database that you can name. There, you will have the option to add metadata about sources and targets to dynamically use them in your activity.

    You can also decide on the cadence and/or the schedule for your task.

2. Define an existing data store or create a new one as your source connection.

3. Detail the selection of objects to copy from the source data store and configure them.

4. Define an existing data store or create a new one as your target connection.

5. Detail the configuration for your targets and configure them.

6. Configure settings such as **Data consistency verification**, **Fault tolerance**, and **logging**, or performance settings such as **Staging**, the number of data integration units, and **Degree of copy parallelism**.

7. Check the summary.

8. Start the activity in a pipeline. On this screen, you have the option to click **Edit pipeline** to switch to the development canvas, where you will see your newly created artifact. You can change settings there or copy the activity and use it in another pipeline, for example.

As mentioned already, the Copy Data Wizard is a nice option for beginners as it even helps you with more complex tasks such as copying several sources in one go. It is quite intuitive and gets you up and running fast. But it is used by experienced developers too when they quickly want to load data from *a* to *b*, or just want to create a Copy activity.

Integrating Spark notebooks with Synapse pipelines

The Copy Data Wizard is not the only supporting function that Synapse Studio offers relating to Synapse pipelines.

In *Chapter 6, Using Synapse Spark Pools*, you saw how to quickly integrate a Spark notebook either with an existing Synapse pipeline or create a new one directly in the dialog.

This comes in handy when you need to add complex transformation logic to a pipeline. Maybe you want to train a machine learning model as part of the pipeline, you want to integrate a complex training run with Azure Machine Learning, or you have other requirements where you need a Spark environment.

Analyzing data ad hoc with Azure Synapse Spark pools

Before you can use a Spark notebook with a Synapse pipeline, you will, of course, need to create one. Synapse Studio allows you to kick-start a notebook with a right-click on a source, for example, from your data lake, your dedicated or serverless SQL pool, or your Synapse Link-attached Cosmos DB:

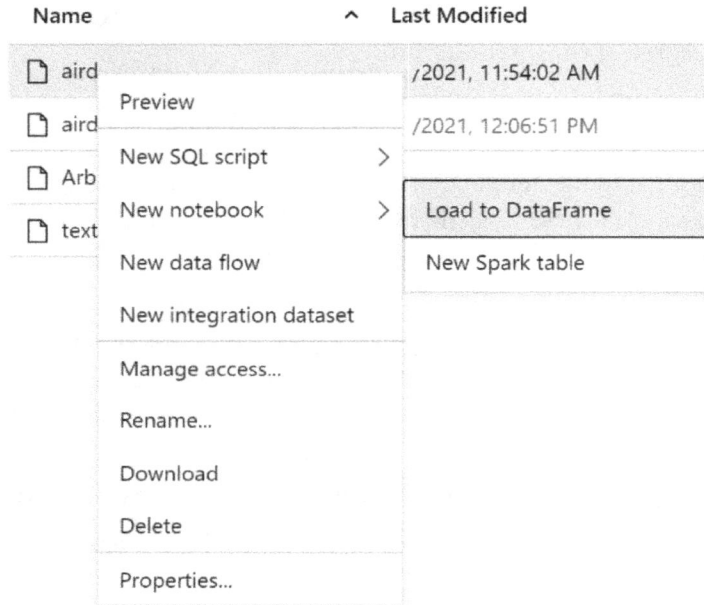

Figure 11.29 – Creating a Spark notebook and loading data to a new DataFrame

You have seen this too in *Chapter 6*, *Using Synapse Spark Pools*. The function will start a new notebook, add the code needed to read the selected source, and create a DataFrame from its content. A nice starting point!

Creating Spark tables

Right below the **Load to DataFrame** option shown in *Figure 11.30*, you'll find the **New Spark table** entry. Again, a notebook is displayed that has code added to create a Spark table from the data that was selected. You have also seen this option already, in *Chapter 9*, *Integrating Azure Cognitive Services and Machine Learning*. You used this option when you integrated the sentiment analysis cognitive service.

This option additionally comes in handy for another option that you find when you right-click an existing Spark table. See the next section, *Enriching Spark tables*.

Enriching Spark tables

In *Chapter 9*, *Integrating Azure Cognitive Services and Machine Learning*, you used a Spark table with an existing cognitive model. But when you right-click a Spark table, you find an additional option there that will take you to another wizard: **Enrich with new model**.

This will take you through a wizard that will create a notebook that will use the automated machine learning option of Azure Machine Learning.

You will need to select an Azure Machine Learning workspace from an Azure Machine Learning-linked service, if available, in your Azure Synapse workspace. You'll configure an experiment name, the best model name, the target column to predict, and an Apache Spark pool from your Synapse workspace. Additionally, you can configure the machine learning model type, such as **Classification**, **Regression**, or **Time series forecasting**, and their options.

The entire step will be coded into a new notebook again and will be made to be displayed immediately.

Do you remember that you can use this notebook now in a Synapse pipeline and orchestrate it into your jobs?

Enriching dedicated SQL pool tables

A similar function to the last one from the *Enriching Spark tables* section is available for tables in dedicated SQL pools too.

When you right-click a table in a dedicated SQL pool, you will again find a **Machine Learning** entry in the context menu. You will again find the **Enrich with existing model** entry:

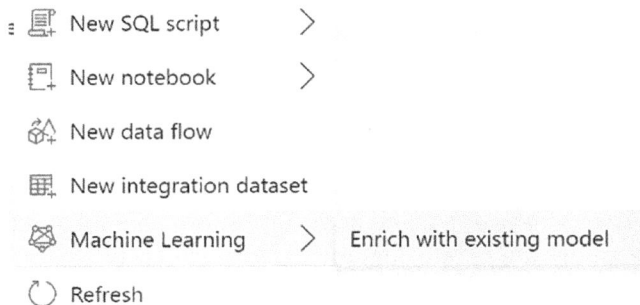

Figure 11.30 – Enriching a dedicated SQL pool table with a machine learning model

This wizard will take you through a sequence where you will select an **Open Neural Network Exchange** (**ONNX**) model from your Azure Machine Learning workspace and create a stored procedure to implement it with a selection of source columns and a target column to predict.

Finally, the wizard will end by importing the ONNX model into a `varbinary(max)` column into a table and the stored procedure that is provided will implement the native `predict()` function to score data that lives inside the dedicated SQL pool or is read via PolyBase. Please check the *Further reading, Native predict with dedicated SQL pools* section for more information.

Creating new integration datasets

Another menu item that you can see in *Figure 11.31* is **New integration dataset**. This entry is a convenient way to quickly create a dataset that can be used as a source or a sink in Synapse pipelines.

This feature is available for the following:

- Tables in dedicated SQL pools
- Files in the attached storage accounts

After the right-click, you will be prompted to add a name and select the file format in the case of a file, as well as whether and from where to get the schema information.

Starting serverless SQL analysis

Let's not forget the serverless SQL pools and the option to create new SQL scripts in many cases. You can use the **New SQL script** entry of the context menu with the following:

- CSV and Parquet files in your storage accounts and data lakes
- Spark tables in your Apache Spark pools
- Data from Synapse Link attached to Cosmos DB collections

Up to now, we have always used it to create a new SQL script to select the top 100 rows from the data. But if you examine the menu item closely, you will find another option, **Bulk load**. This will support you in bulk loading the selected file or folder into a dedicated SQL pool table.

You will be prompted for some configuration options, such as **File type** and **Field terminator**, and you will then set the dedicated SQL pool to load the data to and whether you want to use an existing table or create a new one. Finally, you can decide whether to instantly start loading or to see the resulting SQL script and maybe keep it for reuse.

Backing up and DR in Azure Synapse

There are several aspects that you need to consider when you are planning backup and DR in your Azure Synapse workspace. You'll need to take care of three main areas:

- Data
- Developed artifacts
- Infrastructure setup

Backing up data

You will first look after your data and take care that you don't lose your biggest assets here. The two areas that you want to consider could be the following:

- Storage accounts/data lakes
- Dedicated SQL pools

As the serverless SQL pools and Spark pools are talking to the data lake, mainly you will cover those by backing up the data lake.

Backing up storage accounts/data lakes

A data lake/storage account can be prepared for DR in different ways. There are the following:

- **Automated redundancy built into the service**. You have already learned about this in *Chapter 3*, *Understanding the Data Lake Storage Layer*. You can decide on the zone or geo-redundant storage and their alterations. These functions will ensure that if parts of the, or even the whole, Azure data center where your data is stored goes offline, you will still have your data in a second place where you can access it.

  The caveat with this functionality is that data is replicated by the service automatically. This means if you delete data accidentally, the deletion will be carried through to the secondary area too and the data is still gone.

- **Immutable options for your data lake**. You can place a **delete lock** on your data lake account. This will avoid the unintentional deletion of the whole account.

- The caveat here is that this does not help you if the data center is partially or entirely not available. And still, data might be lost if an administrator accidentally deletes the data anyway.

- **Write locks on your folders**. Again, the data might be not available if the data center is partially or entirely not available and again, data might be lost if an administrator accidentally deletes it.

At this point in time, you will need to back up the data that is needed using a routing process that you will need to set up yourself.

One recommendation would be to create a second data lake store and a Synapse pipeline Copy activity that differentially copies all the files that you want to be backed up over to the second data lake store.

> **Note**
> There are other features, such as versioning and soft delete, that are already available with storage accounts but not yet with data lake accounts but are on the roadmap. These will help you with the drawbacks mentioned previously as they will avoid unintentional data loss automatically without the need for copying data around. Stay tuned for the arrival of these functions as they can then save you some money, making the second (backup) data lake account obsolete.

Backing up dedicated SQL pools

Synapse-dedicated SQL pools perform snapshot backups without your explicit action or configuration. This is a default setting and is performed by the engine automatically. These automated snapshots will give you a **Recovery Point Objective** (**RPO**) of 8 hours.

The service will give you a retention time on your snapshots of 7 days or a maximum of 42 automatic restore points, whatever comes first. At this point in time, there is not yet a feature for long-term retention times on snapshots.

> **Note**
> When the service is paused, there are no snapshots taken. You may need to take care of taking the snapshots yourself.

You can issue user-defined snapshots either using PowerShell or from the portal. Again, these have either a retention time of 7 days or when the threshold of 42 snapshots is hit, whatever comes first.

If you want to dive into the details on how to issue a user-defined snapshot, please refer to the *Further reading, Dedicated SQL pools user-defined snapshots* section.

> **Important Note**
>
> When a SQL pool is deleted, a final snapshot is issued and kept for 7 days. But be aware that if you delete the entire Synapse workspace, you will lose any data as you can't restore a dedicated SQL pool snapshot when the workspace, and with it the server object, is deleted.

Keeping a snapshot

There is a workaround to enable long-term retention of a certain snapshot or version of your dedicated SQL pool.

Remember the creation sequence of a dedicated SQL pool? On the **Additional settings** tab, you can decide to **Use existing data** and whether you use either **None**, a **Backup**, or a **Restore point**:

Data source

Start with a blank dedicated SQL pool or restore from a backup to populate your new dedicated SQL pool.

Use existing data *   None   Backup   Restore point

Figure 11.31 – Creating a dedicated SQL pool from existing data

You can use a restore point to create a new dedicated SQL pool and instantly pause it. Doing so, you are only charged for the storage that the SQL pool uses and you are able to immediately spin it up if you need data from the restore point.

For more information on backup and restore on dedicated SQL pools, please see the *Further reading, Backup and restore* section.

Monitoring your MDWH

Monitoring your workloads with different services on Azure might be a challenging idea when you consider the isolated service. But as mentioned in other chapters already, all Azure services that we have touched on integrate with Azure Monitor. Their logs and metrics can be sent to Log Analytics and can be analyzed together with other services at a higher level.

When we check the Azure Synapse-dedicated SQL pools, for example, you will have the option to select and push from the following log information to Log Analytics (and two other options; see the following):

- **SQLRequests**: These are insights about the queries run in the pool, especially the distribution of the steps.

- **RequestSteps**: Here, you will find the details about the query steps.

- **ExecRequests**: High-level information about the queries.

- **DmsWorkers**: Insights about the worker nodes that are executing steps during query execution.

- **Waits**: Wait states during query runs; you will also find locks here, for example.

- **SQLSecurityAuditEvents**: Insights about the who, when, and what in your database.

These logs represent the content of the dynamic management views with similar names.

Additionally, you will find an extensive list of metrics that can be provided regarding your dedicated SQL pool, such as `DWHLimit`, `DWHUsed`, and `DWHUsedPercent`. You will find a link to the list of the most important ones in the *Further reading, Dedicated SQL pool metrics* section.

When you enable diagnostic settings for a dedicated SQL pool but also for other services, you will always have the option to define the target to where the information will be sent. You can choose from the following:

- Send to a Log Analytics workspace.

- Archive to a storage account.

- Stream to an event hub.

We have touched on Log Analytics already. This is where you can correlate different logs from different services.

You can additionally archive your logs and even stream them to an event hub from where you can react to different entries. Remember Azure Stream Analytics and what we covered in *Chapter 8, Streaming Data into Your MDWH*? The circle closes. By using Event Hubs and Stream Analytics, you will be able to stream events to a location of your choice and analyze them to create additional insights or maybe even trigger events based on certain log entries.

Understanding security in your MDWH

When you are using Azure services, there are always two aspects regarding security. You can set up access control where you grant or revoke **Role-Based Access Control (RBAC)** roles or **Access Control Lists (ACLs)**.

We have touched on these concepts already in *Chapter 3*, *Understanding the Data Lake Storage Layer*, and in other chapters too when we have set up services and their connections.

The other perspective in the security topic is networking, such as when you want to hide your services completely from the outside world and the so-called public internet. You can peer your on-premise network to Azure Virtual Network. Typically, you will set up a so-called landing zone from where you will route traffic to the target services, such as your data lake, for example, or your Azure Synapse workspace with its computes.

Additionally, you will then implement IP firewall rules for the services that you are securing.

Implementing access control

You will find the access control entry in any Azure service that participates in your MDWH setup. In your data lake, for example, you have implicitly added the RBAC Storage Blob Data Contributor for the Synapse Workspace Managed Identity role. This will enable the Synapse workspace with its services to access the data.

Passing through credentials to the data lake

Anyway, Azure Synapse will pass through your personal credentials to the data lake if you don't use any other authentication method. Therefore, you will need to manage the access for users/groups explicitly. In *Chapter 10*, *Loading the Presentation Layer*, *Figure 10.2* sketches the data lake access from a dedicated SQL pool using PolyBase or the COPY INTO statement. You can take this structure as an example of how the information flow and authentication work. Serverless SQL, as well as Spark pools, will work similarly.

Other storage components will work in a comparable manner. You'll need to set up access control for users/groups in a "normal" storage account, for example, as well.

Implementing access control in the Synapse workspace

The Azure Synapse workspace itself implements another layer of access control. This works similarly to what you already know from the Azure portal. Azure Synapse adds additional workspace-level RBACs to the system:

- Synapse Administrator
- Synapse SQL Administrator
- Synapse Apache Spark Administrator
- Synapse Contributor (preview at the time of writing)
- Synapse Artifact Publisher (preview at the time of writing)
- Synapse Artifact User (preview at the time of writing)
- Synapse Compute Operator (preview at the time of writing)
- Synapse Credential User (preview at the time of writing)

When you are adding role assignments, you will find another option to explicitly control access to workspace items. Here, you can control RBACs for the following:

- Apache Spark pools
- Integration Services
- Linked services
- Credentials

For these items, you can grant either Synapse Administrator, Synapse Contributor, or Synapse Compute Operator for the upper two and Synapse Administrator or Synapse Credential User for the lower two.

The combination of Storage Access Control, Synapse Workspace Access Control, and Synapse Workspace Item Access Control determines the access options that a user can be granted.

For a detailed overview of all the Azure Synapse RBAC roles and their detailed access rights, please check the *Further reading, Synapse Workspace RBAC roles* section.

Understanding SQL user access control

Both dedicated SQL pools and serverless SQL pools implement the typical user/group access rights. Again, please check *Chapter 10, Loading the Presentation Layer, Figure 10.2*, for an overview of SQL rights in the example of creating external PolyBase tables.

Please refer to *Further reading, SQL authentication* for a detailed overview of both serverless and dedicated SQL pools.

Implementing networking

Do you remember the provisioning sequence of your Azure Synapse workspace? In the **Networking** blade, you had the **Allow connections from all IP addresses** and **Enable managed virtual network** options. These two settings will allow you to route your traffic in a nicely controlled way through your virtual networks.

> **Note**
> You can still configure the IP firewall settings after the workspace is created.
> But you can't switch on the managed virtual network if you didn't do so during the creation process.

Let's examine together how you would set up networking between your Azure Synapse workspace and your data lake store:

1. First, please go to your Azure Synapse workspace with managed virtual network enabled and navigate to the **Management** hub. The **Managed private endpoints** menu item in the **Security** section should be available.

Here you will find two predefined private endpoints. Please click + and start creating a new one. In the dialog that is displayed on the right, please select **Azure Data Lake Storage Gen2**:

New managed private endpoint (Azure Data Lake Storage Gen2)

ⓘ Choose a name for your managed private endpoint. This name cannot be updated later.

Name *

AzureDataLakeEndpoint

Description

Account selection method ⓘ
🔘 From Azure subscription   ⚪ Enter manually

Azure subscription ⓘ

Select all   ⌄

Storage account name *

⌄ ↻

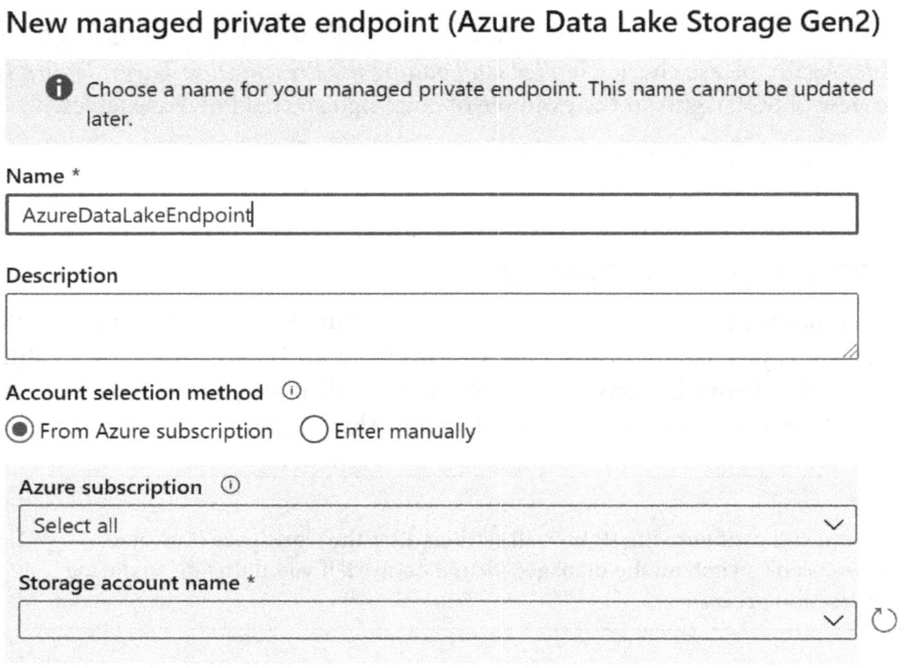

Figure 11.32 – New managed private endpoint dialog

After that, you can now select your Azure subscription and the storage account name where your newly created private endpoint should be used. When you click **Create**, your new private endpoint will show up but will need approval from the administrator of the data lake. It will be displayed in the list, too, with the **Pending** approval state:

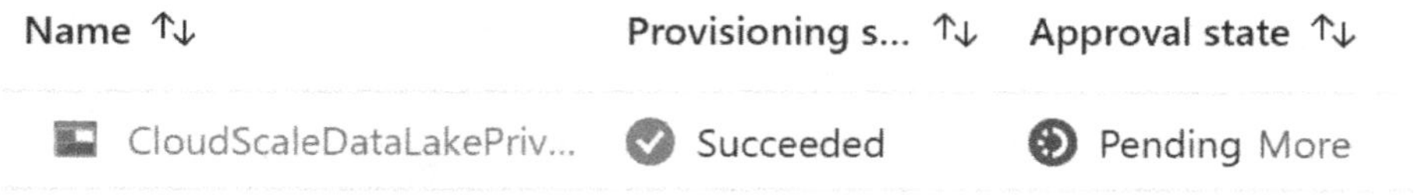

Name ↑↓	Provisioning s... ↑↓	Approval state ↑↓
🖼 CloudScaleDataLakePriv...	✅ Succeeded	🔵 Pending More

Figure 11.33 – Pending private endpoint

2. Now, please navigate to your data lake store in your Azure portal and search for the **Networking** entry in the **Navigator** blade.

3. On the **Firewalls and virtual networks** tab, please click **Selected networks**. After saving the settings, your data lake will not be accessible from anywhere else.

4. Please now approve your newly created private endpoint. Please switch tabs in the top area of the blade to **Private endpoint connections**. You will see the private endpoint that was issued from your Azure Synapse workspace, too, in the **Pending** connection state:

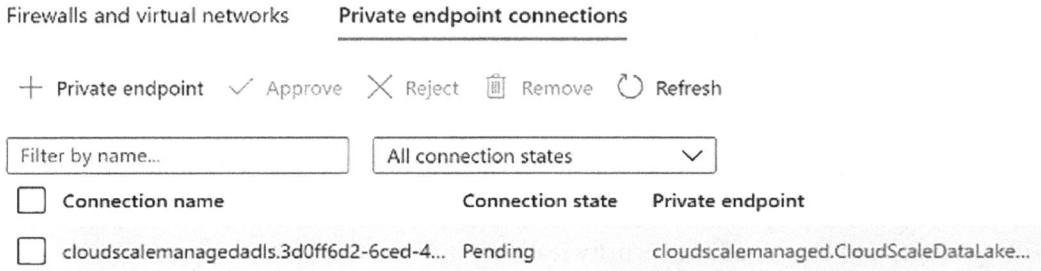

Firewalls and virtual networks   **Private endpoint connections**

$+$ Private endpoint $\checkmark$ Approve $\times$ Reject   🗑 Remove   ⟳ Refresh

Filter by name...	All connection states ⌄

☐ Connection name	Connection state	Private endpoint
☐ cloudscalemanagedadls.3d0ff6d2-6ced-4...	Pending	cloudscalemanaged.CloudScaleDataLake...

Figure 11.34 – Data lake networking: Private endpoint connections

5. Please select the entry and click **Approve** in the ribbon above the list. After entering a description, you can finish the approval.

6. Please switch back to the **Firewalls and virtual networks** tab and search for the **Resource instances** section somewhere in the lower middle of the page. Please select `Microsoft.Synapse/workspaces` for **Resource type** and next to that field, in the **Instance name** field, select your Azure Synapse workspace.

7. Finally, please click **Save** to store your changes.

Your data lake is now privately connected to your Azure Synapse workspace and can no longer be reached from elsewhere.

> **Important Note**
> If you don't have access to your files after setting up the IP firewall, the private endpoint, and the exception for your workspace and you are completely sure that you have set up the access control correctly for your user, you might want to follow this script: `https://docs.microsoft.com/en-us/azure/synapse-analytics/sql/develop-storage-files-storage-access-control?tabs=user-identity#querying-firewall-protected-storage`.

As you have seen from the list of data services on Azure, you can now also derive and transfer your knowledge to other components that might participate in your analytical estate.

Summary

This chapter took you through the integration between Azure Synapse and Azure DevOps. You learned how to connect Azure Synapse to Azure DevOps and how to automate deployments when creating new, or updating existing, artifacts.

You gained some insights into developer productivity with Synapse Studio and how to benefit from supporting functions of Synapse Studio.

Afterward, you examined how to implement backup and DR and how to monitor your Azure Synapse environment.

Finally, you saw and examined the security features of Azure Synapse.

In the upcoming chapter, *Chapter 12, Distributing Data*, you will learn how to connect your Power BI service with Azure Synapse and create Power BI reports directly in Synapse Studio.

Further reading

- Continuous integration: https://docs.microsoft.com/en-us/azure/synapse-analytics/cicd/continuous-integration-deployment

- Native predict with dedicated SQL pools: https://docs.microsoft.com/en-us/azure/synapse-analytics/sql-data-warehouse/sql-data-warehouse-predict

- Dedicated SQL pools user-defined snapshots: https://docs.microsoft.com/en-us/powershell/module/az.sql/new-azsqldatabaserestorepoint

- Backup and restore: https://docs.microsoft.com/en-us/azure/synapse-analytics/sql-data-warehouse/backup-and-restore

- Dedicated SQL pool metrics: https://docs.microsoft.com/en-us/azure/synapse-analytics/monitoring/how-to-monitor-using-azure-monitor

- **Synapse workspace RBAC roles:**
- a) https://docs.microsoft.com/en-us/azure/synapse-analytics/security/synapse-workspace-synapse-rbac-roles
- b) https://docs.microsoft.com/en-us/azure/synapse-analytics/security/synapse-workspace-understand-what-role-you-need
- **SQL authentication:** https://docs.microsoft.com/en-us/azure/synapse-analytics/sql/sql-authentication

12
Distributing Data

After you have brought your data into your analytical environment and massaged it into the required shape, you may have even created machine learning models to acquire deeper insights into the circumstances that your data describes. Then, after loading it into the right storage for optimal access, you, of course, are now looking for ways to distribute the data in the right way to the right target groups.

This chapter will show you ways in which to create **data marts** to distribute insights in your modern data warehouse with Power BI. You will see how to use Power BI data models and the options to visualize and publish their content and even use the data with other tools.

Additionally, you will explore the options that you have to create a data model with **Azure Analysis Services (AAS)** aside from Power BI, that is, other options available to you to form insights from your data and make them available.

Finally, we will examine Azure Data Share as another option for providing datasets to others.

You will find these topics covered in the following sections:

- Building data marts with Power BI
- Creating data models with Azure Analysis Services
- Distributing data using Azure Data Share

Technical requirements

To follow the exercises in this chapter, you will need the following:

- A Power BI workspace and Power BI Desktop, which you can download for free from `https://powerbi.microsoft.com/en-us/desktop/`
- An Azure subscription where you have at least contributor rights or you are the owner
- The right to create an AAS instance
- A Synapse workspace with a data lake attached
- The right to create an Azure Data Share instance

Building data marts with Power BI

In our modern data warehouse architecture in *Chapter 2, Connecting Requirements and Technology*, you were introduced to Power BI as the central reporting tool if you are either planning for small-sized implementations or planning for an enterprise-wide analytical data estate.

Power BI offers a toolset to do the following:

- Get data and wrangle it into the Power BI data model if required.
- Store your data in a columnstore database for fast visualizing and reporting.
- Equip you with a data analysis language called **Data Analysis Expressions (DAX)** to implement the business logic that you need for your reports and dashboards.
- Develop all the required artifacts.
- Publish datasets, reports, and dashboards and collaborate with your co-workers.

Understanding the Power BI ecosystem

If you examine the Power BI ecosystem, you will find different components that will serve the functions mentioned:

- **Power BI Desktop**: This is the central development tool. You develop the data acquisition, data model, business logic, and all the visualization options of your reports on the desktop. It offers the complete functionality for developing with Power BI. Power BI is a free tool. You can download it and use it for free.
- **The Power BI service**: `www.powerbi.com` delivers a portal and workspaces for each participant and workgroup. This is the place where you publish reports with their datasets. The Power BI service offers you a rich surface for the creation of dashboards and collaboration on your data.

- With **Q&A**, the service adds a query option where you can interact with your data models using your own natural language. You don't need to use a SQL query or another programming language to get data and the service even suggests visualizations that best suit the data queried.

  Streaming datasets complete the functionality of the service. You can stream data directly to `www.powerbi.com` and create live dashboards that show up-to-date information about your company.
- **Power BI mobile**: This is a mobile app to be used on mobile devices to interact with your reports and dashboards.
- **Power BI Embedded**: When you are planning to embed Power BI into your application, this is the service to hold the data models for it. You can embed visuals within your application and connect them to data models.
- These visuals can be controlled from within your application. You can inject a user context and filters from your code and control the appearance and what is displayed.
- **Power BI Report Server**: This is a kind of legacy application that originates from SQL Server Reporting Services. You can install this service on-premises and create a reporting environment on your servers. This option will not offer you the complete functionality of the Power BI portal. For example, you cannot use the Q&A feature for natural language querying, but it is a good opportunity to start a reporting ecosystem if you are not allowed to move to the cloud.
- **Synapse Power BI**: Synapse Studio integrates a Power BI authoring option, too. With this, you don't need to leave the Synapse environment for report authoring.

Do you have Power BI Desktop already installed? If not, please find the link to the free download here: `https://powerbi.microsoft.com/en-us/desktop/`. Install it now as you will need it to follow the exercises in the next section.

Understanding Power BI object types

Power BI uses the following object types to fulfill its purpose:

- **Datasets/data models**: Will be built from imported or connected data sources and can combine data from many different data sources.
- **Visuals**: Are used to display data from the datasets.
- **Reports**: Combine visuals that belong together into one semantical collection.
- **Dashboards**: Are created in the Power BI portal and can display visuals from different reports.
- **Workspaces**: Collaborative spaces for workgroups. Each user has their own private workspace for their own work.

Understanding Power BI offerings

Power BI offers different pricing tiers according to the availability of functionality. It is important to understand the availability of a free tier, which allows you to explore the whole Power BI experience:

- **Power BI free**: You can use Power BI Desktop and the Power BI service for free for your personal usage. You won't be able to share reports and dashboards with others with this option.

- **Power BI Pro**: Enables you to share your work with other Power BI Pro license holders, for example, or in a premium capacity. You will have a certain volume available for your datasets on the portal and you are limited, for example, to a data refresh interval of a maximum of eight times a day. The Pro license is required for every Power BI developer who needs to publish reports to a broader audience.

- **Power BI Premium**: This edition comes in two flavors: a per-user and a per-capacity version. You have a possible refresh interval of 48 times a day, and 100 GB of data model size is available for the per-user version and 400 GB on the per-capacity model. Remember the columnstore compression? The ratio is 1:10 on average. So, the 400 GB of the premium model translates to 4 TB in raw row-oriented data.

To understand the detailed differences between the offerings, visit `https://powerbi.microsoft.com/en-us/pricing/`.

Acquiring data

Once you have installed Power BI Desktop, you can start to build your first report. To do so, you will need to get data into your Power BI file.

You can connect Power BI to 120+ different sources. From SQL Server to Oracle to Sybase and Teradata, there is a multitude of database connectors available as well as file formats such as Excel, CSV, XML, JSON, and Parquet.

There is also much more, including application sources such as SAP or Salesforce and OData feeds or folders. Another option would be, for example, to write a Python or R script that will produce a data frame and hand it to Power BI. We distinguish the data sources into the following groups:

- Databases
- Files
- Power Platform

- Azure

- Online services

- Others

Please check the *Further reading, Power BI data sources*, section for an exhaustive list of available connectors.

Connecting to a dataset

When you start Power BI Desktop, you will be prompted to get data to create visuals. On the dialog that pops up, you'll find additional information and videos with examples. Let's start with a simple example and use the Parquet file that you created with the Synapse serverless SQL pool in *Chapter 10, Loading the Presentation Layer*:

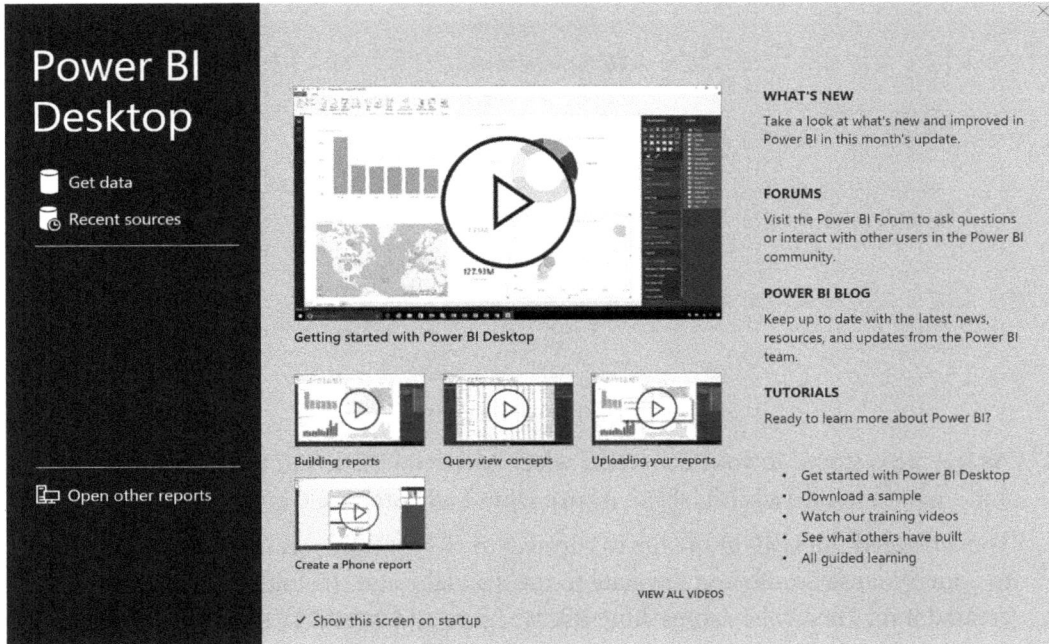

Figure 12.1 – Starting Power BI

Figure 12.1 shows the splash screen that is displayed when you open Power BI Desktop:

1. To start with a new dataset, click on **Get data**. The **Get Data** dialog will show up:

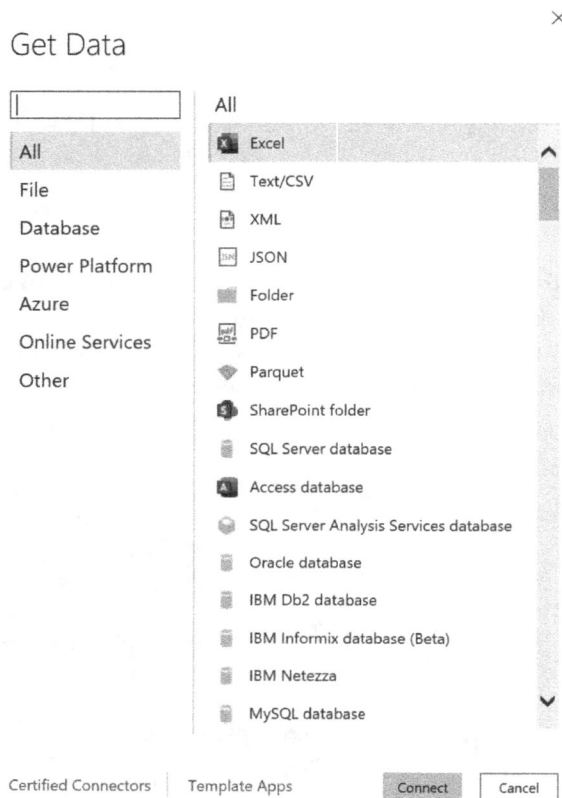

Figure 12.2 – The Get Data dialog

2. As the file is stored in your data lake, select **Azure** on the left in the navigator view, and then, in the details list, select **Azure Data Lake Storage Gen2**.

3. You will need the URL of the file to connect to. To get this, you will need to browse to your Synapse Studio and navigate to the data lake and the folder that you have created it in. This will be something like `https://[YOURFILESYSTEMNAME]@[YOURSTORAGEACCOUNTNAME].dfs.core.windows.net/[YOURPATH]/SQLServerlessTarget/myParquetfile.parquet`.

    You can right-click the file and select **Properties** and then copy the path from the URL property in the properties dialog.

4. With the URL in your cache, switch back to Power BI Desktop with the started **Get Data** sequence. In the **Azure Data Lake Storage Gen2 URL** dialog, paste the URL that you got from the data lake in *step 3* and then click **OK**:

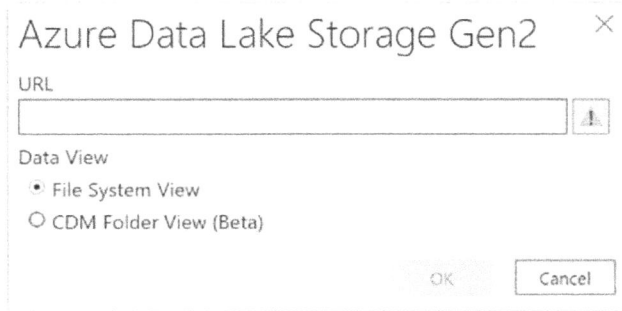

Figure 12.3 – URL prompt for your data lake file

5. You'll now be prompted to authenticate against your Azure account. Once you are logged in, you will be presented with the following view where you will see some file metadata in a row:

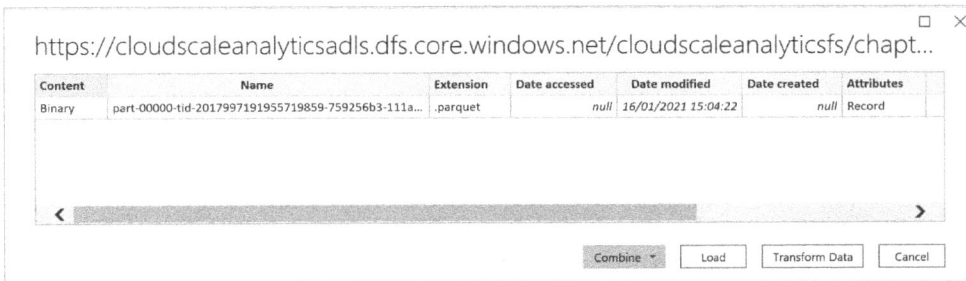

Figure 12.4 – The Transform Data dialog

Click **Transform Data** and go on to the Power Query Editor to start transforming the input that comes from the file:

Figure 12.5 – Power Query Editor

Look familiar? Yes! You have seen a similar view already in *Chapter 5, Integrating Data in Your Modern Data Warehouse*. You have a similar option to wrangle data in your Synapse pipeline/Azure Data Factory wrangling data flow.

6. Up to now, you have only seen one row that contains metadata about your file. This is due to the binary character of the Parquet file. You haven't seen the content yet. A CSV file would be shown directly with all columns in this editor.

    If you check the columns, you will find the first one with the header **Content** and the word **Binary** representing a link in the value field below.

    There are two options to resolve the content of your file so as to be able to work with it. You can click on either the link in the value field or the transform button next to the column title.

    Clicking on the link will only display the content of the file. If you use the button in the header, there will be a dialog showing the content first. Then, hit **OK**. This will give you the advantage of adding the filename to the data, too.

    Click on the link for now and take the short version:

Figure 12.6 – The transform button to the right of the column name

7. The content will display. From here on out, you have all the options that Power Query offers. Maybe you'll create a datetime field from the content in the file.

    We have a year, a month, and a day. Select these three and click **Column from Examples**. In the input field, enter the date from the values that you have in your three columns. So, if you have the year 1987, month 10, and a day such as 23, enter 23.10.1987 in the input field and hit **OK**. A text field with the intended date is displayed with the name **Merged**. However, it is still a text field.

    If you want to convert it into a date, you may now either select the column and choose **Parse** from the **Date** menu in the **Transform** or **Add Column** ribbon (a new column will be created). Alternatively, you can click the type icon next to the column name and select **Date** from the context menu that will be displayed:

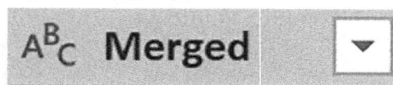

Figure 12.7 – The type icon to the left of the column name

Finally, you want to rename the newly created column. You can click on the name and type in a new one. Perhaps name it something like `ad_date`.

8. Maybe you want to convert the `ad_year`, `ad_month`, and `ad_dayofmonth` columns to number formats, too. Try it by clicking the type icon and selecting **Whole number** from the drop-down menu.

9. If you have finished your work on the data import, go to the **Home** ribbon and click **Close & Apply**. The dataset will now be stored in your Power BI file and you will be taken to the Power BI editor, where you can start building business logic now and later visualize your data on the Power BI canvas.

Modeling your dataset

Once you are taken to the Power BI canvas, you can, for example, switch to the Model view. There is a tool navigator on the left side of Power BI Desktop:

Figure 12.8 – Tool navigator with the Model view selected

This will give you a view of your imported tables/queries in an entity relationship-like diagram. All imported sources will show up here and you can either view them with their relation to each other or you can even create new relations if they are not yet present.

You can add data from many different sources and establish relationships between them. The Model view will help you to display and sort this and keep an overview of your data model:

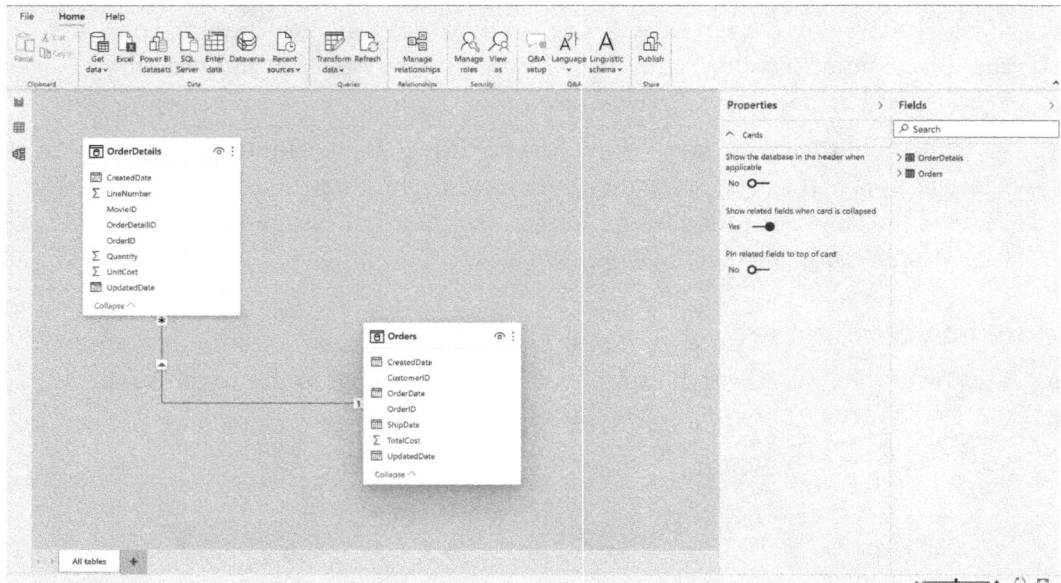

Figure 12.9 – Model view with two tables and their relationships

Relationships can be created as follows:

- Many to one
- One to one
- One to many
- Many to many

Power BI will try to detect relationships between tables based on column names at the time of the import. If this is not possible, you will have the option to create relationships on your own.

You can either drag and drop columns between tables to do so or you can use the **Manage relationship** menu option from the displayed ribbon. In the dialog, you can select one or more columns from each table that constitutes the relationship.

> **Note**
> Power BI will always examine the columns and the cardinality between the two tables. When you have a *one* side in your relationship, Power BI will ensure that the column values in this table are unique. You will receive an error message if this condition is not met.

This view additionally allows you to add all kinds of additional information to the model. You can rename columns, add synonyms to columns, hide certain columns, and influence their data type and format. You can also create hierarchies here from columns in your tables. This adds nice navigation features to your model.

> **Note**
> Date columns will automatically be extended with date hierarchies. You will find this handy when you are using them in your visualizations.

Optimizing the columnstore database in Power BI

Power BI will keep all data in columnstore format. We have talked about this format already in *Chapter 4, Understanding Synapse SQL Pools and SQL Options*.

Power BI uses a similar technology that is also used in the AAS database. You can read more about that in the *Creating data models with Azure Analysis Services* section.

As you have already seen, the columnstore technique benefits a lot from the elimination of redundancies and you can help the database to optimize its performance a lot if you follow certain rules. Here are the two most popular ones:

- Only use columns that you really need for either visuals or calculations. All data that is not needed that you stress the database with will do exactly that: it will stress and slow down the database. Get rid of it.

- Split columns that contain data such as date and time or first name and last name. If you keep date and time in one column, you can calculate the distinct values in 1 year depending on the precision of the time portion (31,636,000 entries for time precision to the second). If you split the time portion into a second column, you will only have 365 entries per year in the date column. For the time column, your number of entries will vary depending on the time precision stored. So, if you store the time down to the second, this column will hold 86,400 entries. A lot of entries avoided, right?

There are many more optimization rules to follow. Please see the *Further reading, Data model optimization*, section for more details.

Let's now move on and explore the options and capabilities of Power BI when it comes to business logic and flexible calculations that you can only run with a significant effort on your presentation-layer database.

Building business logic with Data Analysis Expressions

DAX is a language that is quite easy to learn. It offers an Excel-like notation and the tabular character of Power BI datasets makes it easy to understand.

The purpose of DAX in your Power BI reports is to help you create so-called **measures** to reflect your business logic calculations. These measures can then be used in the Power BI visuals and are always evaluated in the context they are displayed in.

Such a measure, for example, can be created to show the correct summary of a value from a table if you are displaying the very details of the table or displaying a hierarchy field together with the measure.

Say we have the following myTable table:

Family	ParentChild	Member	Screenhours
Miller	Parents	Dad	2
Miller	Parents	Mum	.5
Miller	Children	Lucy	3
Miller	Children	Franky	5.5

Figure 12.10 – Example myTable table

We have the following formula:

```
SumScreenhours = calculate(sum(myTable[Screenhours])
```

It will display as follows when used in different contexts. For example, it is displayed together with the ParentChild field as follows:

ParentChild	SumScreenhours
Parents	2.5
Children	8.5

Figure 12.11 – Measure used with the ParentChild field

This is displayed together with the Family field as follows:

Family	SumScreenhours
Miller	11

Figure 12.12 – Measure used with the Family field

And, of course, the details when displayed with the `Member` field are as follows:

Member	Screenhours
Dad	2
Mum	.5
Lucy	3
Franky	5.5

Figure 12.13 – Measure used with the Member field

As you can see, the measure will show the correct sum in the different contexts that you can use it. Handy, isn't it?

Let's now try this on the data that you imported in the *Connecting to a dataset* section.

If you have followed the steps in that section, you will find yourself in Power BI Desktop like this:

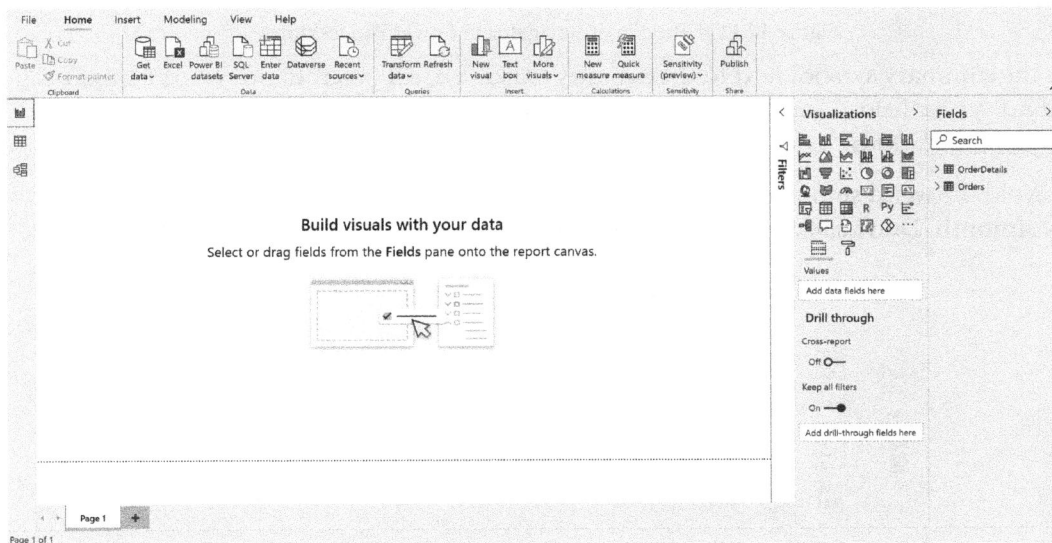

Figure 12.14 – Power BI Desktop after connecting to a dataset

> **Note**
>
> When you are importing/connecting to data, Power BI tends to present you with all numeric fields as possible summary fields that can be summed up or used in calculations. In your case, the `ad_year`, `ad_month`, and `ad_dayofmonth` fields representing date portions are presented as numeric fields and if you don't reconfigure this, Power BI will display arbitrary results that you didn't intend to.

You can derive this information from the display of the columns in the right-side **Fields** navigator:

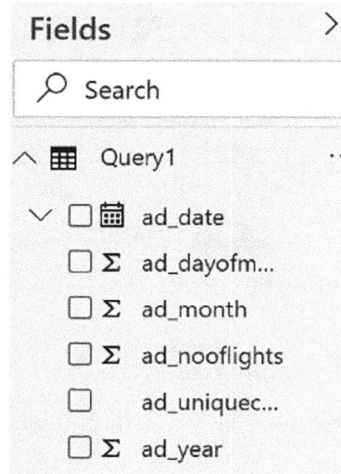

Figure 12.15 – The Fields navigator in Power BI Desktop

See the summary symbol next to the `ad_dayofmonth`, `ad_month`, `ad_nooflights`, and `ad_year` fields? The only one that really should be configured for summation is `ad_nooflights`.

To avoid the weird summarization of the other fields, you can reconfigure them. Click **ad_dayofmonth**. Don't click the checkbox in front of it; just click the name that is marked:

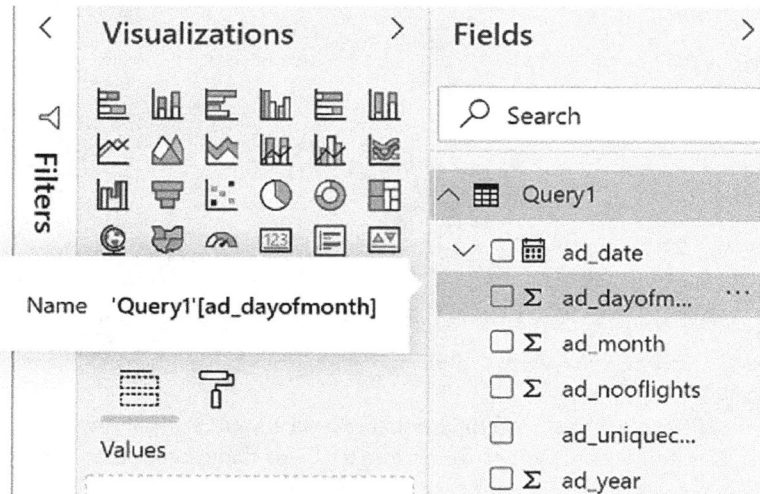

Figure 12.16 – The ad_dayofmonth field selected

The top ribbon now changes to **Column tools**:

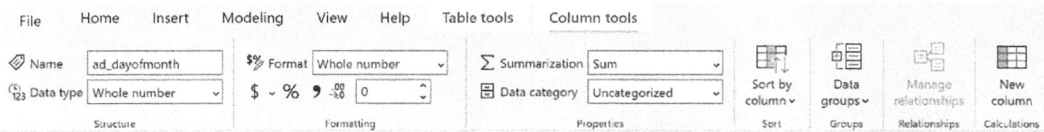

Figure 12.17 – The Column tools ribbon when a column is selected

Right in the middle of the ribbon, you see the **Summarization** drop-down box. For the ad_dayofmonth column, select **Don't summarize** from the options:

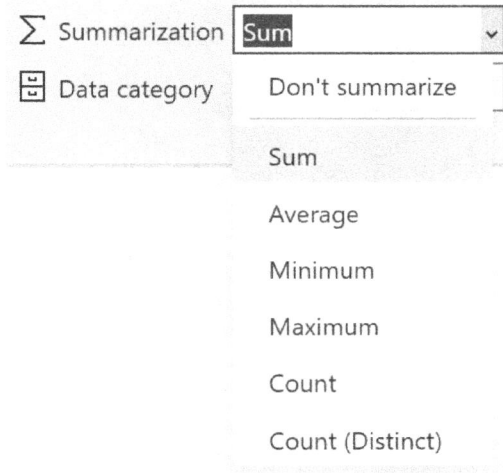

Figure 12.18 – The Summarization drop-down box and its options

As you can see from *Figure 12.18*, you have several options on how you can configure the standard summarization for numeric fields.

When you have selected **Don't summarize**, you can see the immediate result of your action: the summarization symbol disappears.

Repeat this action now for the other time portion columns in your dataset. Your **Fields** navigator should look like this finally:

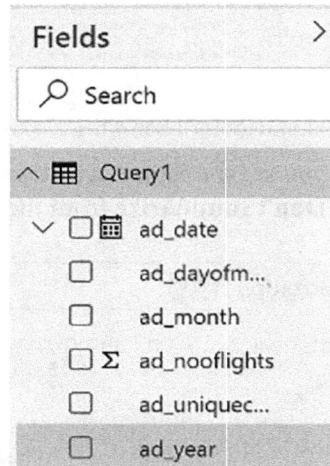

Figure 12.19 – Final column list

If you now click the checkbox in front of the `ad_year` and `ad_nooflights` columns, they will get selected and displayed on the visualization canvas:

ad_year	ad_nooflights
1987	448620
Total	**448620**

Figure 12.20 – First display of selected columns

But what happened now? You didn't yet create a measure yourself and this one seems to already do the job.

Yes, this is one of the supporting functions of Power BI. The summarization configuration of the `ad_nooflights` column helps you to now display it as a sum. If you now select the `ad_dayofmonth` column, you will see the values in detail again.

But let's create a real measure now. To do so, you can, for example, go to the table view on the left side:

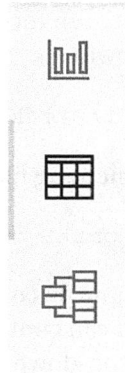

Figure 12.21 – Tool navigator on the left with the table tools selected

In the **Table tools** menu ribbon, you will find the **New measure** option. Click it and start entering the following formula:

```
FlightsSum = calculate(sum(Query1[ad_nooflights]))
```

Here, `Query1` represents the table name that is required for this formula.

The `ad_nooflights` column is placed in brackets here to follow the formula notation.

The `sum()` function represents the intended calculation that you use on the column.

The `calculate()` function wraps the inner function and can offer some additional filters. You could replace the whole calculation with the `sumx()` function, for example, if you only want to simply sum up a value. `calculate()` extends your options massively. Refer to the *Further reading*, *The calculate() DAX function*, section for an overview of the options and their usage.

Maybe you can, together with the documentation, create another DAX measure that always displays the sum for a certain day in the month? Maybe the first? (See the *Further reading*, *DAX formula to filter ad_nooflights for the first of the month* section.)

Creating complex calculations

You can nest measures in measures. This comes in very handy for more complex math on your data. Given the measures that you have created, you might, for example, calculate a ratio between the two values.

First, you would calculate the difference between the actual day calculated against the first of the month. This would look like the following:

```
Difference = [FlightsSum] - [FirstofMonthSum]
```

And, of course, the ratio would then be something like this:

```
Ratio = [Difference]/[FirstofMonthSum]
```

For the `Ratio` measure, you want to additionally configure another format to be displayed. As this is a percentage value, you can configure it as such during its creation in the **Measure tools** ribbon in the **Format** drop-down box:

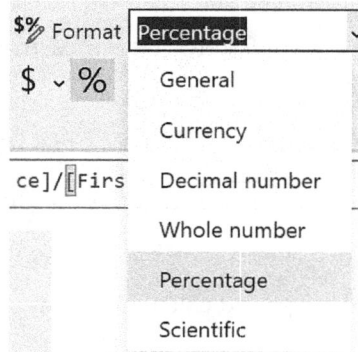

Figure 12.22 – Format options for measures

The measure will display as a percentage value everywhere you use it in your visualization:

Ratio
0.00%
-0.12%
-8.32%

Figure 12.23 – Ratio measure formatted as a percentage

You have a lot to discover when it comes to the creation of business logic using the DAX language. For example, we didn't handle the total case for the Ratio measure. At the moment, it will show an absolutely unusable value. You want to handle these cases too and, for example, different cases for intermediary sums and more.

However, this certainly falls outside the scope of this book. Please refer to the *Further reading, DAX books and readings*, section for more information.

There is not only business logic that you can implement using DAX. Imagine you have a dataset that contains data for several subs and you don't want to create as many reports as you have subsidiaries. You only want to create exactly one report that should display data depending on the user that is logged in and views the report.

You can achieve this using a feature that is called **Row-Level Security** (**RLS**). Please check the next section, *Implementing Row-Level Security*, for an introduction.

Implementing row-level security

RLS will help you to limit access to your data. You can use RLS to implement advanced security based on DAX expressions. With the features provided, you will be able to control who can see what from your datasets.

Additionally, implementing RLS will help you in the use case described previously. You don't need to create different views or even reports or similar to be able to individually provide data to your recipients, such as a region report for example. This will reduce your effort for creating and reports their amount drastically.

Imagine a report that generally shows all rows from a certain table but uses RLS to narrow down the data the logged-in user can see. You would create just one report and let RLS do the trick of displaying the right data to the right person.

If RLS were to not be available, you would need to create one report for each different user group to display the selected data for that particular group, such as the regional sales teams, for example.

Creating roles

First, you will need to define the roles that you want to assign to your recipients. With these roles, you will then add DAX filters that select the data that the different roles can access:

1. Navigate to the **Modeling** ribbon. Here, you will find an entry called **Manage roles**. When you click it, the **Manage roles** dialog will be displayed:

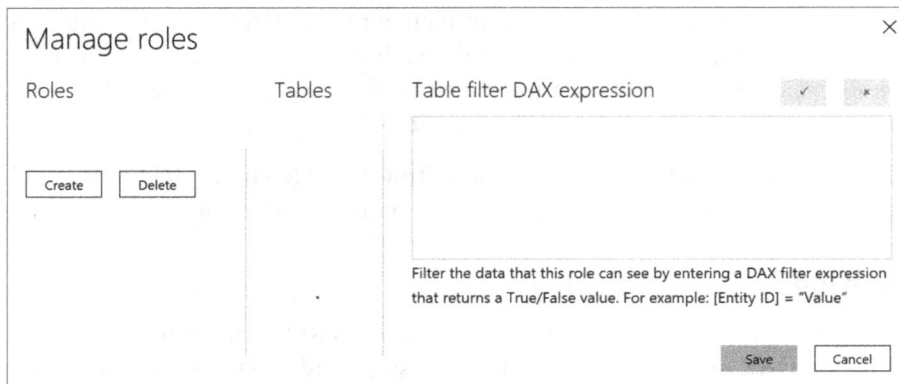

Figure 12.24 – The Manage roles dialog

2. When you click **Create**, a text entry field will appear where you can enter the name for the new role.

3. In the **Tables** section, you will now find all tables that are available in your dataset. If you click the three dots next to the table name of the one that you want to use in your filter criteria, you can browse to the fields of the table by selecting **Add filter....** The fields are listed and you can select one that will be added to the DAX expression field. You can repeat this step as often as necessary. If you check the menu closely, you will find another entry: **Hide all rows**. Using this will prevent the role from accessing any row of that table and the ones related.

4. Click **Save** once you are ready to proceed.

5. To test the roles that you have created, you can now click **View as** next to the **Manage roles** button. A selector will be displayed (see *Figure 12.25*) where you now can pick a role that will be used for viewing. You will see everything impersonating the role that you have selected:

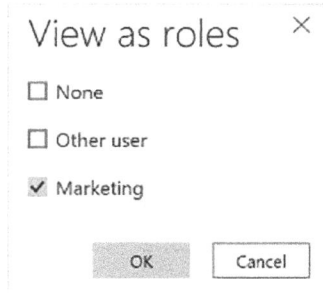

Figure 12.25 – The View as roles dialog

6. The last step, where you will add users to the defined roles, will be implemented after the report with the dataset published to the www.PowerBI.com service. Please check the *Publishing insights* section for information regarding the publishing process.

    You will need to browse to the dataset in your workspace and, from the **More options** menu, select **Security**. In the following dialog, you can add the email addresses of users that you want to add to your created roles. These can even be external users that can be invited to your tenant. In general, you would add users or groups from your Active Directory.

Visualizing data

Once you have modeled all the sources and their relationships and created all the calculations and you are sure that you can display everything as intended, you can move on to the **Report** view. Here, you can start creating the visualizations that will display your data in the most suitable way. You can create filter controls, charts, lists, and many more visuals that can dynamically change following the selections that your users will do on the report.

But let's start by creating a simple visual first. Have you kept the Power BI file open where you have imported the data from your data lake?

1. Navigate to the **Report** view:

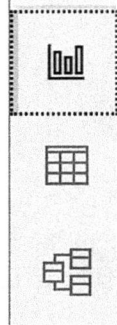

Figure 12.26 – The tool navigator on the left with the Report view selected

2. You will find the **Visualizations** palette and the **Fields** list on the right of the view:

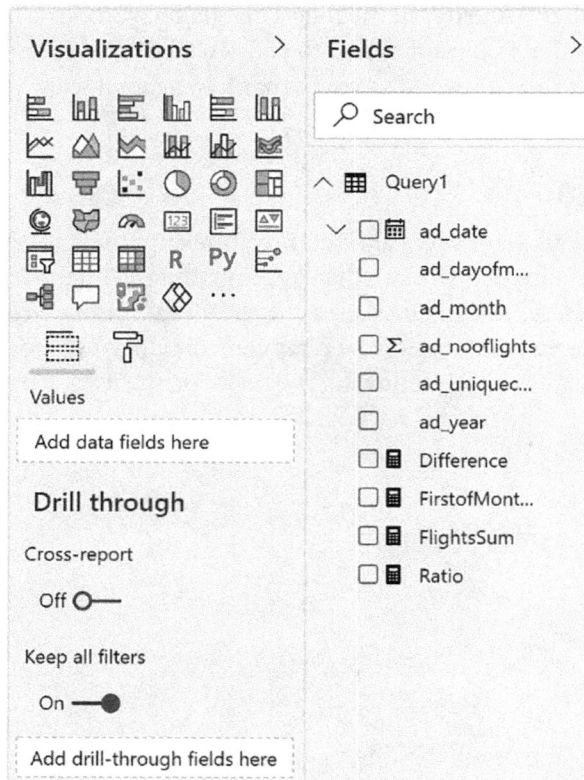

Figure 12.27 – Visualizations and Fields on the right of the Report view

3. Navigate to and select the **ad_uniquecarrier** and **ad_nooflights** fields. The data will display as a table on the canvas.

4. To change the table view to something else, now select, for example, **Clustered Column Chart** in the top row in the **Visualizations** sections. It is the fourth one from the right:

Figure 12.28 – Clustered Column Chart selected in the Visualizations palette

5. You should receive a visual as in *Figure 12.29*:

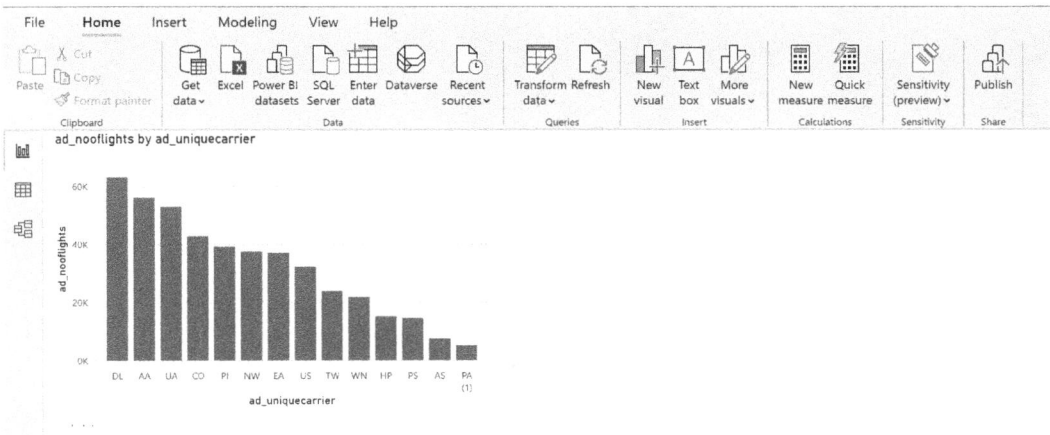

Figure 12.29 – Your first Clustered Column Chart visualization

Congratulations! You have created your first visual on Power BI. You can now, of course, examine the options of the visual that you have created in the properties that are displayed below the **Visuals** palette when the particular visual on the canvas is selected:

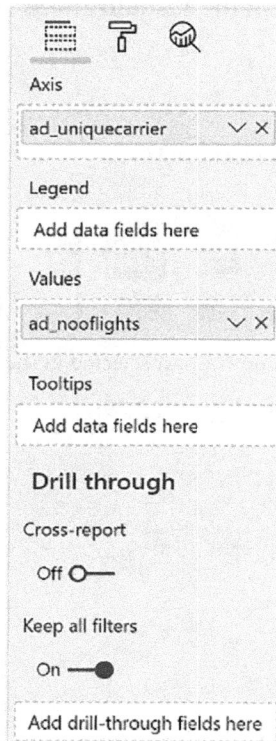

Figure 12.30 – Visuals properties

Many visuals have several properties pages and you can click through them with these buttons:

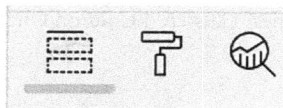

Figure 12.31 – Different options for visuals

The paint roller symbol will take you to the **Format** options of the individual visual, and the magnifier will give you access to **Analytics** functions that the individual visual can offer.

By default, you will find different columns and lines, areas, scatter and pie charts, value cards, and maps that can display geodata and more. Additionally, you will find filter visuals with date functions and sliders, and so on.

It gets interesting when you examine the R and Python visuals. Here you can write your own R or Python code to plot any R or Python visual. You can add plotting libraries to your report and use them here.

Understanding the dynamics in reports

One thing that all the visuals have in common is that they interact with the data model and they influence and are influenced by the filter context, which again affects the data that is selected from the underlying database and is displayed.

Let's add some more visuals to the canvas and examine the behavior. Maybe you want to see the number of flights overall. You can use a card for this and this time, we're first going to place the visual and then select the columns or measures to be displayed:

1. First, click somewhere in the white space to deselect any visual to avoid it being reconfigured when you select another visual from the palette. Now, examine the **Visualizations** palette, look for the **Card** visual (see *Figure 12.32*), and then click it. It will be placed on the canvas below the visual that you have created already:

Figure 12.32 – Card visual

2. Now, with the empty visual still selected, click the **ad_nooflights** column. The overall total for the number of flights is displayed. Maybe reduce the size and put it right next to the column chart that you already have on the canvas:

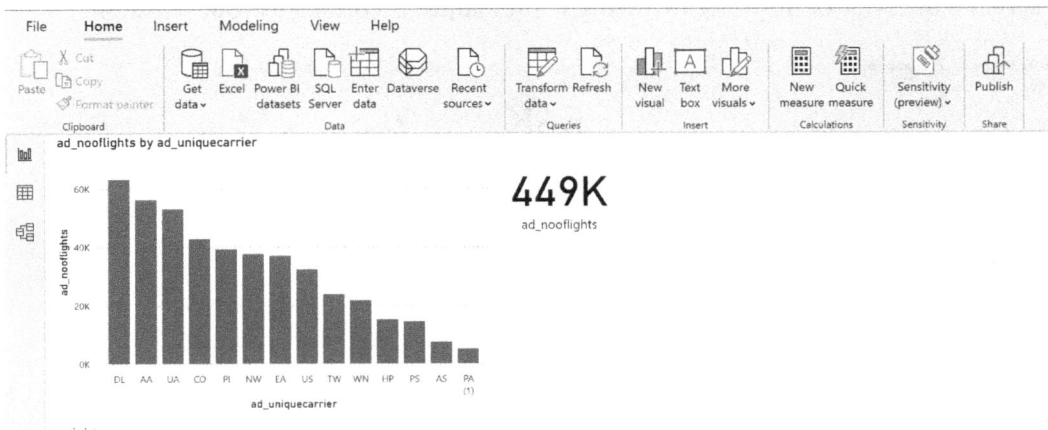

Figure 12.33 – Card visual added

3. Maybe you want to see the number of flights over the course of a month as a line. You can put a line chart on the canvas and configure it accordingly. You can find a line chart as the first visual in the second row of the **Visualizations** palette.

    Once again, first, click somewhere in the white space on the canvas to deselect any visual. Otherwise, you would just be changing an existing one by clicking on another one in the palette.

    Now, click and put the line chart on the canvas and first add `ad_dayofmonth` and then the `ad_noofflights` column. After rearranging it, you might have a view like this:

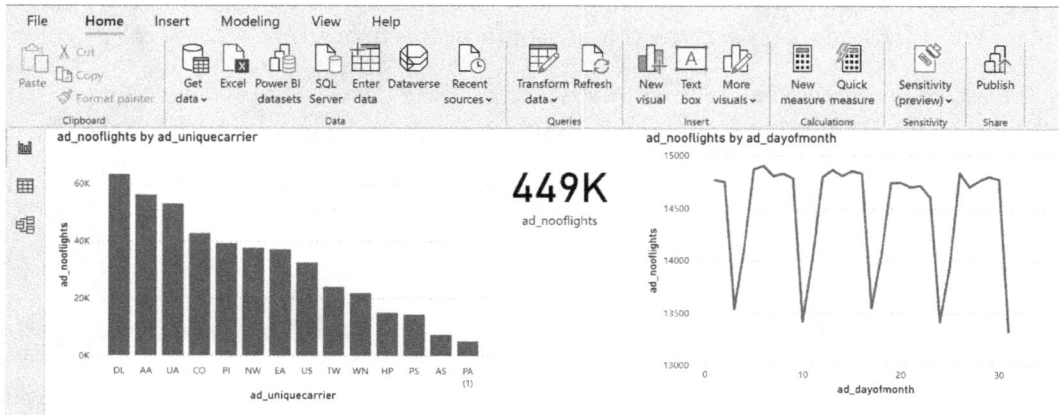

Figure 12.34 – Line chart visual added

Congrats! You have created your first small report. You can now start examining the behavior of the visuals. Perhaps click into the column chart on one of the columns and observe the behavior of the other two visuals. They adjust according to your selection:

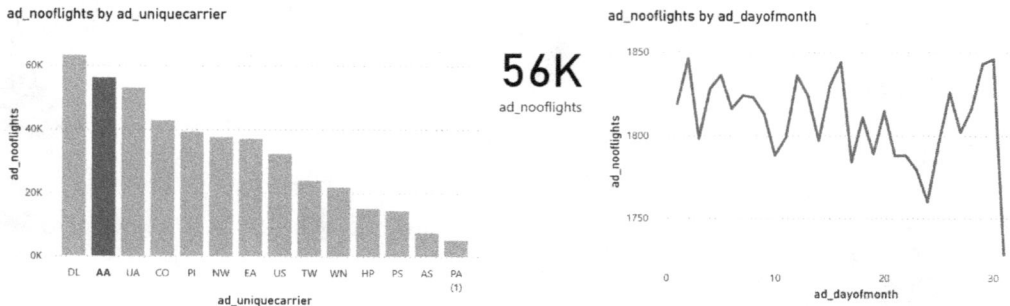

Figure 12.35 – Visuals with a certain column selected

Maybe you add another filter option to the report on the canvas. Try and add the slicer visual from the palette and add the `ad_date` field to it. This will add a date filter option to your report. You can either use date pickers for the start and end date or use the displayed slider to adjust the date range that you want to display:

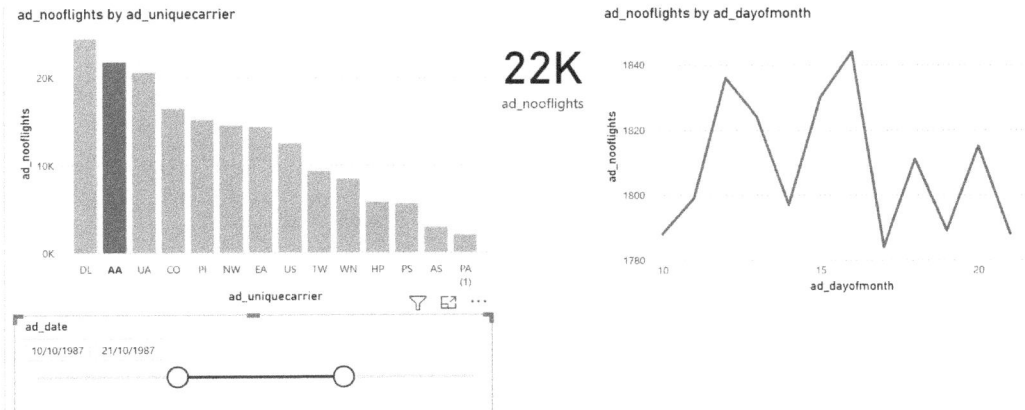

Figure 12.36 – Date slicer added to the report

Why not experiment with your data and your ideas? Have some fun. Try to examine the different settings and additional features for the visuals that you use.

Using the visuals gallery

The default visuals available can already help you build quite comprehensive and versatile reports. But still, you will find yourself in situations where the available palette does not offer enough options.

However, there is a way to extend the palette with a wide collection of additional visuals from a continuously growing gallery.

At `https://appsource.microsoft.com/en-us/marketplace/apps?product=power-platform%3Bpower-bi-visuals&page=1`, you can find 320+ additional visuals, some provided by Microsoft, others created by Microsoft partners or enthusiasts. Most of them can be downloaded free of charge. Some give you a preview and can be purchased for full functionality:

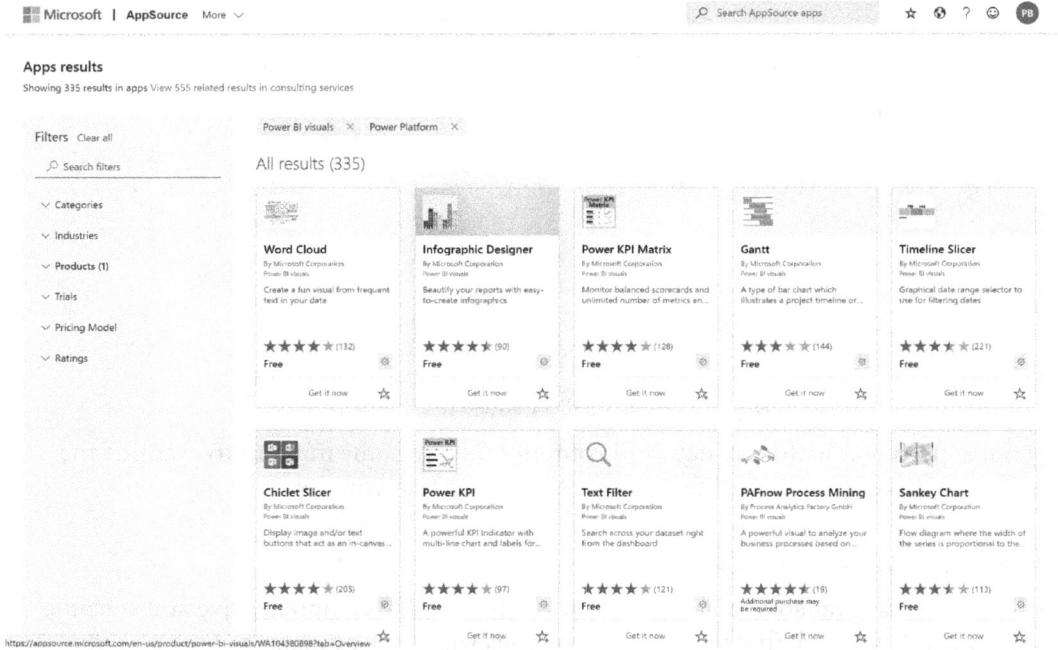

Figure 12.37 – Power BI custom visuals gallery

After the download, you can import a visuals file by clicking the three dots in the **Visualizations** palette. Once the visual shows up in the palette, you can use it throughout your report and it will also be deployed to the www.`PowerBI`.`com` service.

Browsing the Data Stories Gallery

Another rich source for visualizations is the Data Stories Gallery. This can be seen more as a source of inspiration of how others have used Power BI before. At `https://community.powerbi.com/t5/Data-Stories-Gallery/bd-p/DataStoriesGallery`, you will find a ton of interesting Power BI data stories.

Browsing the Power BI Desktop functions

Take a moment and browse through Power BI Desktop and examine the menus. You will find a lot more functions. Give them a try and experiment with them.

For example, you might want to check the options that the **View** menu has to offer. Did you find the mobile layout option? You will be taken to an editor that allows you to arrange all the visuals that you have created on the standard report into a layout that would be pulled when the report is used on a phone or tablet. Useful, right?

Publishing insights

Once you are done with developing your datasets and reports that visualize your data, you can publish your work to the Power BI portal.

To do so, you will need to find the **Publish** button in the Report view:

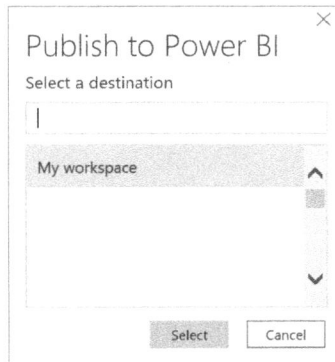

Figure 12.38 – The Publish to Power BI dialog

Before you get to the dialog in *Figure 12.38*, you will need to log in to your www.PowerBI. com portal. If you don't have an account there, you can create one free of charge.

Once the publishing is done, you will be prompted and a link displayed to your Power BI portal that will instantly take you to your newly published report. Give it a shot! It will look and feel exactly like in your Power BI Desktop tool.

Another link will give you the option to instantly invoke **Quick Insights**. Why not give this tool a try? This function will scan your dataset for significance and will suggest a set of additional visuals that will display them. This function is based on statistical and AI functions and can help you to discover new aspects of your data. Maybe you'll find constellations and insights that you didn't even expect.

Building dashboards

A dashboard lives in the Power BI service. It is a collection of visuals that can originate from one or more reports. It can act as an entry point into your data story and can deliver an overview of the most important key metrics. You can use the visuals displayed on your dashboard and jump to the reports that they represent by clicking on them.

To build your first dashboard, navigate to your newly published report and the clustered column chart. Like nearly everywhere here, you will find a pin symbol in the upper-right corner. If you click it, you can pin the visual to a dashboard. If there is no dashboard yet available, you can create a new one in the following dialog:

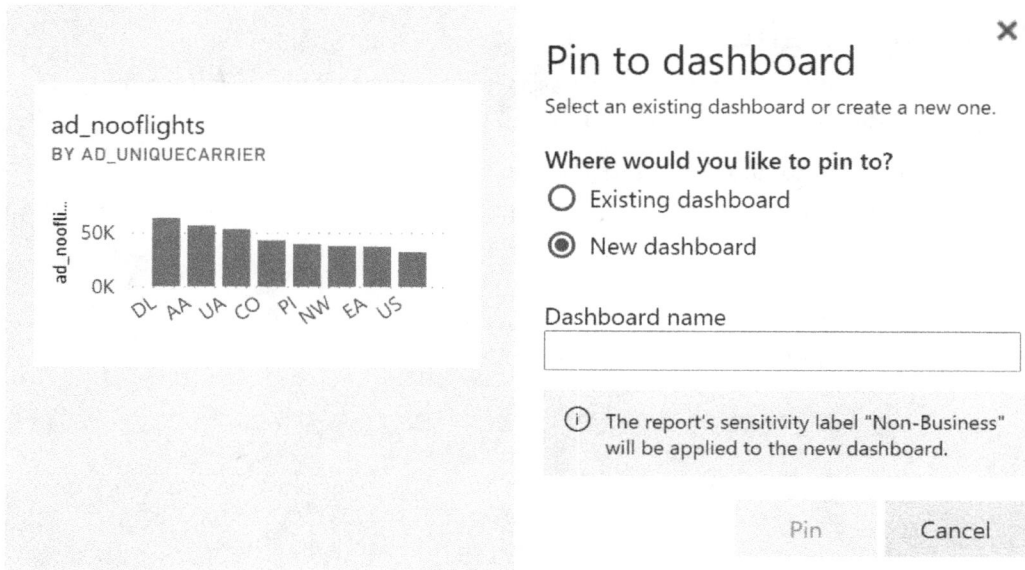

Figure 12.39 – The Pin to dashboard dialog

Again, you can create a mobile view of the dashboard after it has been created. The service will prompt you.

On the left side of your portal view, you will find a navigator. In your workspace, you will have different nodes to browse: **Dashboards**, **Reports**, and **Datasets**. Check the **Dashboards** node. Your newly created dashboard will show up here. Browse to it. You will find your pinned visual there. When you click it, you will be taken to your report.

Using Q&A

Another very handy feature of the Power BI service is called **Q&A**. If you check your dashboard, you will find a text entry box in the upper-left corner that says **Ask a question about your data**.

Give it a try and type `What were the FlightsSum per ad_uniquecarrier?` and see what happens.

You'll receive a card visual that displays the number of flights for the carrier UA.

What you just did is you used the natural language query function of Power BI to analyze the dataset that you have published with your report. Now, how useful is that?

You can deliver insights into your report and dashboard users and you can even enable non-technical users to create new insights into the data. If you check the upper-right corner, you will find the **Pin visual** option again. So off you go and put the newly created card on your dashboard.

Exporting data

When browsing reports and dashboards in the Power BI portal, the user can export data to CSV files if needed and if they are allowed to. This comes in handy for users if they want to further investigate the data that sits behind a visual.

If you click the three dots in the upper-right corner of a visual, you will receive the following menu:

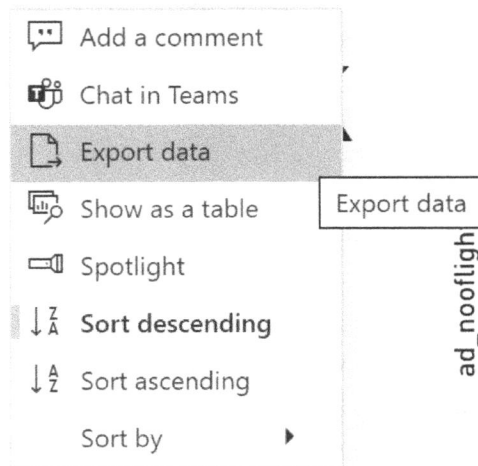

Figure 12.40 – Context menu of a visual in the browser

In the next step, after clicking **Export data**, you can decide whether you want to export summarized data or the underlying details behind the visual. If you aren't allowed to export the details, you will receive an error:

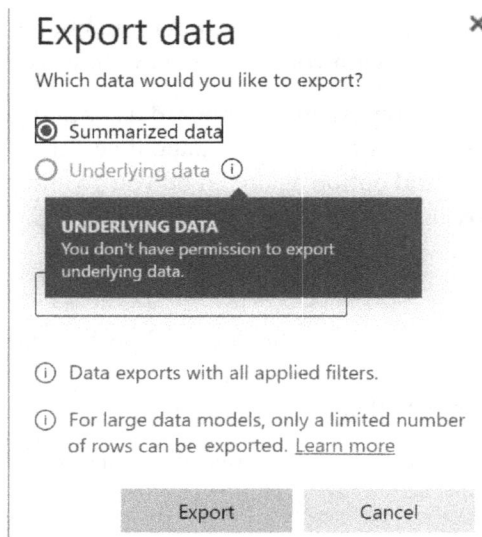

Figure 12.41 – The Export data dialog with an error

You can configure either Excel or CSV as the target format. When you then hit **Export**, the download will start and you will receive a file with the same name as the visual in your download folder.

This was just a brief look at the most important Power BI features. There is a lot more to discover. Please check the *Further reading, Discovering Power BI*, section for a link to the Power BI documentation.

You will find in Power BI apps, for example, an option to package datasets, reports, and dashboards and create reusable templates out of them.

Additionally, there are options to create alerts based on data content and constellations. You can be notified when numbers hit thresholds or things are changing.

There are workbooks to discover on the www.powerbi.com service and you can benefit from the rich sharing capabilities on the portal.

Maybe you want to examine the Power BI streaming datasets that can be populated by Azure Stream Analytics to build near real-time or real-time dashboards.

These are only some examples of the additional capabilities of the Power BI ecosystem.

Creating data models with Azure Analysis Services

AAS is a sibling of Power BI that can be provisioned from your Azure portal. Compared to Power BI, AAS doesn't offer the same visualization functionality, however. You would use Power BI or Excel or another tool that can talk DAX or **Multidimensional Expressions (MDX)** to consume the data model that you can build with AAS.

That is exactly the purpose of AAS: to act as a columnstore database and host a DAX data model just like you would create it with Power BI. AAS will give you a similar set of features and DAX functions when creating a data model that you have seen in this chapter already.

Up to now, the advantage of AAS over Power BI has been the higher amount of data that you could host with AAS. But with the current and further releases of Power BI and the Premium offering, these limits will vanish.

Microsoft has announced that Power BI Premium will be the superset of all the functions of both Power BI and AAS in the future. At the moment, there are a few functionalities of AAS that haven't yet made it into Power BI Premium. For example, you can't yet implement translations in Power BI, but you can in AAS. This is a feature that enables your dataset to display metadata of the dataset in the language of the client that is logged in. For a comprehensive overview of all the differences and to help decide whether to use Power BI or AAS, please visit the *Further reading, Differences between Power BI and Azure Analysis Services*, section.

Developing AAS models

To develop a data model to be hosted in AAS, you will use **SQL Server Data Tools (SSDT) for Visual Studio**. You will find similar options there to what you have seen in Power BI Desktop. They might be placed a little differently from the ones in Power BI, but they are there. Even the **Get data** experience with Power BI is available.

If you want to explore the options there and don't have Visual Studio installed, check the current Visual Studio Community edition (`https://visualstudio.microsoft.com/downloads/`).

To install SSDT, you will need to run the Visual Studio installer and modify the installation. In the installer, you will first need to look for **Data storage and processing**. On the **Installation details** tab, search for **SQL Server Data Tools** and tick the checkbox. Then you can start the installation by clicking **Install**.

Once SSDT is installed, you need to proceed to the **Extensions** page within your Visual Studio and select **Manage Extensions**. From **Marketplace**, you can then install the extension for **Analysis Services**.

Distributing data using Azure Data Share

Up to now, we have examined options to consume data from your modern analytical setup using the typical reporting interfaces. But what about consumers that need data transferred to you as files? Or if you want to deliver outputs of your analytical efforts directly into a database that lives in another Azure subscription? Maybe it is even created in a subscription of your customer.

If you are thinking of sharing data, you might then also consider topics such as an agreement on the terms of usage of the data or regular updates if the shared dataset is updated on your side. And, of course, security will play a central role in your setup.

One big asset that Data Share offers is the ability to control your shares and always know who has access to your data – who receives what data when and in what location.

Azure Data Share will allow you to share data from the following sources with consumers with an Azure subscription:

- Azure Storage accounts
- Azure Data Lake Storage Gen2
- Azure SQL Database
- Azure Synapse Analytics (standalone SQL pools)
- Synapse dedicated SQL pools
- Azure Data Explorer

When you create a Data Share instance, you invite your consumer to use the functionality in the Azure portal.

After setting up the service, you can add datasets from your sources, such as Synapse, for example, and set up a schedule for the snapshot updates.

Finally, you will add consumers/recipients to your share and invite them.

On the consumer side, when the invitation is received, the consumer will accept the invitation (they can decline too) and will need to name a storage target in their Azure subscription. This is where Data Share will put the data. Please refer to the *Further reading, Data Share source-to-target mapping*, section for an overview of compatible source and target options.

If you have set some terms of usage before you issued the invitation, your consumer will need to agree to them when they accept the invitation. Otherwise, the data will not arrive at the target.

Please check the *Further reading, Azure Data Share*, section for a link to the documentation, which contains some interesting ideas about scenarios and tutorials regarding Azure Data Share.

Summary

In this chapter, you have learned about Power BI and the options to create and deliver datasets, reports, and dashboards to your consumers. You have seen how to publish reports with their datasets to the Power BI service, where you can share and collaborate on your analysis.

You have seen how to build DAX calculations and the options that you have for RLS implementation on your models.

We have also discussed the creation of visuals and how they interact with each other and the data model.

You have learned how to publish your reports to the Power BI service and how to create dashboards from reporting visuals. With the created dashboard, you have investigated Q&A and seen how to export data for further usage.

Furthermore, you have learned about AAS, Power BI, and Excel as the clients for the columnstore database.

Finally, you have read about Azure Data Share and the options for automating the exchange of data between Azure subscriptions.

In *Chapter 13, Introducing Industry Data Models*, you will learn about the options in the Azure Industry Data Workbench and how you can leverage predefined data models in your modern data warehouse architecture.

Further reading

- Discovering Power BI: `https://docs.microsoft.com/en-us/power-bi/`

- Power BI data sources: `https://docs.microsoft.com/en-us/power-bi/connect-data/desktop-data-sources`

- The `calculate()` DAX function: `https://docs.microsoft.com/en-us/dax/calculate-function-dax`

- DAX formula to filter `ad_nooflights` for the first of the month:

```
FirstofMonthSum = calculate(sum(Query1[ad_nooflights]),
Query1[ad_dayofmonth]=1)
```

- DAX deep dive readings:

  a) Alberto Ferrari, Marco Russo, and Daniele Perilli: `https://www.sqlbi.com/`

  b) Chris Webb's BI blog: `https://blog.crossjoin.co.uk/category/dax/`

  c) Adam Saxton, aka *Guy in a Cube*: `https://guyinacube.com/tag/dax/`

  d) DAX reference documentation: `https://docs.microsoft.com/en-us/dax/`

- Data model optimization: `https://docs.microsoft.com/en-us/power-bi/guidance/import-modeling-data-reduction`

- Differences between Power BI and Azure Analysis Services: `https://powerbi.microsoft.com/en-us/blog/power-bi-premium-and-azure-analysis-services/`

- Data Share source-to-target mapping: `https://docs.microsoft.com/en-us/azure/data-share/supported-data-stores`

- Azure Data Share: `https://docs.microsoft.com/en-us/azure/data-share/overview`

13
Introducing Industry Data Models

A challenge that you already might have come across when modeling your analytical environment is to create not only the model but also all its details. When you need to integrate data from different source systems, you don't want to forget the necessary details when you create your target model.

There is nothing worse than identifying missing attributes or wrong data types in a target object when you are already two-thirds of the way through your implementation. Adjusting data mappings and cleansing routines and formulas at a late stage in your development process can be a cumbersome situation that slows down your development and extends your delivery time. Industry data models that reflect years of experience and best practices can help you accelerate your development.

This chapter will give you an overview of the industry data models that you can leverage using Microsoft's **Common Data Model** (**CDM**). You will discover the metadata model that is used as the foundation for CDM and the available predefined schemas that you can use for your implementation.

In the second part, we will examine a new service on the Azure platform called **Industry Data Workbench**. At this point in time, it is only in preview but is a very promising service that allows you to leverage a multitude of predefined industry data models and extend them to your needs.

You will find the following sections covered in this chapter:

- Understanding Common Data Model
- Examining and leveraging predefined entities
- Discovering Azure Industry Data Workbench

Understanding Common Data Model

In its initial version, CDM provided predefined entities and their data types in a standardized notation. In the second generation, Microsoft extended it to be able to reflect complex semantic contexts such as relationships.

CDM is not just targeted at analytical use cases. You can use its predefined entities, language elements, rules, and structures in any application.

If you examine services such as Azure Data Factory/Synapse pipelines, for example, or the **Modern Workplace power suite** with apps such as Power Apps, Flow, and Power BI, you will find connectors that will be able to use the models that you define with CDM.

CDM is reflected in a collection of JSON documents (`*.cdm.json`) that follow a certain schema that can be seen as the language of CDM. Let's dive into its elements.

Examining the basics of the SDK

The top-level container that collects all artifacts of a CDM construct is called the **corpus**. It provides path definitions to all objects that are used in your model. It labels and categorizes the paths at the same time into different storage types, such as the standard model components that you will instantiate from the CDM GitHub. Another example is Azure storage types that might hold the data that is described in the model. These files are called partition data files.

And, of course, you will add files that describe the data and the relationships between the different objects.

Understanding solutions and the manifest file

The manifest file represents a map of a certain solution. It can contain the following:

- Versioning
- Imports
- Entity lists
- Relationships
- Submanifest lists

But let's first see an example to get an overview and then explore the different sections:

```json
{
    "manifestName":"MyAirdelay",
    "jsonSchemaSemanticVersion":"0.9.0",
    "imports":[
        {
            "corpusPath":"cdm:/foundations.cdm.json"
        }
    ],
    "lastFileStatusCheckTime":"2021-04-18T17:12:01Z",
    "lastFileModifiedTime":"2021-04-18T17:12:01Z",
    "lastChildFileModifiedTime":"2021-04-18T17:12:01Z",
    "entities":[
        {
            "entityName":"Aiport",
            "entitySchema":"AirportData/Airport.cdm.json/
Airport",
            "dataPartitions":[
                {
                    "location":"AirportData/Airports.csv",
                    "exhibitsTraits":[
                        {
                            "traitReference":"is.partition.format.
CSV",
                            "arguments":[
                                {
                                    "name":"columnHeaders",
                                    "value":"true"
                                },
                                {
                                    "name":"delimiter",
                                    "value":","
                                }
                            ]
                        }
                    ]
                }
            ]
```

```json
                    }
                ]
            },
            {
                "entityName":"Airdelay",
                "entityPath":"AirdelayData/Airdelay.cdm.json/
Airdelay",
                "dataPartitions":[
                    {
                        "location":"AirdelayData/Oct/airdelays.csv",
                        "exhibitsTraits":[

                        ],
                        "arguments":[
                            {
                                "month":[
                                    "Oct"
                                ]
                            }
                        ],
                        "lastFileStatusCheckTime":"2021-04-
18T17:12:01Z",
                        "lastFileModifiedTime":"2021-04-18T17:12:01Z"
                    }
                ],
                "dataPartitionPatterns":[
                    {
                        "name":"AirdelayByMonth",
                        "rootLocation":"AirdelayData/",
                        "regularExpression":"(\\\\w{3})/airdelays.
csv)",
                        "parameters":[
                            "month"
                        ]
                    }
                ],
                "lastFileStatusCheckTime":"2021-04-18T17:12:01Z",
```

```
            "lastFileModifiedTime":"2021-04-18T17:12:01Z",
            "lastChildFileModifiedTime":"22021-04-18T17:12:01Z"
      }
   ],
   "relationships":[
      {
            "fromEntity":"AirdelayData/Airdelay.cdm.json/
Airdelay",
            "fromEntityAttribute":"ORIGIN",
            "toEntity":"AirportData/Airport.cdm.json/Airport",
            "toEntityAttribute":"AirportId"
      }
   ],
   "subManifests":[
   ]
}
```

Examining the imports section

The `imports` section establishes a connection to the `cdm:/foundations.cdm.json` file. This adds all the necessary basic types, structures, and definitions. You can interpret this import to get access to all the base classes of CDM. Please check the *Further reading, CDM object types and definitions* section for an overview.

Examining the entities section

If you then read further down to the `entities` section, this is where you will define your data artifacts, your entities.

The section defines the following for your objects:

- Name
- Schema
- Format
- Data partitions (if available)

If you look closer, you will find the `exhibitsTraits` section, for example, as a sub-section in the `dataPartitions` section. `dataPartitions` themselves represent the files that hold the described data.

Traits are object types that you will find in `foundations.cdm.json` and the referenced structures there. They describe the properties of your entities, for example.

In the `Airports` entity, you can find `"traitReference":"is.partition. format.CSV"` and `arguments`, which describe the details of the CSV format of the `Airports` object.

Another section that comes in handy when you describe your solution and the file structures is `dataPartitionPatterns`. Did you check the regular expression path pattern?

Understanding the relationships section

This section enables you to define relationships between the structures that you have defined earlier in the manifest.

> **Note**
> At the moment, it is only possible to create relationships on one attribute. There is no option to have multi-attribute relationships yet.

Please find additional information in the *Further reading, Entity relationships* section.

Understanding the subManifests section

In the `subManifests` section, you can add additional manifests that can be seen as child objects to the solution that you describe in your manifest file. These can be complete child solutions again that will even "know" the solution that they are used in.

Deep diving into CDM

Please check the documentation link in the *Further reading, CDM* section for a deep dive into the options and the complexity that CDM offers.

Examining and leveraging predefined entities

Wait, didn't we talk about predefined entities at the beginning of the chapter? Up to now, it was about the understanding of the structures and the objects. But the major benefit of using CDM is the collection of predefined entities that you can use and extend to support your application and your analytical estate.

The collection of entities is grouped into so-called sub-folders:

- `applicationCommon`
- `industryCommon`
- `operationsCommon`

There you will find a multitude of predefined entities with their structures ready for you to use in your system.

Let's double-click on `applicationCommon`, for example. You will find the following entities there:

Account	Contact	KnowledgeArticleViews	SLA
Activity	Currency	KnowledgeBaseRecord	SLAItem
ActivityParty	CustomerRelationship	Letter	SLAKPIInstance
Address	Email	Note	SocialActivity
Appointment	EmaiSignature	Organisation	SocialProfile
Article	Fax	Owner	Task
ArticleComment	Feedback	PhoneCall	Team
ArticleTemplate	Goal	Position	TeamMembership
BusinessUnit	GoalMetric	Queue	Territory
Connection	KnowledgeArticle	QueueItem	User
ConnectionRole	KnowledgArticleCategory	RecurringAppointment	

Figure 13.1 – Entities of applicationCommon

You will find another sub-folder too: `foundationCommon`. This holds 3 more sub-folders with 22 entities related to **Customer Relationship Management (CRM)**, 5 entities related to financial topics, and 2 more product-related ones. Additionally, `foundationCommon` offers another 27 entities.

If you look at the details of the `industryCommon` sub-folder, you'll find sub-folders for `automotive`, `education`, `financialServices`, and `nonProfit`. Microsoft has added a `Solutions` folder too that takes you deep down into the structures of Microsoft Cloud for Healthcare. You will find entity collections for the following topics:

- `Administration`
- `Clinical`
- `Diagnostics`
- `Financial`
- `Foundational`
- `Medication`
- `Workflow`

Each of them will have another layer of sub-folders holding entities to complete an application in the healthcare sector.

The `operationsCommon` sub-folder is also divided into more sub-folders like the `Entities` one:

- `Commerce`
- `Common`
- `Finance`
- `HumanResources`
- `ProfessionalServices`
- `SupplyChain`
- `System`

Folders over folders and entities over entities. Please find detailed documentation in the *Further reading, Available entities in CDM* section.

Finding CDM definitions

In addition to the documentation, Microsoft has created a GitHub repository, `https://github.com/microsoft/CDM`, where you will find all the entities as CDM notation JSON files. The repository has the CDM object model also available in C#, Java, Python, and TypeScript for your convenience. You will find samples and schema documents and, of course, comprehensive documentation.

Using the APIs of CDM

As mentioned, the object model of CDM is available to you in the GitHub repository and you will find the APIs to use in your applications well documented in the link in the *Further reading, The CDM APIs* section.

Introducing Dataverse

Microsoft introduced a service called Common Data Service in 2019. This was one of the first services that implemented CDM. It was renamed Dataverse in 2020.

You can think of it as an abstraction layer that uses Azure storage services in the background. It comes with the entities of CDM available to be set up as tables and columns and connect to other services. You can also create your own tables and structures on top or adjust CDM tables to your needs.

Microsoft Dynamics 365 and the Power Platform (the Microsoft low-code/no-code platform) with Power Apps, Power Automate, Power Virtual Agents, and Power BI are tightly connected with Dataverse.

With Dataverse, you can easily connect, for example, Dynamics 365 content to your Power Platform and build apps to integrate with other data or build workflows to react to events and much more.

Please find detailed definitions and insights about Dataverse in the *Further reading*, *Dataverse* section.

Discovering Azure Industry Data Workbench

Recently, Microsoft has launched a beta version of the so-called Industry Data Workbench. This is a new application that will give you a graphical interface to industry data models for 10 different industries:

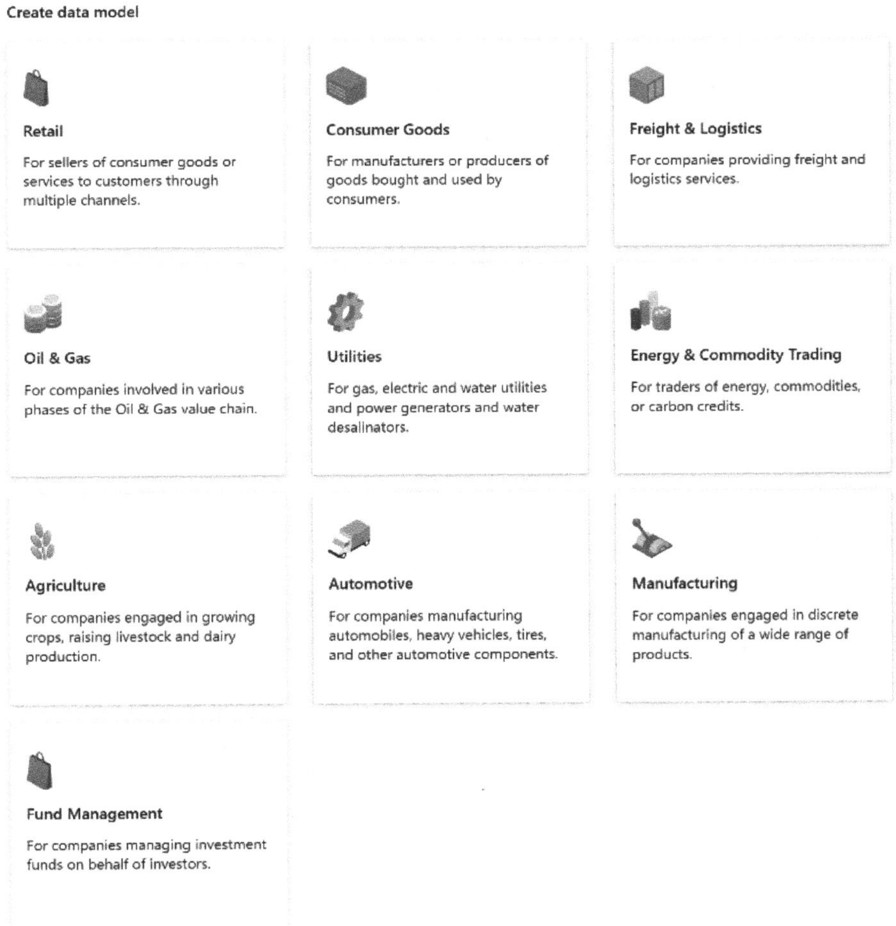

Create data model

Retail

For sellers of consumer goods or services to customers through multiple channels.

Consumer Goods

For manufacturers or producers of goods bought and used by consumers.

Freight & Logistics

For companies providing freight and logistics services.

Oil & Gas

For companies involved in various phases of the Oil & Gas value chain.

Utilities

For gas, electric and water utilities and power generators and water desalinators.

Energy & Commodity Trading

For traders of energy, commodities, or carbon credits.

Agriculture

For companies engaged in growing crops, raising livestock and dairy production.

Automotive

For companies manufacturing automobiles, heavy vehicles, tires, and other automotive components.

Manufacturing

For companies engaged in discrete manufacturing of a wide range of products.

Fund Management

For companies managing investment funds on behalf of investors.

Figure 13.2 – Available industry models of Industry Data Workbench

When you click on one of the tiles, you are taken to a development canvas. You can choose to browse the whole collection graphically by browsing to the **Select entities by enterprise model** drop-down box in the upper-left corner and checking the whole group:

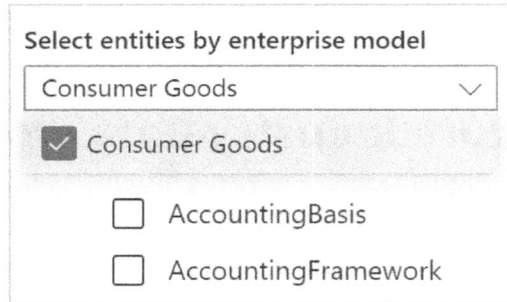

Figure 13.3 – The whole group for Consumer Goods selected

The canvas will display all the tables and their relationships and you can zoom in and out as you wish:

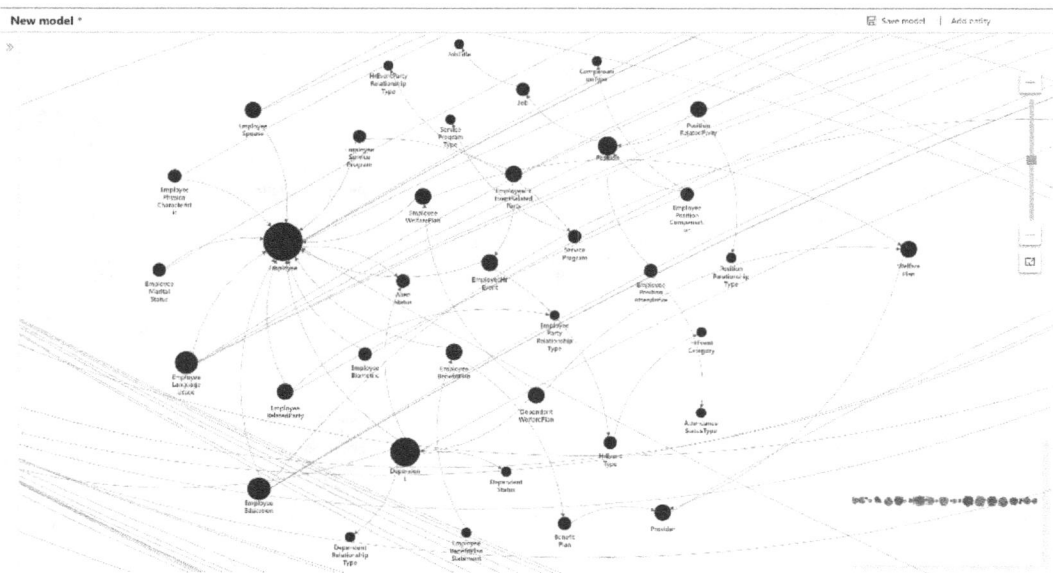

Figure 13.4 – Graphical representation of the Consumer Goods model

In the lower-right corner, you will find a small navigation map that will help you in moving quickly through the model.

Now, when you click a table, you will get the details of that table displayed below the canvas. Let's examine the `Employee` table:

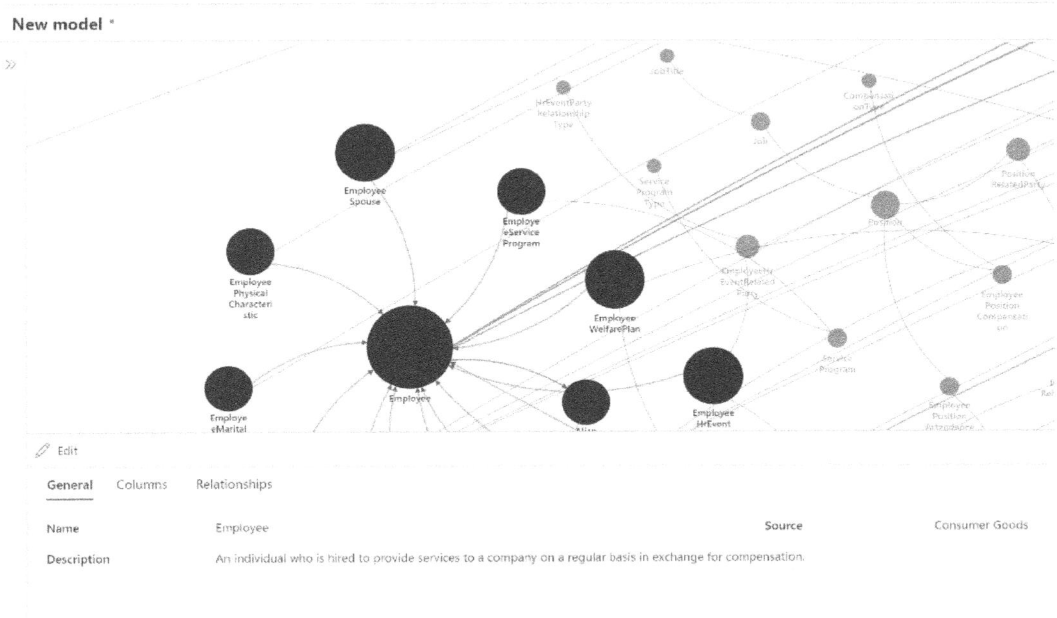

Figure 13.5 – Employee table selected with the General tab displayed in the details

You can now examine the **Columns** and **Relationships** tabs. For all the details that you display, you can click the **Edit** button and modify the entity/table as you wish:

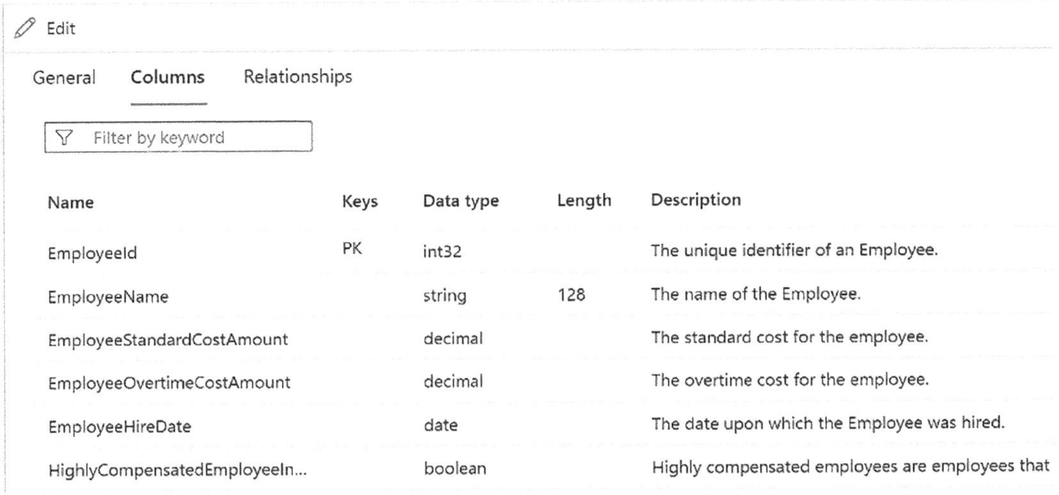

Edit

General Columns Relationships

Filter by keyword

Name	Keys	Data type	Length	Description
EmployeeId	PK	int32		The unique identifier of an Employee.
EmployeeName		string	128	The name of the Employee.
EmployeeStandardCostAmount		decimal		The standard cost for the employee.
EmployeeOvertimeCostAmount		decimal		The overtime cost for the employee.
EmployeeHireDate		date		The date upon which the Employee was hired.
HighlyCompensatedEmployeeIn...		boolean		Highly compensated employees are employees that

Figure 13.6 – The Columns view of the Employee table

You can add or remove columns and relationships and you can even add new tables to the model:

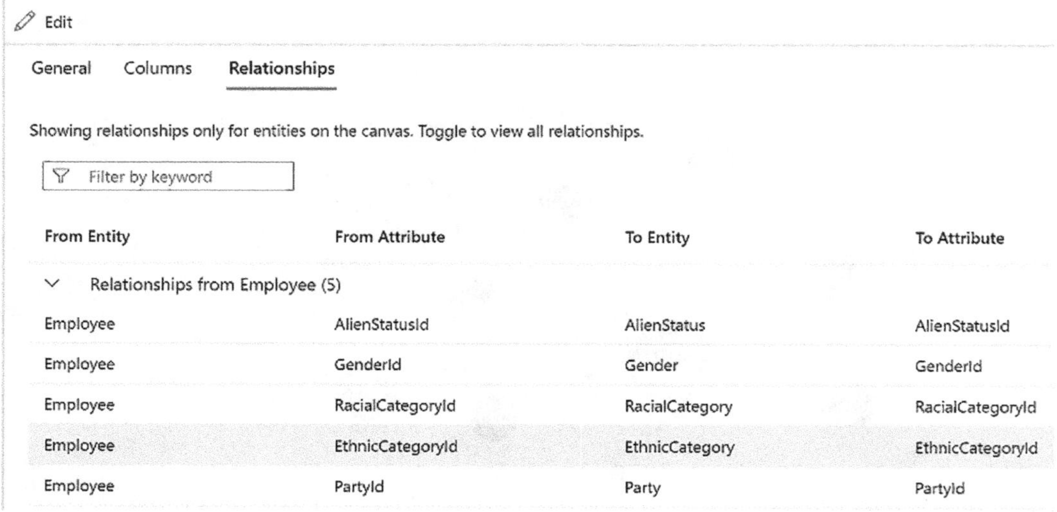

Figure 13.7 – Relationships view of the Employee table

If you check the upper-right corner of the development canvas, you will find the **Add entity** button. This will display a blade on the right where you can start creating your new table and add it to your model:

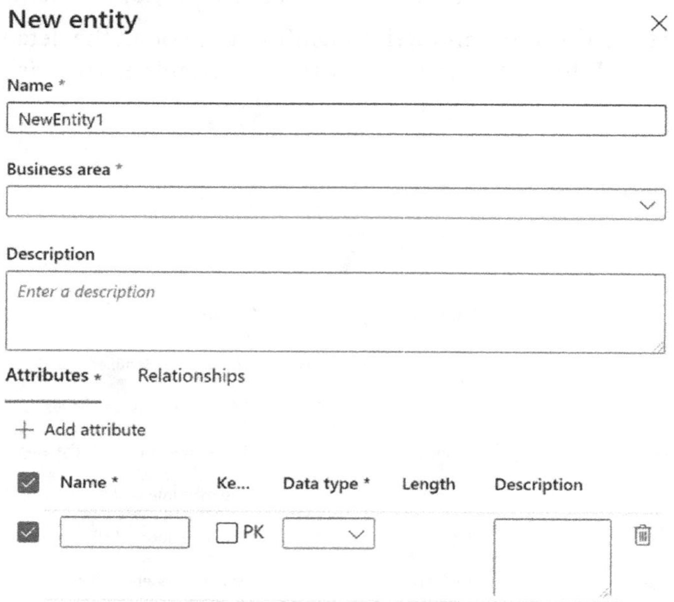

Figure 13.8 – Creating a new entity

Once you have finished your target model, you can save it:

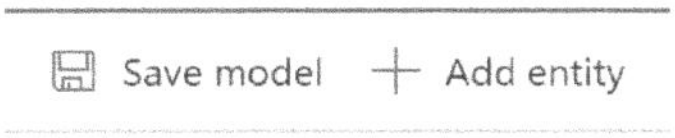

💾 Save model + Add entity

Figure 13.9 – The Save model and Add entity buttons

The model will be saved in the CDM format that was discussed in the *Examining the basics of the SDK* section. You will receive a JSON file in the notation that you used previously. You can use entities with Dataverse, for example, and all the tools that can use the CDM format, including the Power Platform tools and Microsoft Dynamics 365:

```json
{
    "apiVersion": "2021-01-01-preview",
    "properties": {
        "baseModel": {
            "name": "ConsumerGoods",
            "version": "0.1.0",
            "format": "CDM"
        },
        "entityDefinitions": {
            "existingEntities": [
                {
                    "name": "BalanceSheet",
                    "baseEntityReference": {
                        "name": "BalanceSheet",
                        "path": "/BalanceSheet.cdm.json"
                    }
                },
                {
                    "name": "CashFlowStatement",
                    "baseEntityReference": {
                        "name": "CashFlowStatement",
                        "path": "/CashFlowStatement.cdm.json"
                    }
                },
                {
                    "name": "Currency",
                    "baseEntityReference": {
                        "name": "Currency",
                        "path": "/Currency.cdm.json"
                    }
                },
                {
                    "name": "DepreciationMethod",
                    "baseEntityReference": {
                        "name": "DepreciationMethod",
                        "path": "/DepreciationMethod.cdm.json"
                    }
                },
                {
                    "name": "FinancialStatement",
                    "baseEntityReference": {
                        "name": "FinancialStatement",
                        "path": "/FinancialStatement.cdm.json"
                    }
```

Figure 13.10 – The model file as saved from Industry Data Workbench

Summary

In this chapter, you have learned about Microsoft CDM. You have seen the language elements of its SDK and you had a glance at the depth of the available predefined entities that you can use from CDM for your applications.

In the second part of the chapter, you had a sneak peek into the available beta version of Industry Data Workbench and you were able to build a first idea of how this tool will make your life easier when you're developing applications and analytical systems in the future.

In the final chapter, *Chapter 14, Establishing Data Governance*, you will see how you can use Azure Purview to identify and classify sensible data to secure and organize insights and catalog your data estate to actively govern it.

Further reading

- CDM: `https://docs.microsoft.com/en-us/common-data-model`
- CDM object types and definitions: `https://docs.microsoft.com/en-us/common-data-model/sdk/logical-definitions`
- Entity relationships: `https://docs.microsoft.com/en-us/common-data-model/sdk/manifest#entity-relationships`
- Available entities in CDM: `https://docs.microsoft.com/en-us/common-data-model/schema/core/overview`
- The CDM APIs: `https://docs.microsoft.com/en-us/common-data-model/1.0om/api-reference/api-reference`
- Dataverse: `https://docs.microsoft.com/en-us/powerapps/maker/data-platform/`

14
Establishing Data Governance

In the modern data warehouse architecture with its various options to land and store data, you will no longer have the one database where your single version of the truth resides. This will make it more complex to keep track of the content, relationships between entities, and their sensitivity, for example.

Actual regulations such as the **General Data Protection Regulation (GDPR)** require you to be able to locate your customer data and all the related information in case the customer asks for a report or, more importantly, when the customer requires deletion.

But it's not only regulative requirements that force you to gain insights into your data. By adding more and more different datasets in a more and more diverse business, you will need additional tools – tools that enable you to find your way through your data lake, your data warehouse, and for sure the sources where you extract the data from to create all the analytics, insights, and predictions that you need to run your business. Tools that help you and your colleagues to identify the right datasets for your analysis and to plan data access and security.

Data lineage is another big topic when it comes to establishing control over your data estate. There is nothing worse than missing related entities or attributes to an object that you need to change. You may break the whole data integration pipeline.

This chapter will take you through the options that the Azure Purview preview offers for scanning your data and qualifying it. You will see how you can benefit from predefined and custom search patterns and how Purview helps you to find information in your data estate.

You will see how to integrate with other Azure services such as Azure Synapse Analytics or Data Factory and how your data engineers, data scientists, and others working with data will benefit from using the data lineage feature and the other integration features.

The sections covered in this chapter are as follows:

- Discovering Azure Purview
- Classifying data
- Integrating with Azure services
- Using data lineage
- Discovering Insights
- Discovering more Purview

Technical requirements

To be able to give Azure Purview a try, you will need the following:

- An Azure subscription with the rights to provision a Purview service.
- A Synapse workspace where you own the `System Administrator` RBAC role (see *Chapter 4, Understanding SQL Pools and SQL Options*).
- An Azure Data Lake Storage account, perhaps from previous chapters such as *Chapter 3, Understanding the Data Lake Storage Layer*. You will need at least reading permissions on the data lake.

Discovering Azure Purview

With the Azure Purview preview, Microsoft introduces the first wave of data governance tools. You will find modules for the following:

- Data scanning and cataloging and search
- Data classification and glossary
- Data lineage including Data Factory/Synapse pipelines and Power BI
- Metadata insights

Purview integrates with Synapse in a way that you can use the Purview search functionality within the Synapse workspace. The search results can be used by a Synapse developer, for example, to start a SQL script or a Spark Notebook just in the same way as you would do from the Data Lake browser in the Synapse workspace. Remember what you saw in *Chapter 10*, *Loading the Presentation Layer*?

Microsoft is not only targeting data governance; it also aims to improve developers' productivity and experience.

Provisioning the service

But let's start with the service and provision it first:

1. Please go to the Azure portal and type `Purview` into the search bar at the top of the portal. From the search results, please select **Purview accounts**:

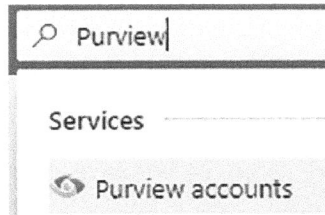

Figure 14.1 – Search results in the Azure portal

2. In the Purview accounts collection, please hit **+ Create** and start the provisioning sequence.

3. In the following blade, please choose your subscription and a resource group (or create a new one), name your new Purview account, and select a location. When you're finished, please either click **Next: Configuration** to continue configuring the service or **Review + Create** if you want to start the provisioning instantly:

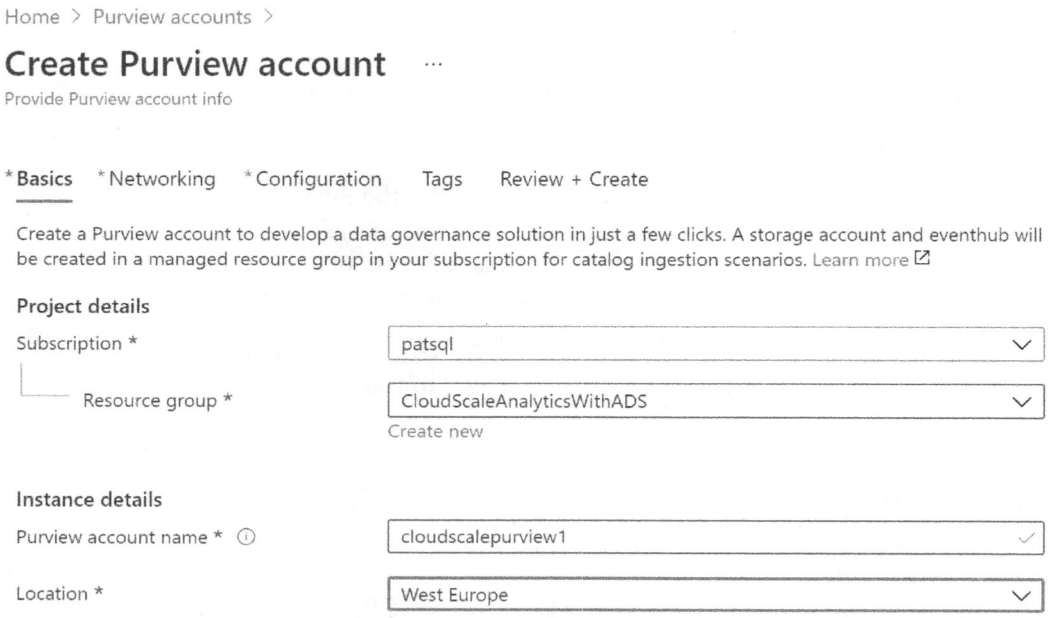

Home > Purview accounts >

Create Purview account ···
Provide Purview account info

* **Basics**   * Networking   * Configuration   Tags   Review + Create

Create a Purview account to develop a data governance solution in just a few clicks. A storage account and eventhub will be created in a managed resource group in your subscription for catalog ingestion scenarios. Learn more ☑

Project details

Subscription *

| patsql   ⌄ |

Resource group *

| CloudScaleAnalyticsWithADS   ⌄ |
Create new

Instance details

Purview account name * ⓘ

| cloudscalepurview1   ✓ |

Location *

| West Europe   ⌄ |

Figure 14.2 – The Basics blade when creating a Purview account

4. If you have continued the configuration, you will be taken to the **Networking** blade. Like other services, Purview too will give you the option to configure it to be available either from all networks or via private endpoints only:

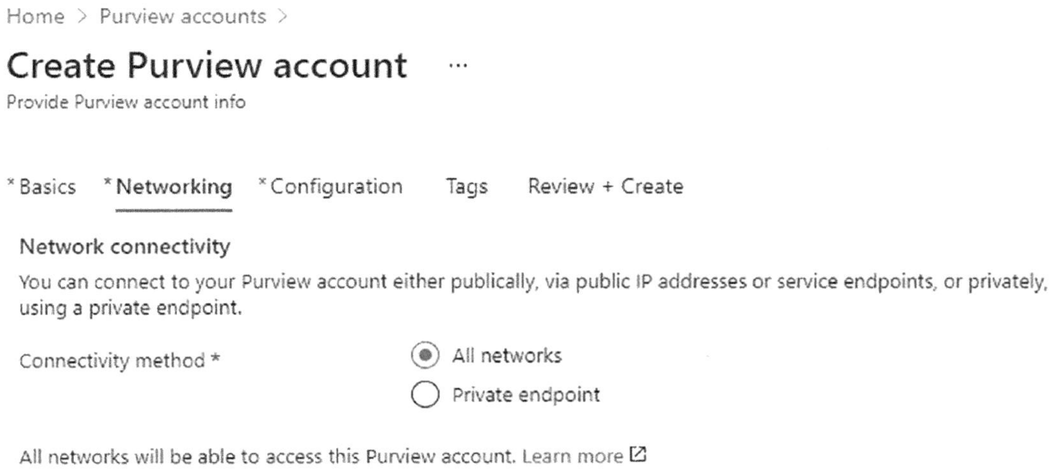

Home > Purview accounts >

Create Purview account   ...

Provide Purview account info

*Basics   *Networking   *Configuration   Tags   Review + Create
            ‾‾‾‾‾‾‾‾‾‾

Network connectivity

You can connect to your Purview account either publically, via public IP addresses or service endpoints, or privately, using a private endpoint.

Connectivity method *   ◉ All networks

                                   ◯ Private endpoint

All networks will be able to access this Purview account. Learn more ☒

Figure 14.3 – The Networking blade creating a Purview account

Please hit **Next: Configuration** to continue when you have configured the networking.

5. The **Configuration** blade will let you configure the capacity settings for the Purview account:

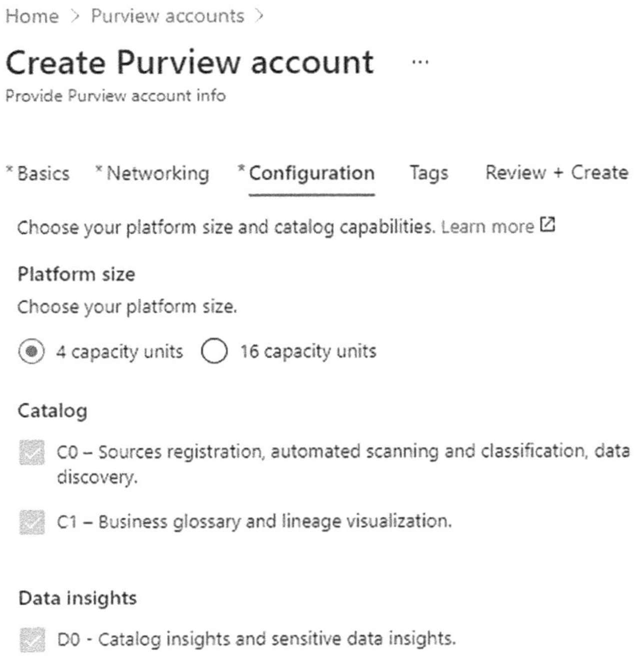

Home > Purview accounts >

Create Purview account   ...

Provide Purview account info

*Basics   *Networking   *Configuration   Tags   Review + Create
                          ‾‾‾‾‾‾‾‾‾‾‾‾‾

Choose your platform size and catalog capabilities. Learn more ☒

Platform size

Choose your platform size.

◉ 4 capacity units   ◯ 16 capacity units

Catalog

☑ C0 – Sources registration, automated scanning and classification, data discovery.

☑ C1 – Business glossary and lineage visualization.

Data insights

☑ D0 - Catalog insights and sensitive data insights.

Figure 14.4 – The Configuration blade when creating a Purview account

6. For the **Tags** blade, you already know what's waiting for you. Like other services too, Purview allows tagging for cross-charging and so on.

7. You have seen the **Review + Create** blade too throughout the book. You can check the configuration of the service that you are about to create here before you start the provisioning.

8. When you're finished with the configuration and the review, please hit **Create** and create the service.

9. Once the provisioning is done, you will be able to directly jump to the newly created service from the provisioning results and derive the ARM template for further automated deployments.

Ready with the service? Let's dive into it and start examining and cataloging your data estate.

Connecting to your data

Once your service is deployed, you will find the following tiles on the **Overview** blade:

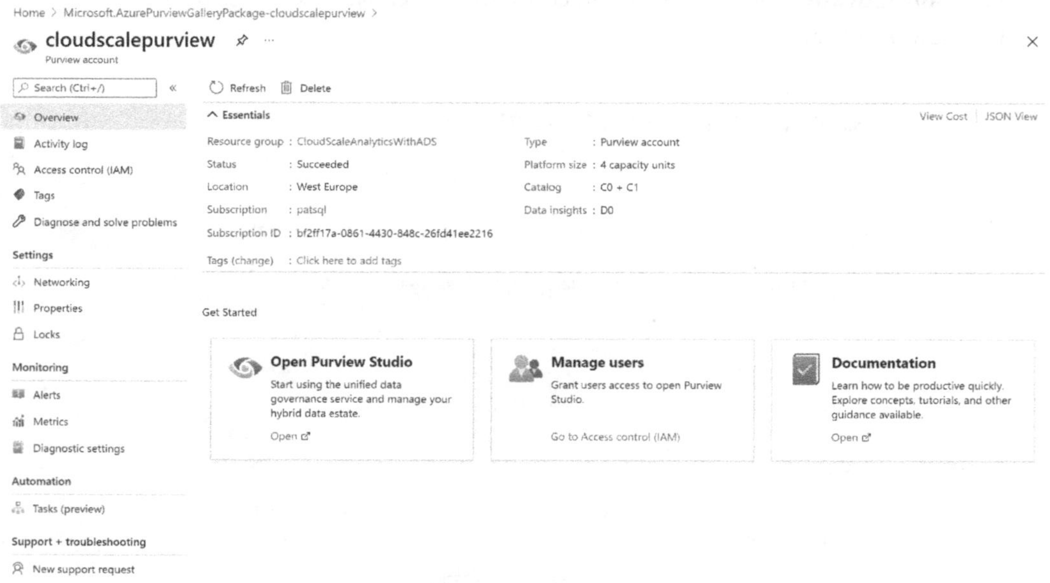

Figure 14.5 – The Overview blade of the Purview service

You will reach Purview Studio via the first option, **Open Purview Studio**. The second tile, **Manage users**, will take you to the **Access control (IAM)** section of the service, where you can add users and groups to your service and control their RBACs for the service. I'm sure you can derive where the third tile will take you.

Let's first open Purview Studio and examine our options there:

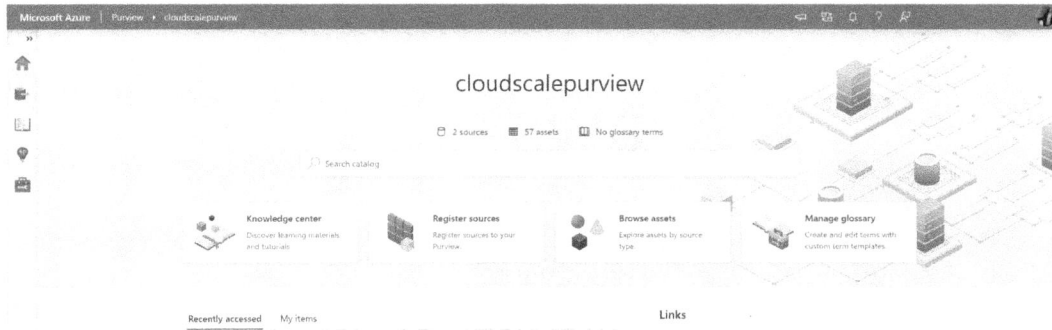

Figure 14.6 – The Home blade of Purview Studio

Let's connect our Data Lake account to prepare it to be cataloged in the next step:

1. Please click on the **Register sources** tile. You are taken to the sources map view with an empty canvas.

2. Sources are grouped into collections. So, your first step is to create one by clicking **+ New collection**. On the blade that slides in from the right, please enter a name, maybe something such as DataLakes. As this will be the top parent, you don't need to add another parent in this case. Please hit **Create**:

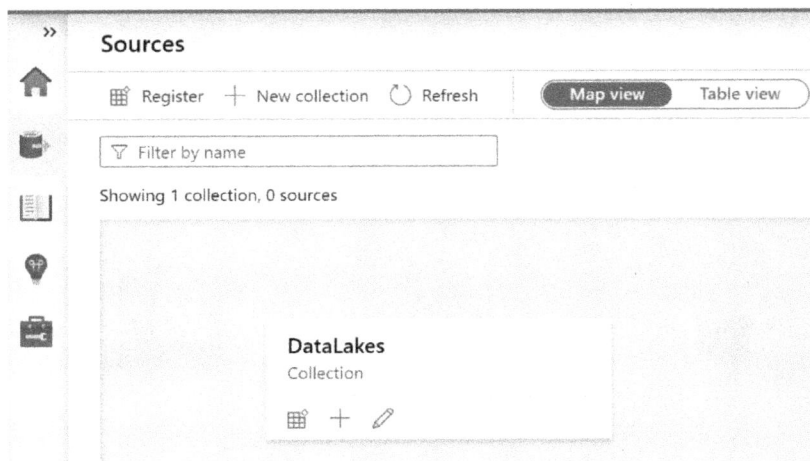

Figure 14.7 – Sources map view with an empty collection

3. Now you can start to register your data lake. Please click **Register**. The following blade will be displayed on the right:

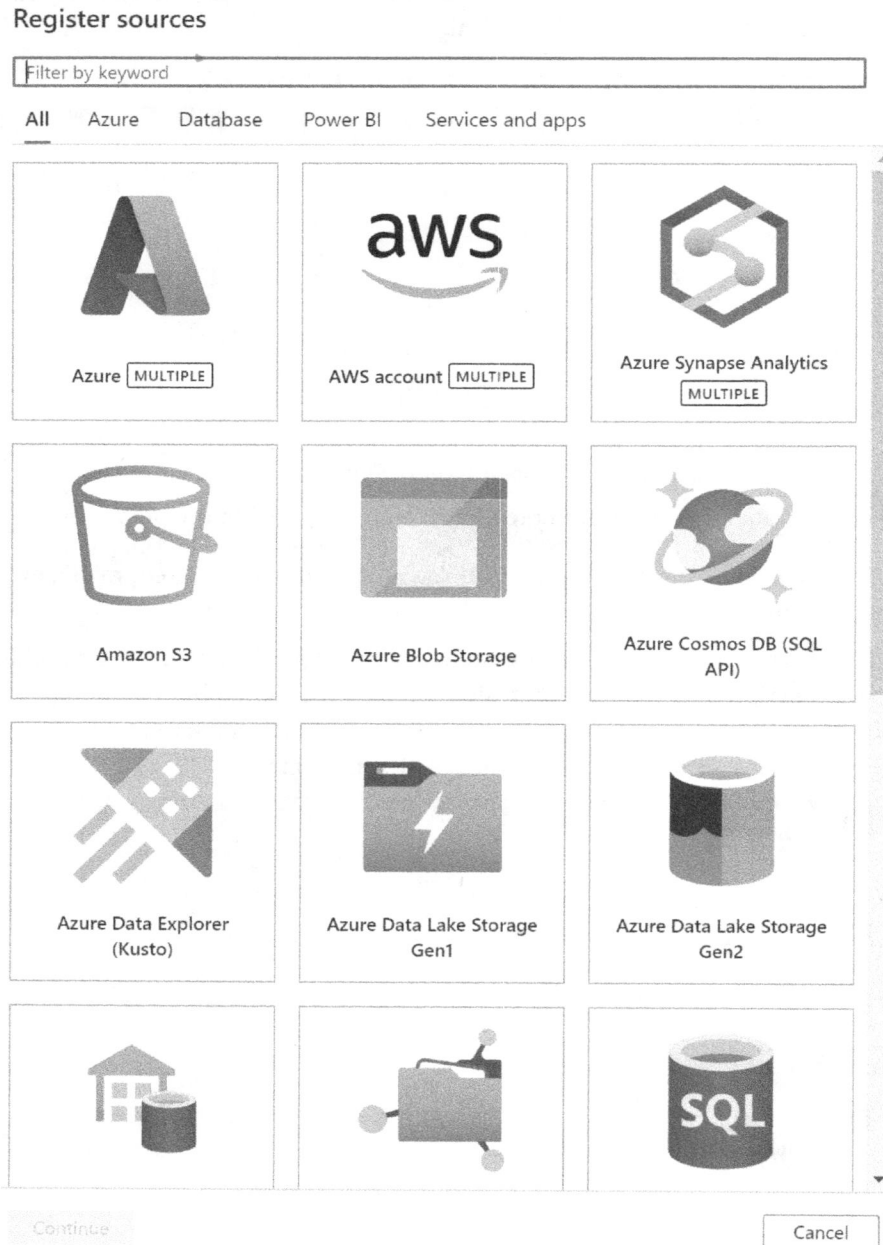

Figure 14.8 – Register sources selector

You can choose from a growing collection of source connectors such as Azure Storage accounts and Data Lake stores, for example, or Amazon S3 buckets. You will find databases such as Azure SQL databases, SQL Server, Oracle, Teradata, SAP HANA, and others.

Power BI is another option that you can select here and will play an additional role when it comes to data lineage.

The **Azure**, **AWS account**, and **Azure Synapse Analytics MULTIPLE** tiles at the top will even give you access to the contents of the particular accounts and their metadata.

Please choose **Azure Data Lake Storage Gen2** and proceed to the next step by clicking **Continue**.

4. On the **Register sources (Data Lake Storage Gen2)** blade that comes in from the right, please fill in or select the necessary information such as a name, your Azure subscription, the Storage account name (**Endpoint** will be filled in automatically), and finally, in the **Select a collection** field, choose the collection that you created in *step 2*, where your data lake will be displayed below as a node:

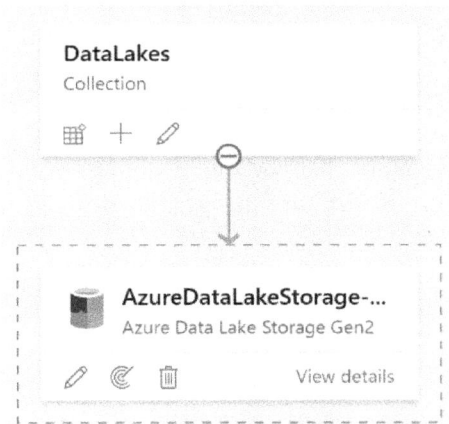

Figure 14.9 – Data Lake Storage node below the new collection

There is one thing that is now left to do before you can start scanning. You will need to set up access rights for your Purview managed identity in your Data Lake account. Please refer to *Chapter 3*, *Understanding the Data Lake Storage Layer*, to check again how to set this up. You will need to add your Purview managed identity to the Blob Storage Data Reader RBAC role in your data lake.

> **Note**
>
> When setting this up, you will find the Purview managed identity with the same name that you chose for your Purview service.

You have now finished the fundamental setup to start the discovery data estate. You will now be able to find sensitive information in your files, for example, and automatically classify this content.

Scanning data

To set up your first scan, please click on the **New scan** (blue Pacman-like) symbol ⦅ in the Data Lake node. A dialog blade is displayed on the right:

1. Please fill in a name and leave the rest of the settings as their defaults. But remember, you can select another integration runtime here in case you need to access an on-premises source, for example. This is the same concept that the Data Factory/Synapse pipelines use.

2. Additionally, you can set up other credentials than the Purview managed identity's ones to react to individual security settings of your source.

3. Please click **Continue** once your connection test is successful.

4. The second step of the sequence asks you to scope your scan. In the tree view of your Data Lake account, you can now individually check the boxes for the folders that you might want to scan or that you might want to avoid being scanned. You might leave this with the default settings and click **Continue**:

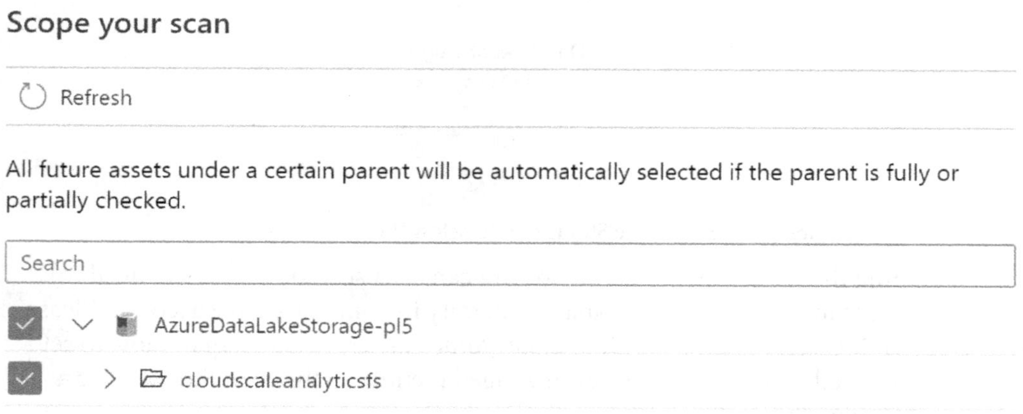

Scope your scan

⟳ Refresh

All future assets under a certain parent will be automatically selected if the parent is fully or partially checked.

Search

☑ ⌄ ▥ AzureDataLakeStorage-pl5

☑ ﹥ ▱ cloudscaleanalyticsfs

Figure 14.10 – Scope your scan tree view of your Data Lake account

5. On the **Scan rule set** blade, you can now either select an available scan rule set or choose directly from the blade to create a new one. Scan rule sets are collections of classification patterns and file formats to be searched for in your data lake. We will examine them in the *Classifying data* section. Please proceed by clicking **Continue** with the default **AdlsGen2** selected.

> **Note**
>
> The default scan rule set selected contains around 100+ built-in classifications already and will indicate information such as people's names, birthdates, addresses or bank accounts, IDs, healthcare, and other patterns that can be found in your data. Please refer to the *Further reading, Supported classifications* section for a link to the detailed collection of predefined classifications available with Purview.

6. In the next step, you can choose **Set a scan trigger** for your scan. Let's choose **Once** for now (all the calendar options will disappear) and click **Continue**:

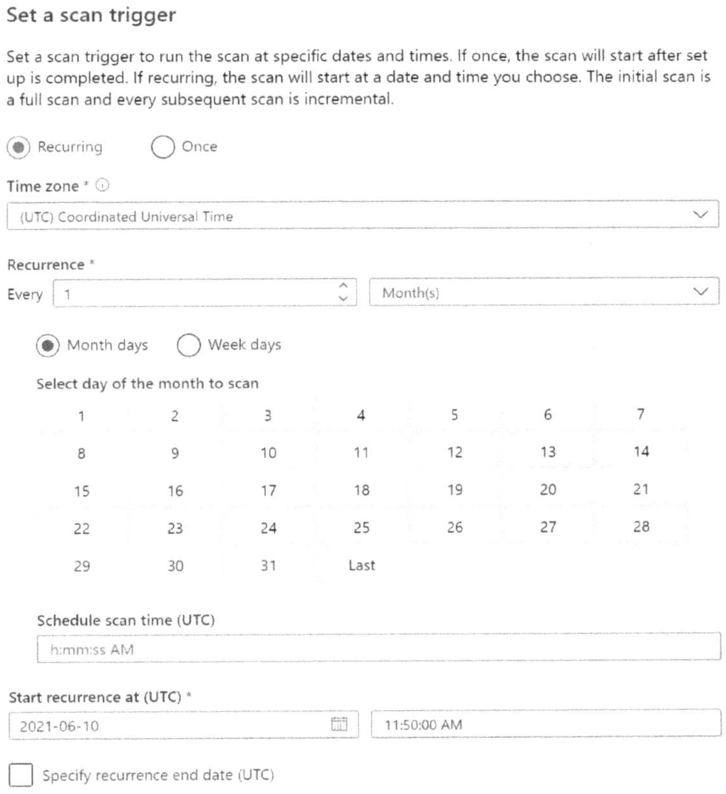

Set a scan trigger

Set a scan trigger to run the scan at specific dates and times. If once, the scan will start after set up is completed. If recurring, the scan will start at a date and time you choose. The initial scan is a full scan and every subsequent scan is incremental.

◉ Recurring   ○ Once

Time zone * ⓘ

| (UTC) Coordinated Universal Time | ⌄ |

Recurrence *

| Every | 1 | ˄˅ | Month(s) | ⌄ |

◉ Month days   ○ Week days

Select day of the month to scan

1	2	3	4	5	6	7
8	9	10	11	12	13	14
15	16	17	18	19	20	21
22	23	24	25	26	27	28
29	30	31	Last			

Schedule scan time (UTC)

| h:mm:ss AM |

Start recurrence at (UTC) *

| 2021-06-10 | 🗓 | 11:50:00 AM |

☐ Specify recurrence end date (UTC)

Figure 14.11 – Set a scan trigger

7. You will get a **Review your scan** display and the option to correct settings on the final blade. When you click **Save and Run**, the scan will start. You're taken back to the data map view. Please allow 10 to 15 minutes to finish the scan.

8. You can follow the status by clicking the **View details** link in the Data Lake node that you see in the data map:

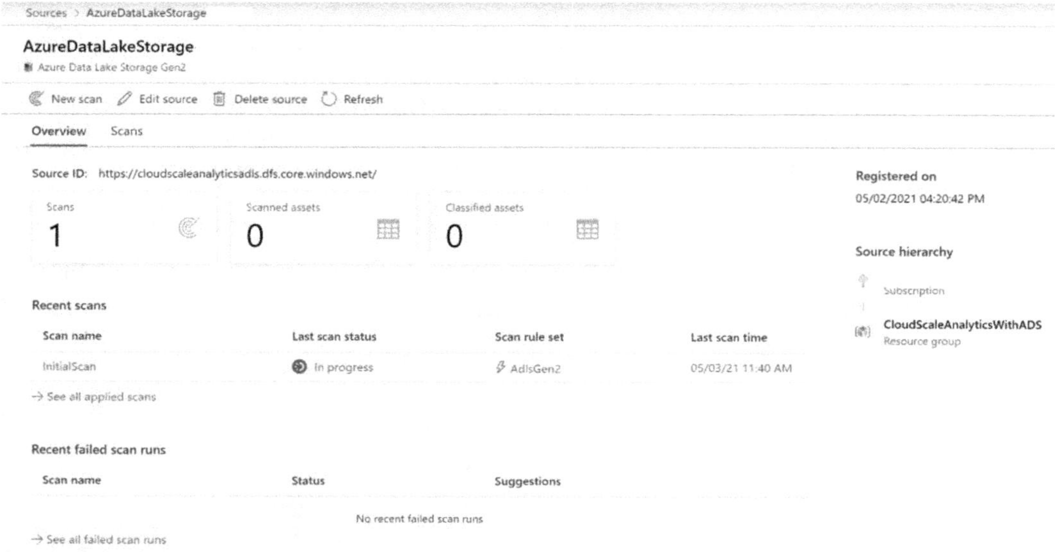

Figure 14.12 – Overview of the Data Lake node

During the scan, the different tiles displayed here will show the counts of assets that are scanned and classified. You can refresh this view from time to time if you want to follow the progress.

Searching your catalog

Now you can start to discover the data that you have stored in your data lake and you have different options to do so.

One of your options is the search field in the top ribbon of Purview Studio:

Figure 14.13 – Purview search

You can use this option when you are looking for certain keywords or parts of them. Just type in your search term and give it a go. Purview will display a quick result instantly below the search:

Figure 14.14 – Quick results of the Purview search

If you hit *Enter* or click the **View search results** link in the quick results, you will be taken to the **Search results for [keyword]** display:

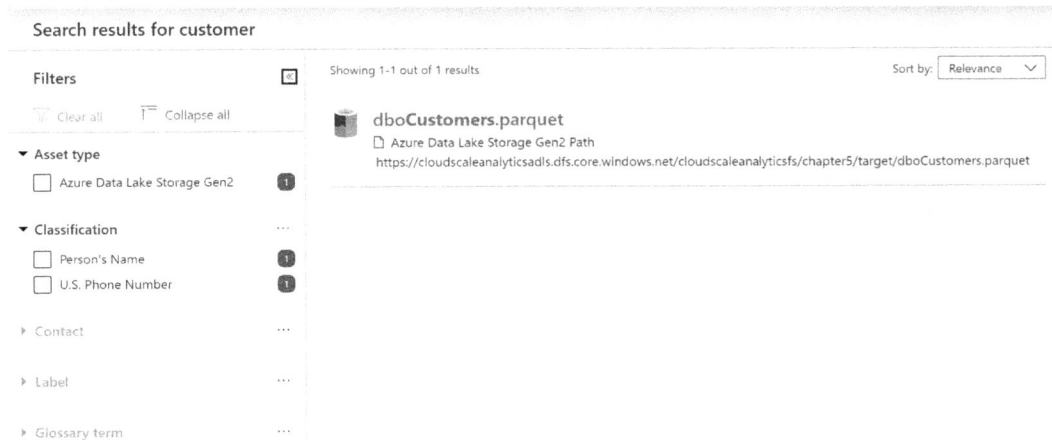

Figure 14.15 – Search results

On the left side of this view, you have the **Filters** pane. You can narrow down the search results by checking the boxes of the displayed entries here. The available filters are determined by the entities that are available in the search results. You will find the following depending on their availabilities with the assets found:

- Asset types
- Classifications
- Contacts
- Sensitivity labels
- Glossary terms

We'll check these later.

Browsing assets

Another option to find assets, especially when you want to gain an initial overview, is to use the **Browse assets** tile on the home screen. The registered data sources are displayed here as tiles.

> **Note**
> Please don't let the displayed resources confuse you when you find an Azure Storage account that you haven't created yourself. This one gets created with your Data Lake source as the Data Lake Storage account is based on the Azure Storage account. You can ignore this one.

When you now select your Data Lake account that you registered in the *Connecting to your data* section, you are taken to another view that displays the name of the Data Lake account as a link. This one will take you to a browser-like overview of your Data Lake. You can browse this just like you would browse your storage explorer and get an overview of what is available there and dive into the folders and files. Once you have found what you were searching for, you can click the asset and dive into its details.

But now let's examine an asset and its details.

Examining assets

Once you have found what you are searching for, you will find the details of the asset on the following different tabs:

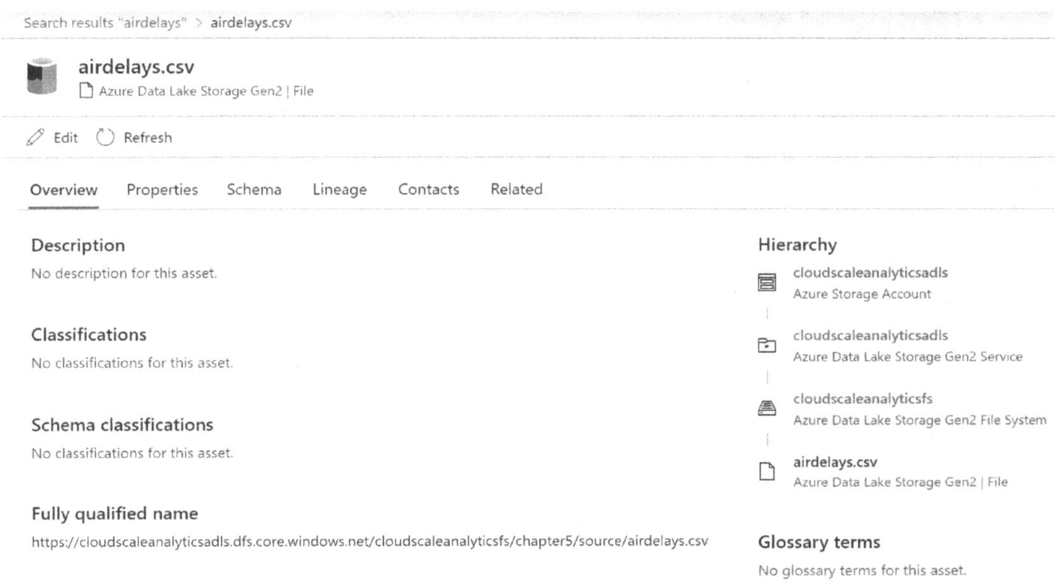

Search results "airdelays" > airdelays.csv

airdelays.csv
Azure Data Lake Storage Gen2 | File

✎ Edit   ⟳ Refresh

Overview   Properties   Schema   Lineage   Contacts   Related

Description
No description for this asset.

Classifications
No classifications for this asset.

Schema classifications
No classifications for this asset.

Fully qualified name
https://cloudscaleanalyticsadls.dfs.core.windows.net/cloudscaleanalyticsfs/chapter5/source/airdelays.csv

Hierarchy

📄 cloudscaleanalyticsadls
Azure Storage Account

📁 cloudscaleanalyticsadls
Azure Data Lake Storage Gen2 Service

📦 cloudscaleanalyticsfs
Azure Data Lake Storage Gen2 File System

📄 airdelays.csv
Azure Data Lake Storage Gen2 | File

Glossary terms
No glossary terms for this asset.

Figure 14.16 – Asset details

From here, you can reach the following tabs:

- **Overview**: This will give you a first glance at **Description**, asset-wide classifications, schema classifications, the fully qualified name, the hierarchy, and the related glossary terms.

- **Properties**: Here you will find all the describing properties of the asset, for example, **contentType**, the **isFile** flag, **path** if you're looking at a file, and more.

- **Schema**: This tab lists the schema with the column names of the artifact, **Classifications, Data type, Sensitivity label, Glossary terms**, and descriptions of the particular columns.

- **Lineage**: A very interesting and maximal informative view on the data lineage of your asset if you have connected Data Factory/Synapse pipelines and a Power BI account. You will find the whole path and relations between assets, pipelines, and Power BI datasets.

- **Contacts**: You can use this area to name experts and owners of the dataset. These are contacts from your Azure Active Directory.

- **Related**: This tab displays all assets from your Purview catalog that somehow might have a relationship to the displayed asset. This might be that they have been found in the same folder or database tables that have primary and foreign key relationships in the same database or similar.

Classifying data

Please examine the `airdelayspredict.csv` file that you created in *Chapter 9, Integrating Azure Cognitive Services and Machine Learning*. You will find a schema classification on the **Overview** tab or in the **Schema** view: `U.S. Stage Name` for `DEST_STATE_ABR` and `ORIGIN_STATE_ABR`. This is one of the predefined classification rules that are available in Purview.

Let's discover where you can find them and how you can add your own classifications and classification rules.

Please navigate to the **Management** hub of your Purview Studio, 🧰 . There are many organizational areas available to you as the admin of your account. For the moment, we will concentrate on **Metadata management** with the two nodes, **Classification** and **Classification rules**.

Please examine the two nodes. You will find the **System** and **Custom** categories in these lists.

The **System** list already shows about 100+ different classifications, such as `Age of an individual`, for example, or different patterns for passport numbers and much more. Some of them are pattern-oriented and implement a regular expression to recognize the pattern within scanned data. Others are dictionary-based, which means that the examined value in your data will be compared to a list of values. This is used, for example, with the `Person's Name` classification rule or `World Cities`.

The **Custom** list is empty when you create your account. This will hold the classifications that you define.

Creating a custom classification

First, you will have to create a classification. This will act as the name for your classification rule. This is decoupled from the classification rule itself to enable you to switch the classification rules easily without breaking the related scanning processes in case you need to update a classification rule or delete one and replace it with another.

Please navigate to **Classifications** in the **Metadata management** node. You create a new custom classification by clicking **+ New** and then entering a name and a description. When you click **OK**, the classification is stored and available to be used in a classification rule.

This custom classification will now show up in the **Custom** tab of **Classifications**.

Creating a custom classification rule

Now let's create the rule that will be applied to the data when this classification is used in a scan:

1. Please navigate to the classification rules in the **Metadata management** node and click **+ New**.

2. In the **New classification rule** dialog that slides in from the right, please enter a name and a description, select the classification name that you created previously from the drop-down field, and decide on the state of your new rule. Finally, you will need to select whether you want to either enter a regular expression or use a dictionary for your classification rule:

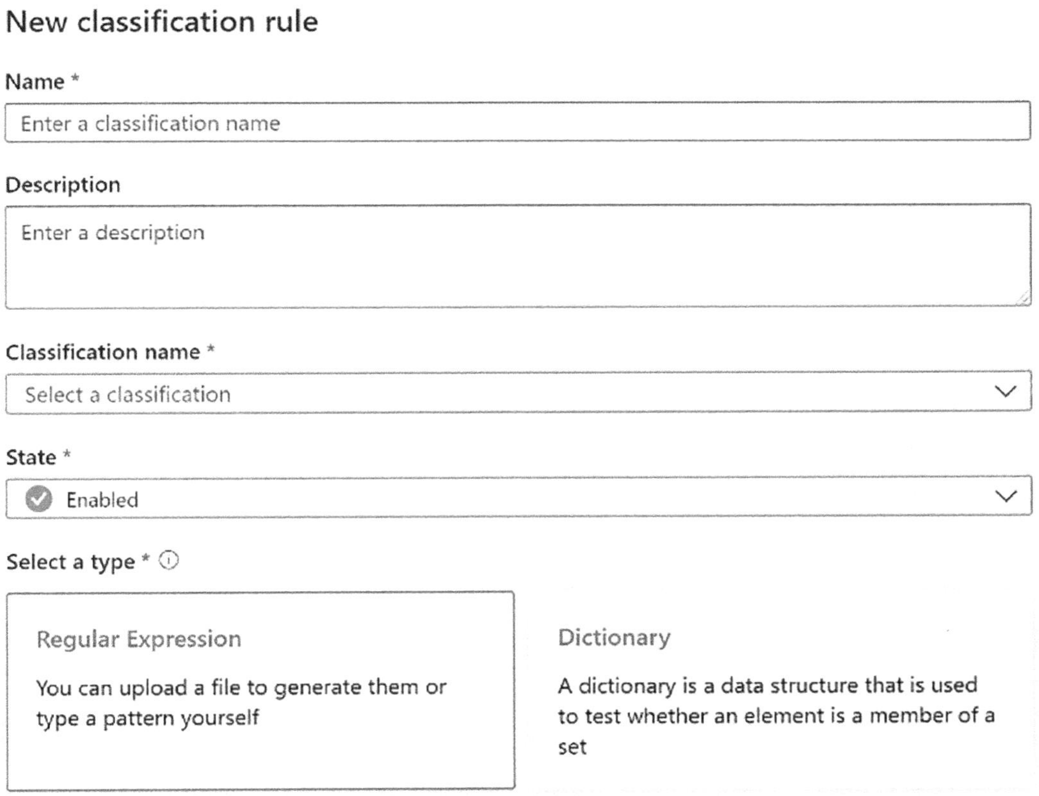

New classification rule

Name *

> Enter a classification name

Description

> Enter a description

Classification name *

> Select a classification ⌄

State *

> ✓ Enabled ⌄

Select a type * ⓘ

Regular Expression	Dictionary
You can upload a file to generate them or type a pattern yourself	A dictionary is a data structure that is used to test whether an element is a member of a set

Figure 14.17 – New classification rule dialog

If you want to use a dictionary, you will be prompted to upload a CSV or a **TSV** (**tab-separated values**) file with the dictionary content that you want to use here. The file should not exceed 30 MB and needs all values that you want to compare against in a single column.

Let's prepare a dictionary about the origin and destination airports from the air delays dataset that you have used earlier in this book. Are you able to create a single-column CSV file from the data that you have in your data lake? For a checklist on how to do so, please check the *Further reading, Creating the airport dictionary* section.

3. Now please select the **Dictionary** tile and proceed with **Continue**.

4. In the upload dialog, please select the dictionary file that you have created. After the content of the file is uploaded, you will get the following display:

New classification rule (Dictionary)

Upload a CSV or TSV file containing all possible values for a classification in one column to generate a dictionary. Please limit the file size to 30MB.

Upload file *

AirportDictionary.csv   🗁

Dictionary generated

Please set thresholds below and click 'Create' to finish setting up the classification rule.

Set thresholds *

Minimum match threshold ⓘ   0%  100%   60% ⌃⌄

Figure 14.18 – Dictionary generated

Use the **Distinct data values threshold** slider to set the number of unique values that you expect before the rule is applied in general.

Minimum match threshold is used to set a minimum amount of values.

Please finish the creation of your classification rule by clicking **Create**.

Creating a regular expression rule

When you want to create a regular expression rule, you would select the **Regular Expression** tile in *step 3* and click **Continue**.

On the following screen, you have the option to either upload a file that contains the regular expression that you want to use here (**Upload file**) or to enter it directly into the **Data Pattern** field.

Use the **Distinct data values threshold** slider to set the number of unique values that you expect before the rule is applied in general.

Minimum match threshold is used to set a minimum amount of values that need to be matched to the regular expression before the scanned attribute will be considered classified.

In the **Column Pattern** field, you have the option to narrow the scan down to only columns that follow the pattern that you enter here:

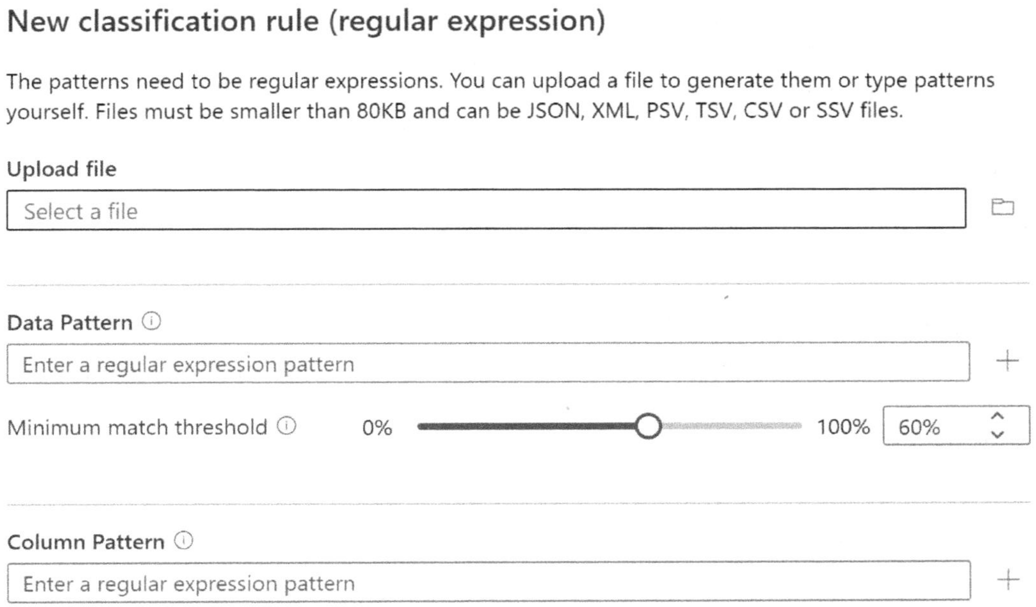

New classification rule (regular expression)

The patterns need to be regular expressions. You can upload a file to generate them or type patterns yourself. Files must be smaller than 80KB and can be JSON, XML, PSV, TSV, CSV or SSV files.

Upload file

| Select a file | 🗁 |

Data Pattern ⓘ

| Enter a regular expression pattern | + |

Minimum match threshold ⓘ   0% ⚫━━━━━━━⭕┄┄┄┄ 100% | 60% ⌄ |

Column Pattern ⓘ

| Enter a regular expression pattern | + |

Figure 14.19 – Regular expression dialog

Again, you would finish your work by clicking **Create**.

Using custom classifications

To use your custom classification in a scan, you will need to create a scan rule set. You can find this node too in the **Management** hub of Purview Studio:

1. Click **+ New** and select the source type on the first blade of the dialog that is displayed on the right. In your case, this will be **Azure Data Lake Storage Gen2**.

    On that same blade, please enter a scan rule set name and a scan rule description. Then hit **Continue**.

2. On the following screen, **Select file types**, you have the option to choose the file types that will be scanned using this particular rule set. You might even add new file types that represent types used in your environment. Please hit **Continue** once you are done with your selection.

3. On the final blade of the dialog, you are now taken to the **Select classification rules** part. The predefined system rules are preselected. When you check the **Custom rules** node, you will find the **Custom classification rules** entry that is expanded and displays all custom classification rules available. Please select your newly created custom classification rule there and finish the configuration of your scan rule set by clicking **Create**.

4. To use this newly created scan rule set, please now go back to your data map and initiate a new scan as you have done in the *Scanning data* section. When you reach *step 3* in the sequence, your new scan rule set will be displayed and you can select it here. Please continue, set up another non-recurring scan, and start it.

Can you find your classification in the search results?

Integrating with Azure services

Purview already integrates with other services, such as Synapse, for example, or Azure Data Factory and Power BI. This integration helps you to leverage the insights that Purview can deliver in different use cases.

Integrating with Synapse

When we examine the integration with Synapse, for example, you will be able to search the Purview catalog from within Synapse Studio. But this is not the only advantage! From the results of your search, you will be able to directly start a SQL script, a Spark Notebook, or a Pipeline data flow just like you would do from a Data Lake folder or a database table or view.

The Synapse integration will be done from your Synapse Studio. You will find the **Azure Purview (Preview)** entry in the **Management** hub of your Synapse workspace:

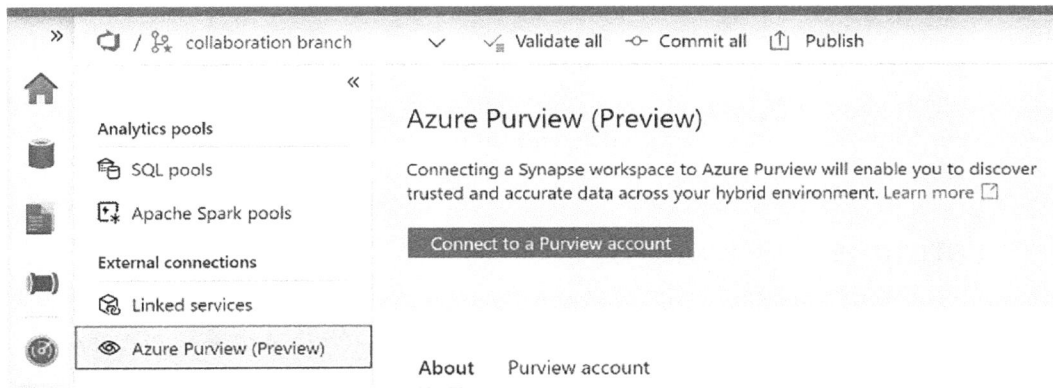

Figure 14.20 – Connect to Purview from Synapse

You will be displayed a selector that will prompt you on whether you want to select from an Azure subscription or enter your Purview connection manually. If you select **From Azure subscription**, you will just need to select the Purview account from the drop-down field and click **Apply**.

When you now enter a search term into the search field at the top of your Synapse Studio screen, you will be connected to your Purview catalog. Why not search for `airdelays.csv`?

When you now access the Purview details of your asset, you will be taken to a very similar view that you already know from Purview Studio, with one extra detail:

Figure 14.21 – Asset details in Synapse

From the **Connect** drop-down field, you can, for example, create the following in your Synapse workspace directly from your found asset:

- A new linked service

- A new integration dataset

When you examine the **Develop** drop-down field, you will find the following:

- New SQL script

- New notebook

- New data flow

A handy extension for your Synapse world, when you don't have to manually browse through your Data Lake folders or your databases to find the right sources for your work here, isn't it?

Integrating with Power BI

Integrating Power BI will add another level of insight to your environment. To add Power BI to your Purview environment, you would simply add Power BI in your data map just as you did with your data lake before.

> **Note**
>
> You will need to add the Purview managed identity to your Power BI tenant to allow Purview to scan Power BI datasets, reports, and dashboards, add them to the catalog, and make them searchable. Additionally, they will appear in the lineage maps together with your data sources, files, and Data Factory pipelines.

Integrating with Azure Data Factory

The available preview of Purview is already capable of displaying lineage information regarding data sources, their origin, and the different steps that they took on their way into your modern data warehouse.

In the future, Purview will also integrate with Synapse pipelines to extract lineage information.

To integrate with a data factory (maybe you still have one left from *Chapter 5, Integrating Data in Your Modern Data Warehouse*), you would navigate to your **Management** hub and go to the **Data Factory** entry in the **Lineage connections** node.

Here you can click **+ New** and in the selector dialog that comes in from the right, you can select a data factory that is available from your subscription.

> **Note**
>
> If you can't access this area, please add yourself to the `Owner` RBAC role in the access control of your Purview service in the Azure portal. You can't do this in Purview Studio. Please refresh the browser window after you have done this.

Using data lineage

Once the data factory is connected, it will send lineage information into your Purview environment for every pipeline that is run. Give it a try and create a Data Factory pipeline that copies data from one folder to another in your data lake. Remember: you are quickest when you use the Copy Data Wizard (or just use the `MyFirstPipeline` pipeline that you created in *Chapter 5, Integrating Data in Your Modern Data Warehouse*, if you used the data factory there).

When you are finished in the data factory, switch back to Purview, repeat your scan (again, this might take a few minutes), and search for the newly created file or the pipeline name, and in the asset details, check the **Lineage** tab:

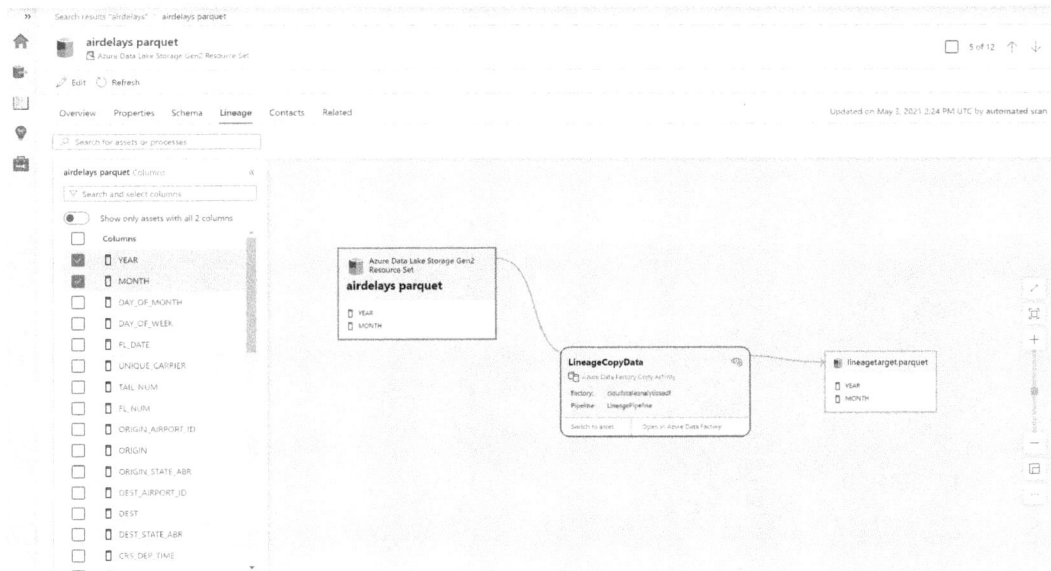

Figure 14.22 – First lineage overview for a Data Factory Copy pipeline

When you check the lineage closely, you will see that you can drill down to the column level and reveal even the column mappings.

Imagine the power of this feature, when you need to make sense of complex pipelines and relationships between sources and targets, spanning over all the levels of jobs in your data integration layer.

As mentioned, data lineage too will be extended in the release version of Purview. You will be able to connect to Synapse pipelines too and get lineage information from there.

Additionally, Purview already exposes the APIs of the Apache Atlas engine that it is built on. This allows you to programmatically add other lineage information to your catalog if you want. Please refer to the *Further reading, Purview REST APIs* section for a link to the documentation.

Discovering Insights

If you examine the **Insights** hub, you will find several sections that will help you to analyze the content of your catalog. You can discover the content of your Purview environment from another perspective. In **Asset insights**, for example, you'll find a tree view and a link to a more detailed grouped overview of all the assets that you have discovered:

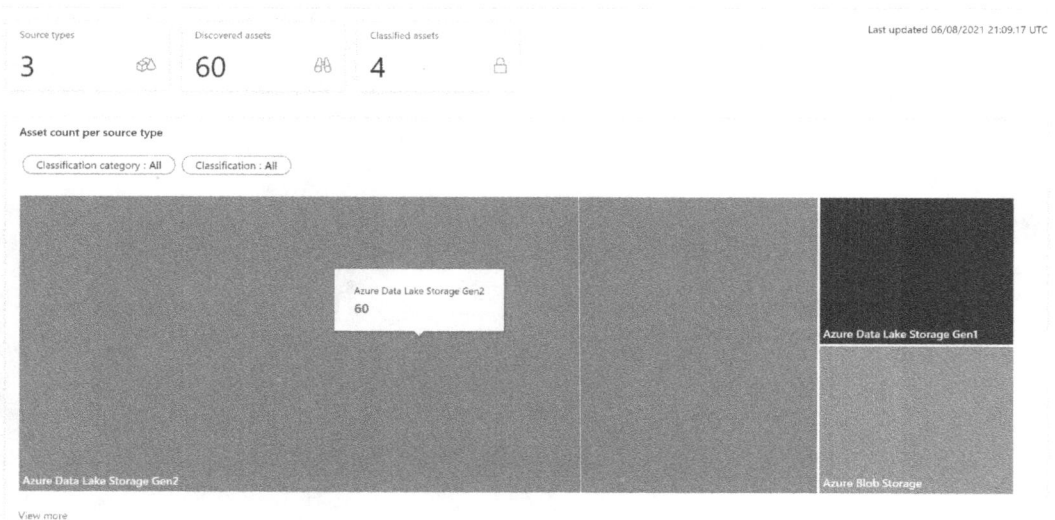

Figure 14.23 – Asset insights

Give it a go and browse the Insights environment.

Discovering more Purview

There is more about Purview that you still can discover. The product is, as mentioned, still in preview status. So, expect changes and new features to come. Maybe you want to examine, for example, a feature that was added recently, such as the usage of sensitivity labels. This means Purview will use sensitivity labels that are extracted from the Microsoft 365 **Security & Compliance Center** (**SCC**). These labels will be applied based on classifications and their combinations found in the scanned data. If you want to find out about sensitivity labels, please refer to the *Further reading, Sensitivity labels* section.

Another interesting concept that was introduced recently is the **Pattern rules** feature and **Resource sets**. These are used to automatically group big amounts of data during your scans. You can find more information about these features in the *Further reading, Pattern rules* section.

Summary

This chapter took you through the Azure Purview preview. You have seen how to connect to data sources and how to set up scans to parse not only your data sources but also your whole modern data warehouse.

You have learned how to use classification rules in your scan rule sets to classify your data. You have seen how to create your own custom classifications and add them to your scans.

In the second part of the chapter, you saw how Purview integrates with other services such as Azure Synapse Analytics, where Purview can help increase productivity by integrating with the Synapse search and the Synapse compute components, such as the serverless SQL engine, the Spark engine, and Synapse pipelines.

You have examined how to integrate Power BI to be able to scan Power BI datasets, reports, and dashboards and how to add this information to the Purview data lineage part.

Finally, you have seen how Purview integrates with Azure Data Factory and can display data lineage from the source to the final Power BI dashboard.

Further reading

- Supported classifications: `https://docs.microsoft.com/en-us/azure/purview/supported-classifications`

- Creating the airport dictionary:

1. Navigate to your Synapse Studio and there, in the **Data** hub, go to the **Linked** section where you have access to your data lake.

2. In your data lake, browse to the `airdelays.csv` file. (Do you have another idea of how to find this using Purview maybe?)

3. Right-click the `airdelays.csv` file and start a new SQL script using serverless SQL in Synapse.

4. Adjust the second line of the query text as follows:

SELECT	
distinct ORIGIN	
FROM	
OPENROWSET(	
. . .	

5. Copy and paste the query text and enter a `UNION` command between the two.

6. In the second query, replace `ORIGIN` with `DEST`. Your query should look like this:

```
1   SELECT DISTINCT
2       ORIGIN
3   FROM
4       OPENROWSET(
5           BULK 'https://[YOURDATALAKEACCOUNT].dfs.core.windows.net/[YOURFILESYSTEM]/[YOURPATH]/airdelays.parquet*.sr
6           FORMAT='PARQUET'
7       ) AS [result]
8
9
10  UNION
11
12  SELECT DISTINCT
13      DEST
14  FROM
15      OPENROWSET(
16          BULK 'https://[YOURDATALAKEACCOUNT].dfs.core.windows.net/[YOURFILESYSTEM]/[YOURPATH]/airdelays.parquet/*.s
17          FORMAT='PARQUET'
18      ) AS [result]
19
```

Figure 14.24 – UNION query to generate the airports dictionary

7. Below the query text, please click on **Export results** and download the results as a CSV file (maybe rename it after you have downloaded it successfully).

- Purview REST APIs: `https://docs.microsoft.com/en-us/azure/purview/tutorial-using-rest-apis`

- Sensitivity labels: `https://docs.microsoft.com/en-us/azure/purview/create-sensitivity-label`

- Pattern rules: `https://docs.microsoft.com/en-us/azure/purview/how-to-resource-set-pattern-rules`

Packt>

Other Books You May Enjoy

If you enjoyed this book, you may be interested in these other books by Packt:

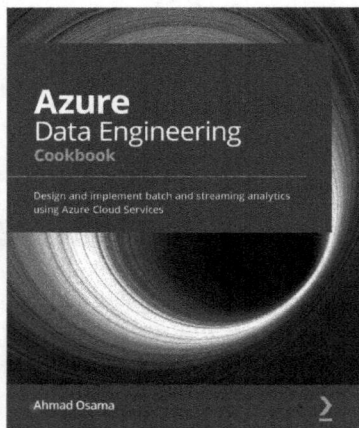

Azure Data Engineering Cookbook

Ahmad Osama

ISBN: 978-1-80020-655-7

- Use Azure Blob storage for storing large amounts of unstructured data
- Perform CRUD operations on the Cosmos Table API
- Implement elastic pools and business continuity with Azure SQL Database
- Ingest and analyze data using Azure Synapse Analytics
- Develop Data Factory data flows to extract data from multiple sources
- Manage, maintain, and secure Azure Data Factory pipelines
- Process streaming data using Azure Stream Analytics and Data Explorer

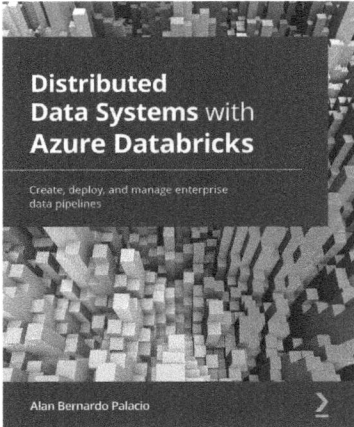

Distributed Data Systems with Azure Databricks

Alan Bernardo Palacio

ISBN: 978-1-83864-721-6

- Create ETLs for big data in Azure Databricks
- Train, manage, and deploy machine learning and deep learning models
- Integrate Databricks with Azure Data Factory for extract, transform, load (ETL) pipeline creation
- Discover how to use Horovod for distributed deep learning
- Find out how to use Delta Engine to query and process data from Delta Lake
- Understand how to use Data Factory in combination with Databricks
- Use Structured Streaming in a production-like environment

Packt is searching for authors like you

If you're interested in becoming an author for Packt, please visit `authors.packtpub.com` and apply today. We have worked with thousands of developers and tech professionals, just like you, to help them share their insight with the global tech community. You can make a general application, apply for a specific hot topic that we are recruiting an author for, or submit your own idea.

Share Your Thoughts

Now you've finished *Cloud Scale Analytics with Azure Data Services*, we'd love to hear your thoughts! Scan the QR code below to go straight to the Amazon review page for this book and share your feedback or leave a review on the site that you purchased it from.

`https://packt.link/r/1-800-56293-4`

Your review is important to us and the tech community and will help us make sure we're delivering excellent quality content.

Index